CONTENTS

FIGURES

ABOUT THE AUTHOR

Joe Hinds has been illustrating works of military interest for more than 40 years. This series, *The Ship Killers,* is a history of the torpedo boat from its earliest development to the fulfillment of its potential in World War II.

Mr. Hinds served as an illustrator in the United States Marines and the United States Navy from 1965 to 1980 and again in the Naval Reserves from 1990 to 1996 in support of Operations of Desert Shield and Desert Storm. During his military career he served on the aircraft carriers *USS Coral Sea* and *USS Enterprise* and later at the Naval Weapons Test Center, Port Hueneme, California. He also worked on the first cruise missile tests in what would become the Tomahawk program.

Mr. Hinds grew up in Needles, California. In this desert river town he was greatly influenced by the National Speed Boat Races. These events featured small, fast boats that were civilian echoes of the torpedo boats of the world's navies. Here in Needles Joe also began to draw at an early age as he grew up admiring the celebrities of the time like Elvis Presley, James Dean and Steve McQueen.

Mr. Hinds has been involved in both technical and editorial illustration. His career includes working in the business community as an independent graphics project manager, illustrator and designer for American Honda Motor Company, British Petroleum/HITCO, Martin Marietta, Grey Advertising, BBDO Los Angeles and others. His art has appeared in magazines such as *Private Pilot, Plane and Pilot, Cycle World, White Mane Publishing,* and others.

In pursuit of historical accuracy, he has been fortunate to work with distinguished authors and historians such as R. Thomas Campbell, Donald Barnhart and Bob Holcombe. Mr. Hinds has also worked closely with Jeff Johnston, Program Specialist and Dr. John Broadwater, Manager of the Monitor National Marine Sanctuary-NOAA. This collaborative effort led to illustrations of the *USS Monitor* and *CSS Virginia* (ex-*Merrimack*). These illustrations, sold as prints, show numerous details of these two historic vessels that have never been shown before. It is believed by all associated with Mr. Hinds that these are the most historically correct renderings to this date.

Mr. Hinds' illustrations of several Civil War vessels appeared In *Civil War Times* magazine from 1999 to 2006. At that same time he began work with the Discovery Channel to produce an accurate illustration of the U.S. Navy Civil War submarine *Alligator.* This illustration was produced with the aid of master model builders and submarine historians, a necessity since no contemporary plans or photographs of the *Alligator* exist. The final product produced by Mr. Hinds was used in television productions, posters and other products.

Mr. Hinds is well underway in producing the first comprehensive treatment of torpedo boats. This monumental research effort on the development of the torpedo boat in the world's navies has produced a series of illustrations and narrative accounts, as well as voluminous research files. One part of the effort, the story of *PT-34*, recently appeared in the magazine *America in World War II*.

By combining his avocation for naval history with his vocation in illustrating, Mr. Hinds continues to make lasting contributions to the history of the world's navies.

Mr. Hinds lives and works in Richmond, Virginia.

ACKNOWLEDGMENTS

This book started in 1999 and the number of people involved is astounding. When I go back to my notebooks of names, web address, site address, phone numbers, I am humbled. This is a team effort to say the least. I will do my best to mention everybody I have ever worked with. That is not possible for me. My memory was damaged several years ago so I must rely on what I am told or old notebooks and my huge collection of yellowing emails. For those who deserve to see their names and don't, I am truly sorry for this omission. I am going to do the best I can.

Joe Judge, Curator, Hampton Roads Naval Museum and the entire museum staff who have given me endless assistance over the years. I do not believe that I could have finished this book on my own. Joe and his staff have gone to great lengths to find me hard to find research papers, photo files, as well as the expense and physical effort to put on a three month art show at the museum in 2004. Joe was all set to talk about an art show for me and then I went and banged my head and the lights went out for two years. Joe, Becky, Marta, Mary, Tom, Susanne, Kathryn, Byron E. 'Jake' Tobin, RADM USN (RET), and the others waited patiently while I spent three years in different therapies reinventing myself and learning how to draw and write again. They have always stood fast by me and given a world of help and hope. I will never forget these lovely people.

I owe a world of thanks to my friend, and naval history author, Richard Worth, for being willing to work with an old guy who can't keep two simple thoughts together for five minutes. Richard has written two naval history books, a novel or two, and is banging away on more while trying to pitch in and help me. That is above and beyond duty but then that is the character of Richard.

Bill Breidenstine, of the Weider History Group, has been another staunch friend for many years. When I bashed my head and could no longer manage the business of selling my prints, Bill, who was a customer, just took over the entire project. No discussion, no deals, he just picked up the slack and pushed on. This was not even a hand shake deal; it was a phone call that lasted about five minutes. Then I went back to being as effective as a head of cabbage and he went to work. He has participated in every way possible in my art shows and symposiums, marketing the prints, getting free ad space in the company magazines and fixing my really dumb writing efforts. He is also a great friend by any standards.

Dana Wegner, Curator, U.S. Navy Ship Model Program, who has sent me well over a hundred emails to answer my questions and send me photo copies of ancient books of how color was made for early ships. Many times he had to really dig through files to find the answers I needed. He also went over the pencil work of many of my Civil War boats before they were painted.

David Grabarek, Interlibrary Loan Specialist, Library of Virginia, has spent years finding books, magazines, and plans for my research from all over the world. To date David has located and received over three hundred books, magazines, reports, obscure papers, and microfilm for my use. It can be said that David has had a piece of every single boat I have done and this book

Jim Kushlan, Editor, *America In WWII,* and past editor of *Civil War Times Illustrated,* who bought my first story and did a terrific job, along with his Art Director Jeff King, in making the layout for my terrible first writing effort really shine.

There can be no particular order for the following names as they are all important to this project, however, these two men must come first. John Lambert of Lambert Plans, United Kingdom and Al Ross of Coastal Forces at Yahoo Groups, United States, have spent years feeding me plans color information, deck details, an encouraging word and a better second guess and the ever final, "What the heck, nobody knows that anyway." Gordon Smith, with his amazing naval history website, has always helped with any question I have ever asked.

From Australia there is Roger Pearson and his wife Deb. Roger has spent hours reading over material and submitting back to me intelligent corrections. Germany is where Michael Emmerich, www.german-navy.de, Hans Frank, Admiral (R), Peter Linao, German naval historian, and Falk

Pletscher, SMML member, have added greatly to my work on *Schnellboote*. In the UK there is Fergus Read, Acting Head Department of Exhibits and Firearms, Imperial War Museum, Chris Huges, a good friend to go out shooting photos of old boats, John Snyder of Short & Snyder Model Paints, who spent way too much time yakking about paint with me, and the lovely Dad and Daughter team of Clive and Caroline who run the finest bookstore in all of England, Aardvark Books,

In America it is, Frank Andruss, Mosquito Fleet; Tomas Tanner; Ken Goldman; Robin Cookson; Archivist for Captured German and Related Records, Modern Military Records, NWCTM; Chip Marshall, model maker, author, historian; David Clark, TV Producer, Discovery Science Channel, who used my work in a TV special; the late Jim Weeks, Editor of *Civil War Times*. From Russia, Andrej Sidenko. From Italy, Captain Umberto Donat, Italian Navy, Louis Henwood, Italian Naval Historian, and a good man named Cristiano who spent many hours translating Italian works for me as a plain favor. I do feel bad about the names that have slipped by and for this I am to be eternally sorry for the oversight.

Fall 2008
Joe Hinds
Richmond, Virginia, USA

DEDICATION

This book is dedicated to my big sister, Camden Classen, who has been my best friend for 62 years, even on those many occasions when I did not deserve her support.

FOREWORD

The U.S. Navy became the greatest in the world with determination, courage and forward thinking. With all its accomplishments, this great navy still needs and welcomes friends. Joe Hinds is such a friend. He is a close student of the service from its earliest experiments with steam power to its modern far-flung operations. Joe has worn the uniform, and in this book he has served the Navy blue in another way.

In these pages Joe enthusiastically records the history of experiments with boats and weapons systems that have had a profound effect on naval development. This history, and the artwork that it supports, advance the discussion about where the Navy has been, and where, perhaps, it needs to go. This is always a worthy discussion. Please join it by enjoying *The Ship Killers.*

Joseph M. Judge
Curator, Hampton Roads Naval Museum

PREFACE

The motor torpedo boat is a large part of the world's naval history. Along the way, in recording world naval history, the small combatant boats have been left out. Their achievements are often ignored and their importance diminished. Previous naval history has always wrapped itself around "Big Ships," "Big Guns" and major sea battles. Historians of the mighty ships have dictated the recording of naval history since the first sea battle. Small boats seldom are mentioned in their esteemed books. Hundreds of books of World War II have been written without a single mention of small craft. This means that there is a hole in history that this book will fill in with facts and illustrations.

Uncorrected history has a nasty habit of becoming a "Fact" in future literary works. When enough time passes the real story is buried deep in achieves and books of misinformation abound. Now errors become facts. A great living historian, Eric Foner, said in his book, *Who Owns History*? "Who owns history? Everyone and no one-which is why the study of the past is constantly evolving, never-ending journey of discovery." It is another way of saying the experts do not own history.

It also means that lost, misplaced, or ignored pieces of historical facts can at any time be re-introduced into its proper place in the history books. So much of our world history has been steadily corrected over the past two decades. That means once a missing fact is uncovered, the books relevant to the subject, need to be corrected.

The most sinful aspect of false history is the irreparable harm it does to forgotten men who fought and died on small boats and their descendants. I have heard many retired WWII naval veteran officers who glibly claim that, "The MTB's never played a major role in anything." They need to tell that to the sons, grandson, wives and mothers of men who died in combat from the deck of a wooden boat while taking on steel cruisers, battleships, destroyers. The only armor plate these craft carried were bits of wood and their shirts. The heart of this book is the stories of the men who manned the frail, craft from 1861 thru 1945. Torpedo boat men are a breed apart.

Take any man from any navy and put him on another torpedo boat and he would be right at home.

You are embarking on a journey of naval history like you have never read it before. While torpedo boats do not seem at first blush to be a "hot" topic, you will be pulled into the simple unvarnished truth that makes for a fascinating new story of naval history.

.

OVERVIEW[1]

The Eighty Years War in Europe[2], or Dutch Revolt, had been going badly for the Roman Catholic forces. The war between Catholicism and Protestantism would continue to be a seesaw affair of bloody military victories and the massacre of entire towns and villages by troops of both sides. The year of 1585 was to be a pivotal moment in Dutch history and would change the map of Europe forever. The Catholic forces were out to crush the Calvinist armies after the fall of Antwerp and met with stiff resistance on the river Scheldt[3] in modern day Belgium. During one critical event in this fight, a magnificent idea was born of desperation. It came from the genius of an Italian Engineer fighting with the Catholic forces known to us as Federico Gianibelli. He needed to destroy the bridge over the river Scheldt just outside of Antwerp. It was not just a bridge but a formidable barrier with 2,000 hardened troops just itching to kill anyone who came close.

The 730-meter bridge was the work of the Duke of Parma.[4] The Calvinist forces were strongly positioned for the bridge to be taken by the troops on hand. Gianibelli loaded two small boats with barrels of gunpowder. He next surrounded the barrels with six-foot gravestone and roofed with the same. On top of this, he made a small pyramid of marble slabs filled with cannon balls, pieces of shot and scrap iron. He placed a clockwork mechanism, with lit wick, to ignite the powder at a preset time. When ready, the clocks were set, the fuses lit, the boats cast off and the stream carried both boats to the barrier over the Scheldt River. Once the device was near the barrier, the powder ignited.[5]

---

[1] *Publishers Note: After this book went to print, we received information about a torpedo design that preceded Gianibelli by three hundred years. The story of Hasan al-Rammah's torpedo is told in Volume 2 of this series.*
[2] Eighty Years War http://en.wikipedia.org/wik/Eighty_Years%27_War, accessed by author 02-06-2007.
[3] *Ibid.*
[4] Siege of Antwerp (1585-1595) http://en.wikipedia.org/wiki/Eighty_Years_War%27 , accessed by author 02-06-2007.
[5] Eighty Years War http://en.wikipedia.org/wik/Eighty_Years%27_War, accessed by author 02-06-2007.

Figure 1. Alexander Farnese, Duke of Parma. Portrait by Vaenius.

Figure 2. Philippe of Marnix, lord of Saint-Aldegonde, fought against the Duke of Parma. Portrait by Jacques de Gheyn II.

A massive explosion followed with pieces of metal and marble blown horizontal in the faces of over 2,000 Spanish soldiers. Eight hundred died and several hundred were wounded in one tragic instant. The blast left a gap in the barrier over 200 feet wide. The effort was a success and had to be quite a surprise for the enemy to lose so valuable a position and so many men to a couple of small boats.[6] This would mark the beginning of great and powerful men humbled by small boats.

The next man to emulate Gianibelli's work was a Dutchman. He made this new type of warfare his sole occupation for the rest of his life. His enthusiasm and energy impressed the power brokers he approached with his invention. Cornelius Jacobszoon Drebbel[7] was no wide-eyed nut case.

Figure 3. Painted shortly after Drebbel tested his submarine in the Thames River for all of Londoners and King James to witness.[8]

He was a serious man in pursuit of scientific devices that had practical applications. He is credited with the first workable submarine. King James of England took Drebbel under his wing in 1604 and gave him money and a place to work.[9] He is known for many inventions with practical application

---

[6] *Ibid.*

[7] Sleeman, C., *Torpedoes and Torpedo Warfare,* (Griffen & Co., 1889), 330.

[8] http://www.dutchsubmarines.com/specials/special_drebbel.htm

[9] *Ibid.*, 330.

such as the telescope lens, the microscope, an astronomical clock powered by atmospheric changes, the invention of mercury fulminate, canisters of chemically made oxygen for his submarines, and the spar-torpedo. He patented the "Water Petard" spar-torpedo in 1620. In 1626, he received from King Charles a warrant to make devices such as floating mines and spar-torpedoes for small ships boats. He had been assigned a workshop at Minories, a district near the Royal Mint, London. He was paid the huge sum of 100 pounds for each "Water Engine." Drebbel took active part in three separate attacks on the enemy ships of La Rochelle, France, for even more money, 150 pounds per month. Drebbel was one of the few men to make real money from his efforts in "Aquatic Warfare."

This is where the idea of explosives, water and small craft had their first romance. Other inventors made much of experimenting with these "Devilish" devices such as, underwater rockets, forms of early torpedoes (mines) and diving boats. It would be Captain D. Bushnell and his craft, the *Turtle*, in 1775 that would go down in history as the U.S. Navy's first torpedo boat.[10]

Bushnell is credited as the first man to explode a charge of gunpowder underwater in this country. He also started something else. He showed the navy how one man in a small semi-submersible craft could threaten an entire naval force of capital ships. The fear from this new contraption was much larger than its capability. The British Navy in American waters went to great lengths to protect itself from this new menace. They never realized how little a threat Bushnell and his *Turtle* truly were. His craft was cramped, stuffy, always in danger of sinking, impossible to control, and subject to the whim of the tides. The plan of the *Turtle* parking itself next to a large ship while the operator used a large drill to punch a hole in the side of a giant warship, attach the charge, move to a safe distance and pull the lanyard to explode the "Bomb," was never going to work. However, it was the first step in the evolution of the torpedo boat.

---

[10] Brown, Allan D., *Torpedoes and Torpedo Boats*, (Harpers New Monthly Magazine, 1882), 36, 37.

Figure 4. This cutaway illustrations shows the intricacy of Bushnell's invention and how truly precarious it would have been to operate.[11] (Library of Congress)

Robert Fulton started out to be an artist and sailed for England[12] at age twenty-one, carrying a letter of introduction from Benjamin Franklin to the famous American portrait painter Benjamin West, residing in London, England. In a short time, he produced many portraits and miniatures that were highly thought of. His interest in painting could not hold his interest for long, though, because his ever-curious mind became fascinated with the engineering aspects of the English canal system.

Fulton left for Paris, France, in 1797 where he started his experiments on torpedoes and submarines. Much of the testing took place on the Seine River with hundreds of onlookers to witness successes and failures. While

---

[11] http://www.navsource.org/archives/08/08441.htm

[12] Herbert, Frederick D., *Robert Fulton's Original Drawings*, (The Society of Naval Architects and Marine Engineers, New York, 1934), 22.

he had his share of failures, he also added greatly to the body of knowledge in explosives and small craft design. In 1801, he designed and built his famous submarine, *Nautilus,*[13] and proved its worth in Brest, France, July 3, 1801. Fulton stayed underwater for four hours and was the first to use compressed air tanks to do so. His invention of compressed air-tanks was again used in the design of the first U.S. Navy submarine, the *USS Alligator* in 1862.

Figure 5. Fulton called his vessel "a submarine boat," which it truly was. (Library of Congress)

---

[13] Brown, *Torpedoes and Torpedo Boats*, 37.

To Fulton goes the undisputed credit for improving the spar-torpedo boat design. This weapon system would have its difficulties for over one hundred years and Fulton would never receive full credit. Here again, the single small craft would be a death threat to capital ships.

Figure 6. Robert Fulton as a young man. *Duyckinick, Evert A.* Portrait Gallery of Eminent Men and Women in Europe and America. New York: Johnson, Wilson & Company, 1873. (Library of Congress)

Samuel Colt of revolver fame was smitten by the new torpedo concept of naval warfare. Sam's vision was with stationary torpedoes (mines) and insulated copper wire that could fire a charge from miles away. He devoted years to perfecting these projects. He had the ear of his friend, the President of the United States, John Tyler. From him, he had received the sum of $6,000 to test his apparatus. He had written President Tyler in 1841: "Discoveries since Fulton's time combined with an invention original with myself enable me to effect instant destruction of either Ships or Steamers... on their entering harbor."[14]

Figure 7. Samuel Colt first became a rich man by selling the British Admiralty his plans for harbor defense. (Library of Congress)

From 1829 to 1845, he never had an experiment go wrong. He blew up many stationary and moving ships at distances of up to five miles distance. Colt's last test, a 500-ton ship moving at five knots, was a whopping 40 miles away and it completely destroyed the test ship. Colt's vision was to have all American harbors protected by elaborate stationary mine fields with trained operators on duty 24/7. The American government and U. S. Navy were not interested. The British government grabbed on to the idea in a flash. The threat of war with France, just thirty miles away, was always present. Britain had hundreds of harbors and thousands of miles of

[14] Connecticut Heritage Gateway, Samuel Colt (1814-1862) http://www.ctheritage.org/encyclopedia/topicalsurveys/colt.htm; Grant, Ellsworth, *The Colt Legacy: The Colt Armory in Hartford*, 1866-1980, (Mowbray Company, 1982) 1-6.

coastline, so it was no wonder they wanted this type of protection. Colt did receive several thousands of dollars over the years but the American navy dragged its feet on committing to build any such apparatus and Sam went back to guns.[15]

The U.S Navy was never fond of all of Colt's successes. John Quincy Adams called it, "... an unChristian contraption."[16] His string of successful demonstrations appalled the Navy brass. His idea for full protection by his torpedo system for harbors, rivers and cities made the Navy look bad. They were able to thwart the experiments of Robert Fulton. Their trickery worked well and Fulton left the field with many failed experiments. He became discouraged enough to drop all of his work on aquatic explosive devices.

The American lack of interest in torpedoes and torpedo boats would carry through to 1940. This lack of foresight, or just plain stubbornness, has caused major setbacks in naval operations of all wars since the American Civil War. It was noted in The Society of Naval Architects and Marine Engineers's *Historical Transactions 1893-1943* that, "The United States Navy was slow to take up the building of torpedo boats."[17]

American reluctance to take up torpedo boats as an important part of its navy was clearly noted by foreign navies.[18] Lieutenant Armstrong, Royal Navy wrote in his famous book, *Torpedoes and Torpedo Vessels*, 1901, "Both torpedoes and mines first came into existence during the course of the American War of Secession, when no fewer than twenty-eight vessels were sunk by mines and six vessels by various forms of torpedoes. Curiously enough, in spite of the enormous benefits which torpedoes gave to their users on these occasions, they received only the scantiest attention after the conclusion of the war."[19]

---

[15] Grant, *The Colt Legacy,* 1-6

[16] *Ibid.*

[17] *Historical Transactions 1893-1943,* The Society of Naval Architects and Marine Engineers, (New York, 1945), 283.

[18] Armstrong, G.E., Lt. RN, *Torpedoes and Torpedo-Vessels*, (George Bell and Sons, London, 1901), 201-202.

[19] Armstrong, *Torpedoes and Torpedo-Vessels*, 2-3.

The point is illustrated again by the Lieutenant when he writes[20] "At the end of 1890 the different Powers possessed the following numbers of torpedo boats built, building or projected: France 210; England, 206; Germany, 180; Italy, 152; Russia, 143; Austria, 61; Greece, 51; Holland, 50; Denmark, 34; China, 32; Norway and Sweden, 31; Turkey, 30; Japan, 24; Spain, 15; Brazil, 15; the United States, 1. During that year the Navy commissioned its first torpedo boat, the *Cushing*. When Robert Whitehead's "Fish" torpedo became operational, the world navies went torpedo boat crazy. No other craft, until the aircraft carrier, garnered so much world attention.

THE AMERICAN CIVIL WAR

The American Civil War started with the U. S. Navy ill prepared and the South with no navy.[21] Within days of the declaration war, it became clear that in spite of wanting war, no one was prepared for it. Ironically, the best assets of the American military were in its navy. Since the early 1840's, industrialization in the world had led to modernization of the armies and navies of the world. Ship builders on both sides of the ocean embraced the paddle-wheel steamer, the screw, twin-screw propellers and steam engines that are powerful.

Forward thinking engineers proved armor for ships was practical. From France came the designs of Paxhans, DuPuy de Lome and Gervasie.[22] In America, it was John Ericsson, *USS Monitor* and the Stevens' brothers advanced designed ship the *Battery*.[23] These men pushed hard to get the Navy to modernize. By the time of the Civil War, many technical advances in engineering had taken place but not adopted by the American or British navy. Floating batteries by the France proved their tactical worth in the Crimean War. Fulton had invented the spar-torpedo boat and friends of the inventor were further testing the design in New York. The submarines by Brutus DeVilleroi[24] and Robert Fulton were far enough along in

---

[20] *Ibid.*, 189.

[21] E.B. Potter & Chester W. Nimitz, *Sea Power, a Naval History* (Prentice-Hall Inc. 1960), 250-251.

[22] *Historical Transactions*, 250.

[23] *Ibid.*, 250.

[24] Maloney, Janet. "The Submarine *Alligator*," *Civil War Times* (Vol. XLIV, No.5)

development to be deemed worthy of military application. DeVilleroi came to America and caught the attention of the U. S. Navy by building and successfully using a small forty-foot version of a design that would become the USS *Alligator*[25] for the U.S. Navy and the first military submarine built in America. Samuel Colt had perfected the detonators (these were used with effectiveness along with his invention of waterproofed copper wire) for underwater explosives as well as working designs for stationary mines and torpedoes. Once again, the U.S. Navy was not interested but Great Britain was. Colt made a great deal of money from the British in setting up shore-controlled torpedo (mine) fields. It was the U.S. Navy that was reluctant—the British navy was all "gung-ho." The Stevens brothers made the biggest splash with radical ship design. They offered up a design that incorporated: (1) armored slopping sides; (2) heavy guns on turntables; (3) balanced rudders; (4) high-pressure boilers; (5) twin-screws and (6) multiple-cylinder, high-speed engines.[26] It was a breath-taking time in marine engineering and weapons development.

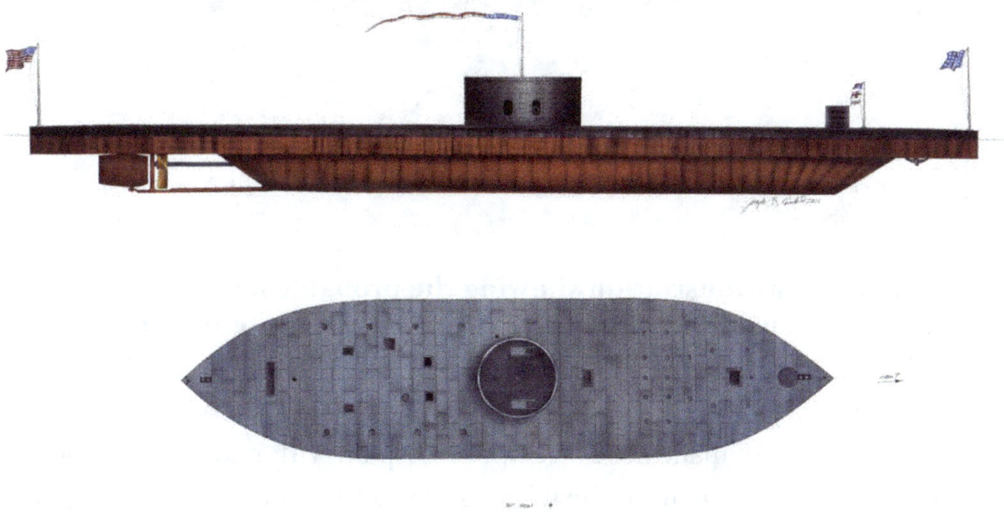

Figure 8. The *Monitor* fought the *CSS Virginia* (ex-*Merrimack*). (Illustrations by author)

PRIMEDIA History Group, 2005, 34.

[25] Bolander, L.H. "The *Alligator,* First Federal Submarine of the Civil War." Naval Institute Press, June 1938.

[26] *Historical Transactions.*

Figure 9. Cutaway illustration showing the probable layout of the interior of the *Alligator*. Illustration is from a contemporary newspaper. (Hampton Roads Naval Museum, Norfolk, Virginia)

The U.S. Navy officer corps included many old and conservative officers who, as a group, had made little attempt to keep up with new technologies. Their training had come from the wars long past with the last war being in 1812. The great weakness in the Navy's promotion system stemmed from the sole determination of rank by age.[27] The other factor was the complete lack of a retirement system. There had been no American wars for fifty

---

[27] *The Union and Confederate Navies*, http://www.civilwarhome.com/unionconfednavies.htm, accessed by author 07-16-2005.

years and this was the knowledge base for much of senior officer training. In the first months of the war, the U.S. Navy's weakness in command personnel became glaring clear. Living fossils were leading a navy trying to modernize.

The Federal fleet on paper had 90 vessels. In reality, the heart of the fleet was twelve first-class screw-sloops of war. The most famous were the *USS Hartford, USS Brooklyn, USS Richmond,* and *USS Merrimack.* Two second-class screw-sloops were the *USS* I*roquois* and *USS Dacotah.* Another ship of the same class but constructed peculiarly was the USS *Pawnee,*[28] which would see years of hard service. Forty others were old and obsolete sailors fit for a big ocean fight against other such ships but completely unfit for this war. Sixteen steamers were docked for repairs or in the ways under construction and nineteen others scattered all over the globe in the Mediterranean, Brazil, East Indies and Japan.[29]

THE CONFEDERATE STATES NAVY

The Confederate States of America did not have a Navy Department until long after the war started. It had no budget, ships, stores, weapons, training academies, recruiting plan or any sign of strategic or tactical war plans. In the rush to war, the Southern leadership never seriously considered the need for a navy. They had envisioned a massive, winner-take-all land battle, which they would naturally win, so that they could then go on to create a country called the Confederate States of America.

On December 20, 1860, the elected officials of the state of South Carolina voted to secede from the United States of America. Six other southern states soon followed. The politicians of these states met at a convention in Montgomery, Alabama, on February 4, 1861. By February 8th, the Southern Confederacy enacted a provisional constitution of the Confederate States of America. There was a new country in the south. America had an enemy within and that enemy was set to destroy them. The

---

[28] *The Union and Confederate Navies,* http://www.civilwarhome.com/unionconfednavies.htm, accessed by author 07-16-2005.

[29] Still, William N., Dr., *The Confederate Navy, The Ships, Men and Organization, 1861-65,* (Naval Institute Press, 1997), 12.

South wanted more land for more cotton and more slaves. It would destroy anyone in trying to get them.[30]

From the first day of forming the new country in-fighting among the Southern states raged. Bickering among government and state officials started immediately. This childish behavior would mark the Southern government for the duration of the war. George Jones of Tennessee said of the Confederate Congress "[it is the] most inefficient branch of government existing in this revolution."[31] Their first priority seemed to be to fight each other and the U. S. military second. The individual states wanted to control their own navies and army so these resources could protect the individual state. Cooperation was beyond the Southern politician's mental powers. The South had already begun to forfeit victory to the United States of America's armed forces.

Figure 10. Scott's Great Snake: how the South was strangled by the U.S. Navy. This map shows the depth of the U.S. Navy's encirclement through waterways.[32]

While the Confederate government pleaded for men, ships, stores and money, the U. S. Navy slowly enacted "Operation Anaconda."[34] This plan

---

[30] Still, 3.
[31] Eicher, David J, *Dixie Betrayed How the South Really Lost The Civil War* (Little Brown and Company, 2006), 140.
[32] http://en.wikipedia.org/wiki/Anaconda_Plan

called for the East Coast, Gulf Coast and Mississippi River controlled by the U.S. Navy. The idea was to put a total stranglehold on all ports and major rivers to force the South into submission. If the Navy had been prepared and the upgrading of ships and training had been accomplished, they might have pulled it off. As it was, the U. S. Navy was weak in leadership and did not have enough vessels to complete the job. Had they acted on the inventions of Sam Colt and Robert Fulton they could have stopped all traffic into the Southern ports with rings of stationary torpedoes as were being built in England.

Confederate President Jefferson Davis appointed Steven R. Mallory, Senator from Florida, to head the Navy department. It would prove to be the best choice of all his cabinet members. Mallory was smart and capable. He came to the Southern cabinet, as one of the few was actually fit for his post. Mallory had recently spent years in the United States Senate as a committee member of Naval Affairs and became its chair in 1855.[35] Mallory tried hard to get old naval officers to accept new technologies for the Navy. Now he had to fight other members of the Confederate Congress who tried several times to have him removed or impeached, or to have his Navy Department folded into the Army Department. The unabated power struggle would continue to the last days of the war. The destruction of the South was a sure thing from within its own government.

Mallory was able to institute a plan to use cotton to buy the best and newest ships from Europe. For the entire war, Confederate agents paid for new, fast, speedy ships to fit the need for the South to have cruisers capable of attacking Federal ships at sea and used as blockade-runners. British and Scottish shipyards outrageously overcharged the Confederate government for these ships and supplies. Later the huge profits realized from this illicit business would be lost in court. After the war, an international court of claims ordered the British to pay the United States $15.5 million in gold to

---

[34] Weisman, John C., *Civil War Chronicles Naval Warfare, Courage and Combat on the Water* (Friedman/Fairfax Publishers, 1997), 30.
[35] Still, 21-23.

offset the damage done to U.S. ships by Confederate raiders built in England and Scotland.[36]

In 1862, the Confederate government weakened its drive for a naval force even more. It brought into being a truly bad idea: "The River Defense Fleet."[37] Secretary Benjamin and the Mississippi delegation maneuvered around Mallory. It called for fifteen vessels to be seized by the Confederate Army under General Lovell at New Orleans. This was to be an Army fleet with Army officers in command of each vessel. It became a failure and a drain on the skimpy resources for the new Confederate States Navy Department. The first use of the River Defense Fleet consisted of eight ships under the command of Captain J. E. Montgomery. The U.S. Navy fleet destroyed three ships by gunfire and captured four. The six vessels commanded by General Lovell at New Orleans were likewise captured or sunk. The technology strapped South turned to other devices as a means of coping with the might of the industrial North. Torpedo warfare (mine warfare) would shock the world not for its brutality but for its effectiveness.[38]

THE CONFEDERATE TORPEDO DEPARTMENT

It became clear to Mallory and others that they could not hold off the U. S. Navy until they had a fighting fleet built in England. They needed to do something now to hold the U. S. Navy at bay. While on the committee for Naval Affairs, Mallory would have been privy to all of applications made by Fulton, Colt, the Stevens brothers, Ericsson and other marine engineers and inventors. During this time, magazines of a technical nature were quite popular and regularly subscribed to by engineers and inventors both North and South. All of the trials of Fulton, Colt and others were not just recorded but illustrated as well. Hundreds of curious readers studied

---

[36] Wideman, John C. *Civil War Chronicles Naval Courage and Combat on the Water* (Metro Books, 1997), 14.
[37] Foote, Shelby. The *Civil War A Narrative, Fort Sumter To Perryville* (Vintage Books, 1986), 361.
[38] Snell, Kelly. The Evolution and Influence of Tactical Warfare in the American Civil War. http://civilwarhome.com/tacticalwarfare.htm, accessed by author 08-21-2006.

Fulton's *Nautilus* submarine, his spar-torpedo boat and the designs of Sam Colt's underwater detonators.

There was enough Southern sentiment in the North to ensure that supplies, detonators, electrical wire and batteries would not be difficult to attain. The Confederates began sending agents north to purchase materials. Materials list were also given to Confederate foreign agents abroad. This would ensure an uninterrupted flow of important materials to the Confederate Navy. Weapons-making material, such as there was already in Southern cities, was rounded up and placed at the disposal of the newly formed Submarine Battery Service, which would become the Torpedo Bureau in 1862.[39]

In the South, as well as the North, interest in submarines and torpedoes ran high for years before the war. One of the great myths of the Civil War is that the South invented the submarine and the *H. L. Hunley* was it. The little-known fact that Lodner Phillips designed and built a submarine for salvage work on the Great Lakes in 1850-55 destroys this myth. Added to this achievement is his underwater cannon used to blast obstructions, which was another important breakthrough.

In the South, there were over twenty private submarines built and many were successful and better built than the Diving Boat, *H. L. Hunley*.

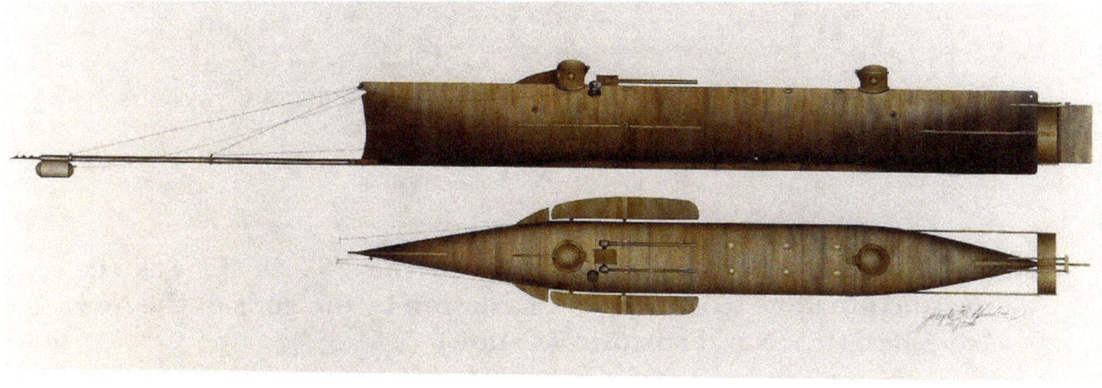

Figure 11. *H.L. Hunley* Diving Boat. (Illustration by author)

The *Hunley* in its short life killed three crews of eight men and managed to kill a few more on a fourth-training mission. The drive to produce successful submarines was not for patriotism alone. The Southern

---

[39] Still, 26.

Congress made an offer that anyone whose invention sank a Yankee ship would receive 50% of the value of the vessel as a reward.[40] Inventors of submarines and torpedoes bombarded the government offices for funds from every, tinker, engineer, and confidence man in the South. The Confederate government spent millions funding one hare-brained scheme after another.

The Confederate Torpedo Bureau came into being in October 1862.[41] It would prove to be more effective at sinking Federal ships than the rest of the entire Confederate States Navy. The damage caused by the CSS *David* and its seventy pound torpedo forced the gigantic warship, *U.S.S. New Ironsides*, completely out of action for the duration of the war. In February of 1862, Federal gunboats on a mission on the Savannah River were turned back due to the threat of submarine torpedoes.

Figure 12. The reason for the creation of the CSS *David* was the destruction done to the city of Charleston by the guns of the *New Ironsides*. (U.S. Naval Historical Center)

[40] *Mine Warfare Nineteenth Century*, http://navy.mil/navydata/cna/n75/Htm/800PopH3.htm

[41] Still, 26.

Figure 13. *CSS David* after the war. (U.S. Navy)

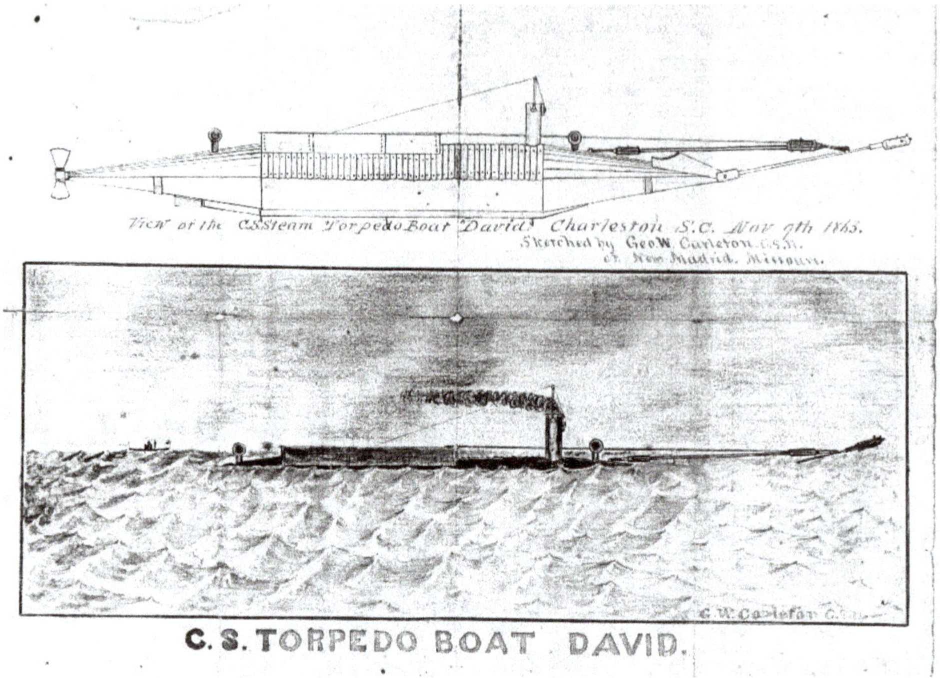

Figure 14. Profile of *CSS David*. The hurriedly built *David* proved successful when it put the mighty *New Ironsides* out of action for the balance of the Civil War. (Museum of the Confederacy, Richmond, Virginia)

Now the U. S Navy was operating on the defensive. While the Federal Navy spent time and resources on defensive measures the Confederates spent time for better and bigger designed torpedoes.[42] The men of the Torpedo Department became more adapt at perfecting explosives and the placement of torpedoes for maximum effect.

The loss of ships to the North was staggering. The first encounter with stationary submarine torpedoes was in the Savannah River, Georgia, February 18, 1862. No ships were sunk but officially the U. S. Navy was now on the defensive. The first U. S. Navy vessel sunk by the new weapons was the USS *Cairo* on the Yazoo River off Haines Bluff, Mississippi. The *Cairo* struck two submarine mines and sank in twelve minutes with great loss of life. 1864 was the blackest year for the river squadrons of the U. S. Navy. They lost 16 ships and hundreds of men. Morale dropped and the hopes of

---

[42] Mine Warfare Nineteenth Century, http://navy.mil/navydata/cna/n75/Htm/800PopH3.htm

the Confederate Navy rose.. In the end, it would be just a respite until the final deathblows on the Confederate Navy fell.

While losing fifty-two ships to torpedoes in the three years between 1861 and 1864, the U. S. Navy leadership at last moved into action to counter the threat from the South by establishing their own U.S. torpedo bureau in, September 1863.[43] The Confederate Torpedo Bureau had placed 123 torpedoes in Charleston Harbor, which held the Federal blockading fleet at bay. In the Roanoke River, North Carolina, they placed 101 torpedoes, which sank five ships alone. The James River near Richmond, Virginia, was peppered with obstructions that contained torpedoes. The Confederates were strewing mines in every bay, harbor and river in their reach. The Federal fleets were hampered for the rest of the war.

The South had contracted with British and Scottish boat builders for two types of spar-torpedo boats during the war, the classic *David* class and the newer and more traditional *Squib* class. A copy of correspondence concerning British firms seems to have been lost along the way. There is a letter, in the Official Records[44] from James D. Bulloch, civilian agent the Navy Department, of July 8, 1864 to Secretary Mallory,

> *"Your letter of April 16, 1864, received Jun 6, encloses description of small marine engine and directs me to have twelve pairs made and sent forward as early as possible. I have given the entire order to Messrs. James & George Thomson, of the Clyde Bank Foundry, Glasgow..."*[45]

He goes on to discuss the problems with the boat plans from a Mr. Williamson as being very inexact and causing problems with their

---

[43] "Unchivalrous" Torpedoes Exhibited at Fort Anderson, North Carolina Department of Cultural Resources, www. Dcr.state.us/news/2004/hs8-504.aswp

[44] Official Records, Series II, Volume 2 Navy Department Correspondence, 1861-1865, With Agents Abroad.. *Navy Department to J.D. Bulloch, Apr.16,* Encloses description of small marine engines and boiler as and orders 12 made as early as practicable and tells of their needs, p. 627 *J.D. Bulloch to Navy Department July 8,* Reports sale of Georgia to E. Bates of Liverpool. Unable to send skilled, metal workers to the Confederacy, given reason thereof. Has order for 12 small marine engines to J. & G. Thomson, pp. 682.

[45] Official Records of the Union and Confederate Navies in the War of the Rebellion, Series II-Volume 2, *Navy Department Correspondence, 1861-1865, with Agents Abroad* (Washington, 1921).

manufacture. It has been reported that earlier in the war three iron-built spar-torpedo boats and their engines had been off-loaded in Charleston. It is known that three complete engines and boilers for torpedo boats did make it through the blockade to be used in spar-torpedo boats under construction.

At war's end the boat yards of England and Scotland would have held several workable torpedo boats. They were paid for but never shipped due to the ending of the war. Undoubtedly, these vessels would have come to the attention of Thorneycroft and Yarrow. Word had been coming back to Europe for three years as to the effectiveness of the spar-torpedo boats and their torpedoes. It is reasonable to propose that is why the French and British were so fast in producing their own spar-torpedo boats after the Civil War.

THE U.S. NAVY TORPEDO BOAT EFFORT IN THE CIVIL WAR

The successful killing caused by Confederate torpedoes and their spar-torpedo boats had taken its toll on the U.S. Navy's river sailors. Operations were moving at less than a snail's pace due to the frequent need to assign ships to minesweeper duties, which were laborious and dangerous. The toll on ship's company morale was noticeable and recruiting became more difficult. Torpedoes were a new and seemingly alien technology. Union sailors hated torpedoes more than any other weapon. They were considered to be, "infernal machines" and "monsters."[46]

The ironclad ships at the time had no watertight compartments and an explosion anywhere below the ship's waterline caused immediate submersion. Many ships sank in less than five minutes. Others disappeared from the force of massive mines placed underwater. Massive torpedoes—now called mines—were filled with a thousand pounds or more of black powder. "... Torpedoes completely destroyed the sailor's myth that their service was safe and that iron guaranteed their security."[47] Fear became crippling to entire crews. Officers became hard-pressed to maintain discipline because in the officer ranks the torpedo fear was just as

[46] Bennett, Michael J. *Union Jacks, Yankee Sailors in the Civil War*,(University of North Carolina Press,2004), 193.
[47] *Ibid.* 193

debilitating. Crews had to contend with boiler explosions that scalded men alive and melted flesh to the bone. Enemy shells caused horrible wounds and tore living men to shreds. The torpedo was a silent killer and gave no warning whatsoever. "The crew of the USS *Sebago* spent the entire night of 14 December 1864 maneuvering around what they thought was a deadly torpedo .... much to their chagrin, the monster that had terrorized them ... was nothing more than an old scow, capsized."[48]

The U.S. Navy turned to the genius of Chief Engineer William W.W. Wood to address counter-measures to the Confederate torpedo boats. The Navy Department had been approached by a young Lieutenant William Cushing who proposed to sink the newest ironclad in the Confederate fleet the CSS *Albemarle*.[49] What would become *Picket Boat No.1* was the creation of Wood and his assistant John Lay. Not only did they modify a common steam launch, they designed, built and tested a superbly functional spar-torpedo packed with enough explosives to seriously damage any ship it struck. A night attack was called for as soon as practical. This date turned out to be on a rainy, dark, night of October 27, 1864.

Figure 15. The success of the spar-torpedo boat and the refinement of the torpedo (mine) spurred inventors, world wide, to race to capture a naval military market and garner vast wealth. Sam Colt, Robert Fulton, John Ericcson and Robert Whitehead all became famous.

---

[48] *Ibid.*, 194

[49] Commander William B. Cushing, USN, (1842-1874). http://www.history.navy.mil/photos/pers-us/uspers-c/w-cushg.htm;

It would be Engineer Wood and John Lay who converted six tug boats, currently under construction, and turn them into the U. S. Navy's first fleet of torpedo boats.[50] At last the Navy Department became serious about countering the Confederate threat. Two spar-torpedo boats saw limited action, as the war ended, but the U.S. was for the moment, a believer in the torpedo boat.

The Wood creations turned out to be formidable vessels. They averaged 207 tons, 84 ft length with 23 officers and enlisted and armor plate of 12 inches on the pilothouse, 5 inches on the hull, and 3 inches over the deck. While they could only make 5 knots, they carried a full 8 days of provisions and water.[51] Their construction was designed to stand the heaviest shelling from Confederate ironclads at close range. With the adaptation of separate steam engine for filling and emptying ballast tanks, the boat could raise and lower itself by 3'-3" in a few minutes. The common practice by the standards of the time called for spar-torpedo boats to carry their charge at the tip of a spar, fully exposed to enemy fire and unhappy, accidental detonation.

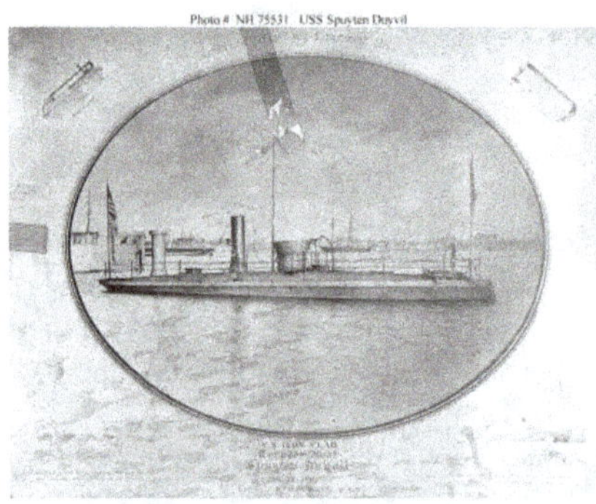

Figure 16. U.S. Navy Picket Boat *Spuyten Duyvil.* (U. S. Navy Archives)

The first of six USN torpedo boats was the the *Spuyten Duyvil* (New York, New York Old Dutch for Speedy Devil), which carried a dozen charges for its two metal spars. The tube-steel spars carried below the waterline and when loaded with charges were run out to their full length (30') by a separate steam engine just for this occasion. Once extended to its

---

50 "The *Spuyten Duyvil,* Engineering", 26 October 1866, http://www.history.navy.mil/danfs/s16/spuyten_duyvil.htm.

51 "The *Spuyten Duyvil,* Engineering", 26 October 1866, http://www.history.navy.mil/danfs/s16/spuyten_duyvil.htm.

full length, the charge released from the spar and allowed to float to the bottom of the enemy ship. A clever mechanism, designed by Woods, released a pin that allowed a steel ball to fall onto a glass container of chemicals that set off a chemical reaction of great heat, igniting the 70 pounds of black powder.

Figure 17. *Spuyten Duyvil.* The world's first underwater torpedo tube—with the world's first underwater torpedo doors—shows clearly in this photo from Pook's Yard. (U.S. Navy Archives)

This boat had twelve chances to destroy an enemy or the ability of targeting twelve ships. All the torpedoes carried by this fleet were of the successful Wood-Lay design, which had sunk the *Albemarle*. Spar-torpedo boat design was making great advances thanks to Engineer Wood.

EARLY EXPERIMENTS AND SMALL WARS, 1861-1900

The end of the American Civil War left an exhausted American nation in its wake. The devastation to cities in the South meant that recovery would not take place for decades. In the North, the factories, shipping firms, and the railroads would be changing their hiring practices for workers. Many thousands of young Northern men were dead or crippled. This sharp decline in manpower led the manufacturers to set aside old prejudices of race and nationality and lure Irish from the East, Chinese from the West and young girls from the farms of the North East to fill the factories by the tens of thousands. Overnight, the two despised races became sought after and women begin to gain importance in the American work place.

The extensive use of the telegraph and railroads during the war became the objects of the acceptance in the commercial world.[52] It also ushered in the practicality of balloon flight for military observation, repeating rifles, and Gatling guns. Communication and railroads would now play a critical role for the world's military war planners. The reports from foreign officers who were guest of both the U.S. Army and Confederate Army would include the stationary torpedo (mine), the spar-torpedo boat, land mines, the improved hand-grenade, the rotating turret, armored vessels, breech loading guns (of all calibers), the Calcium light for ships, shore forts, field artillery, and rockets.

It was after the war that an Austrian Commander, Giovanni de Luppis, came into possession of plans for a small boat, powered by a clockwork engine and guided by wires, which could strike an enemy vessel with a large explosive charge.[53] Luppis did his best to improve the design and then he sought out Robert Whitehead, a brilliant English engineer residing in Italy. Whitehead was intrigued with the crude model boat that Luppis brought with him. His model was called the *Der Kustenbrander* (the Coastal Fireship) from this model and with four years of hard work came the birth of the modern torpedo. Whitehead's new life's work would be the "Automobile Torpedo." It would be the torpedo and torpedo boat that would dominate naval invention, design and construction for the next 30 years.

JOHN ERICSSON AND HIS *DESTROYER*

In 1871 John Ericsson of U.S.S. *Monitor* fame, became an enthusiast for the development of the torpedo boat. His first design, the *Destroyer*,[54] would bear the Ericsson trademark of new inventions, improved equipment designs, and clashes with both the British Admiralty and the

---

[52] Snell, Kelly. *The Evolution and Influence of Tactical Warfare in the American Civil War* http://civilwarhome.com/tacticalwarfare.htm accessed by author 11-10-2006.
[53] Gray, Edwyn. *The Devil's Device, Robert Whitehead and the History of the Torpedo*, Revised and updated (Naval Institute Press, 1991), 34.
[54] Church, William Conant. John Ericsson, The Engineer, *Scribner's Monthly Magazine*, (Vol. II, January-June, Charles Scribner and Son's, 1890).

U. S. Navy. Ericsson had been contemplating underwater weapons since the 1830's. He had made sketches of both turreted ships and underwater artillery weapons. Ericsson built a fantastic ship that could outrun any other vessel anywhere in the world. His new engines were the most powerful of the time. Small in size, they were years ahead of the day's technology. Once again the many, many, ingenious devices throughout the *Destroyer* made it the most modern torpedo boat in the world. Reportedly, she could reach and maintain the unheard speed of 25 knots.[55]

Many of the innovations employed in the boat were heavily influenced by the same innovations built into the U.S. Navy's *Sputen Duyvil,* Torpedo boat No.5. Ericsson's ship would never be an ocean going vessel. It was designed for harbor defense and shallow water operations. Its best defense were its fantastic speed and its low silhouette.

The *Destroyer* has been repeatedly skipped over by past naval historians. It may have seemed to them to be an anomaly and not a terribly practical ship. The *Destroyer* performed well in all trials but received no welcome from the U.S. Navy. It carried an intricate weapons system and equally complicated engines, ballast pumps and other features strange to a Navy ship of the time. It was the British who paid for extensive trials and showed the greatest initial interest. The underwater cannon were complex pieces of machinery that employed both pneumatic (compressed air), and gun cotton. Tests in Great Britain showed some admirable results with the sinking of large target ships, with a 1,500 pound projectile, from a range of 300 to 700 feet in 3 seconds after firing the cannon. It was designed as purely a defensive weapon (like the modern Russian *Shkval*).

When the Admiralty lost interest, Ericsson attempted to sell the craft to Peru, but members of the United States Congress leveled the charge that Ericsson was a traitor for selling "Our Secret Weapon" to a foreign power. The U.S. Navy never offered to buy the craft or reimburse Ericsson his expenses incurred during naval trials. Ericsson was once again bitter and disappointed with the Navy department. Eventually, the *Destroyer* wound up in Peru.

---

[55] Barnard, Charles, Ericsson's Destroyer and Her New Gun, *Scribner's Monthly,* (Vol. XXI, , Issue 5, March, 1881).

Figure 18. In its time, this was the most advanced harbor defense weapon in the world. (Library of Congress)

Figure 19. Ericsson's *Destroyer*. As in the case of the USS *Monitor*, this new boat design ushered in new faster engines, more efficient boilers, and the marriage of hydraulics and compressed air to fire his "Torpedo Gun." (Illustrations by author)

THE HERRESHOFF BROTHERS, TORPEDO BOAT BUILDERS TO THE WORLD

John and Nathaniel Herreshoff were putting together what would become the most famous boat yard in the world, the Herreshoff Manufacturing Company in Bristol, Rhode Island.[56] These two men would soon be leaders in torpedo boat construction for the world's navies. They also built a boat called the *Lighting* (1876) a year before the British and the torpedo boat *Stiletto* (1877) for the U.S. Navy. In 1878, they built Hull No. 44, a spar-torpedo boat for the British Admiralty.[57]

Figure 20. Nathaniel G. Herreshoff (left) and John Brown Herreshoff (right). (Library of Congress)

This was followed in rapid succession by Hull Numbers; 53, 55, 69 for the Peruvian Navy in 1879, and Hull No. 64 for the Russian Navy in 1880. The Peruvian torpedo boats would meet with some success and a few were captured by their arch-enemy, the navy of Chile. These same captured boats in turn sank a few Peruvian ships, much to the dismay of the original owners.

In 1887, USS *Cushing* was laid down by the Herreshoff yards as the first modern steel torpedo boat. This construction included the use of electro-

[56] Simpson, Richard V. *Building the Mosquito Fleet: The U.S. Navy's First Torpedo Boats* (Arcadia Publishing, 2001), 9.
[57] *Ibid.*, 12.

galvanized steel and Herreshoff-built five-cylinder, quadruple expansion engines. This also included the Herreshoff-built fast acting boiler, designed especially for torpedo boat service. The *Cushing's* torpedo room looks amazingly just like the same room of the *Destroyer*. The yard went on to build several more steel torpedo boats for the U.S. Navy. These were the: *Porter, DuPont, Talbot, Gwin* and *Morris*.[58] Two boats, the *Somers* came from the famous German boatyard Schicau and the *Manley* would come from the Yarrow yards of Great Britain. Of the thirty-five boats built by American contractors in this era, Herreshoff would have the lion's share of six boats.

WORLD NAVIES

In 1878, the Italians would launch the *Nibbo*,[59] a unique and beautiful torpedo boat. Built by Thorneycroft and riding low in the water much like the *CSS David*, she carried two forward firing torpedoes and no defensive armament.[60] A new concept for the torpedo boat was the second class torpedo boat designed to be carried on a "mother ship" at night or the end of operations. These mother ships could carry up to six torpedo boats and supply the crews with living quarters. Thorneycroft conceived the idea and built hundreds of second class torpedo boats. These boats showed up in navies all over the world. These came with lifting rings mounted on the iron hull and carried two forward firing torpedo tubes with two torpedoes readied for reload. These were effectively "day boats" and not intended for long duration at sea. The advantage here was that while the boat crew ate, rested and slept, the mechanics could pay uninterrupted attention to the servicing of the boat's mechanical needs and reload her with torpedoes. The next morning, the small boats would be back in the water with a fresh crew and a ready-for-a day's action.

---

[58] *Ibid.*, 20.

[59] Fock, Harald. *Fast Fighting Boats 1870-1945* (Naval Institute Press, 1978), 11.

[60] Fock, 11.

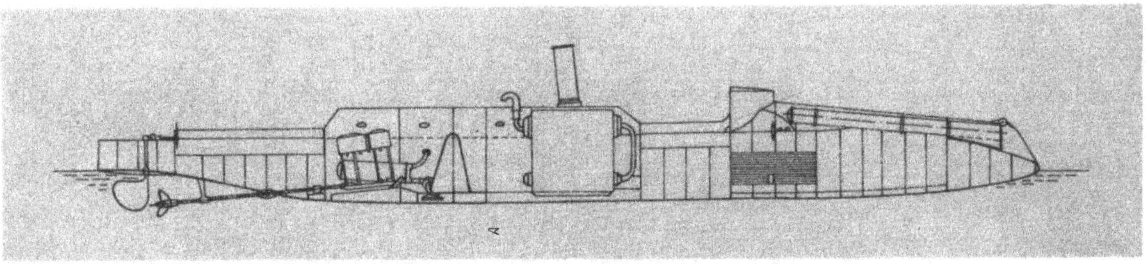

Figure 21. *Nibbo,* Italy's first modern Bow-launching torpedo boat. (U.S. Navy Historical Center)

Torpedo boats designed for use with parent ships went out of vogue in the very early 1900's. They were driven out by the new gasoline engine. The new engines afforded higher speed, lighter fuel loads and longer range. Now a similar size boat could carry liquid fuel in place of coal and have more room for more torpedoes and more armament and a longer hunting range. The internal combustion engine changed the design of the torpedo boat overnight. The Russian Navy capitalized on this by being the first navy to run small boat engines on diesel fuel rather the highly volatile petrol.[61] This extremely intelligent safety measure would be abandoned by the Russians and picked up by the German *Kriegsmarine* in the 1930's.

The Americans

Torpedo boat flotillas were first used in large numbers in the Spanish-American War. There, the U.S. Navy showed it did not clearly know how to use its torpedo boat fleet. The U.S. Navy had no operational doctrine, and the misuse, and lack, of tactical or strategic training was well reported by many young naval officers. In an issue of the U.S. naval magazine, *Proceedings* in March 1903, Lieutenant Edward L. Beach (father of the famous submariner, and himself later a distinguished officer and novelist) wrote, "The Navy is weak in torpedo boat experience because our boats have been so little used ... boats have been put into maneuvers before the officers and crews have had sufficient experience to become expert... There

[61] Greger, Rene. "The Origins of Russian and Soviet MTB's," *Warship* Vol. IX Issue 33 Ed. Andrew Lambert, 39.

Figure 22. The Spanish-American war era torpedo boat *USS Winslow* waiting for repairs to its bow. (U.S. Navy Archives)

Figure 23. Battle damage to *Winslow*. This is where five American sailors died from the cannon fire of Spanish forts in Cardenas Bay. (Library of Congress)

were no torpedo-boat operations worthy of the name in this short war." [63] "... The American boats were used for everything except torpedo-boat work."[64]

Lieutenant Chandler, USN, in the same *Proceedings* article added: "Used with wisdom the torpedo-boat becomes a most powerful weapon; used without practical knowledge of torpedo-boat warfare the torpedo-boat becomes useless to their own flag ..."

This attitude was not unique to the U.S. Navy. It was a common complaint in all navies. Old naval officers who had little or no faith in these small surface combatants were commanding the new boats. This would not change until the young officers with torpedo boat experience began to fill the senior ranks.

The Russians

The Russians had a much different experience. They wholly embraced the spar-torpedo boat. Their war against Turkey for control of the Black Sea and Danube delta was going to be a long one. The daring Russian boat captains met with instant success. In 1877 the Russian Navy had six spar-torpedo boats. Two were of iron construction, the *Shutka* and *Mina* (built by Thorneycroft*),* the others were wooden construction.[66] The four wooden Russian boats attacked and sank the Turkish monitor class ship, th*e Seifi*, on May 27, 1877.

The two iron boats would continue to attack the Turkish fleet until it withdrew from the scene. A few months later, January 26, 1878, Captain Makarov of the ship *Chesma* had a torpedo tube mounted on the keel of his ship.[67] They were going to have a live fire test of a Whitehead torpedo. It was a historical first for the Russians. They would go down in history as the first navy to sink an enemy ship, the Turkish guard ship *Intiba*h, with a Whitehead torpedo. What the Russians learned from this war was that the

---

[63] Beach, Edward L., Lt. USN. Our Torpedo-Flotilla, The Training Needed To Ensure its Efficiency, *Proceedings*, Vol. XXIX No.1, March 1903, 126.
[64] Beach, 121.
[66] Greger, 38.
[67] Greger, 39.

smaller torpedo boats were the most successful in all actions.[68] The small boats were known as *Minonski* or torpedo cutters. Russia (Soviet Union) was still making small torpedo boats up into the 1950's. In December 4, 1904, the Russian bought 10 motor torpedo boats for its Baltic Fleet from the American firm of Lewis Nixon, Perth, New Jersey.[69] The boats were 27 meters long with two engines for 600 hp and a sustained speed of 25 knots. These became the role models for the Russian torpedo boat program.

Figure 24. Early Russian torpedo boats attacking the Turkish Fleet in the Black Sea. (Private collection of Richard Worth)

The Japanese

The Japanese would be the next nation in to adopt the spar-torpedo boat. France had been successful in maintaining good relations with the Japanese since the 1860s. After 1870, the efforts of one dashing Frenchman

---

[68] Greger, 39.
[69] Greger, 39.

named Emile Bertin became the focus for the Japanese to reinvent their navy.[70] The Japanese Navy made great progress in torpedo training during 1870. The officers selected for this mission were ordered to study the Whitehead torpedo. Bertin had been a primary mover in France to change the traditional Navy to a new Navy made up of smaller faster vessels and particularly torpedo boats. He envisioned the torpedo boats working in masses of twelve or more boats per squadron and sending out three or four squadrons at a time.

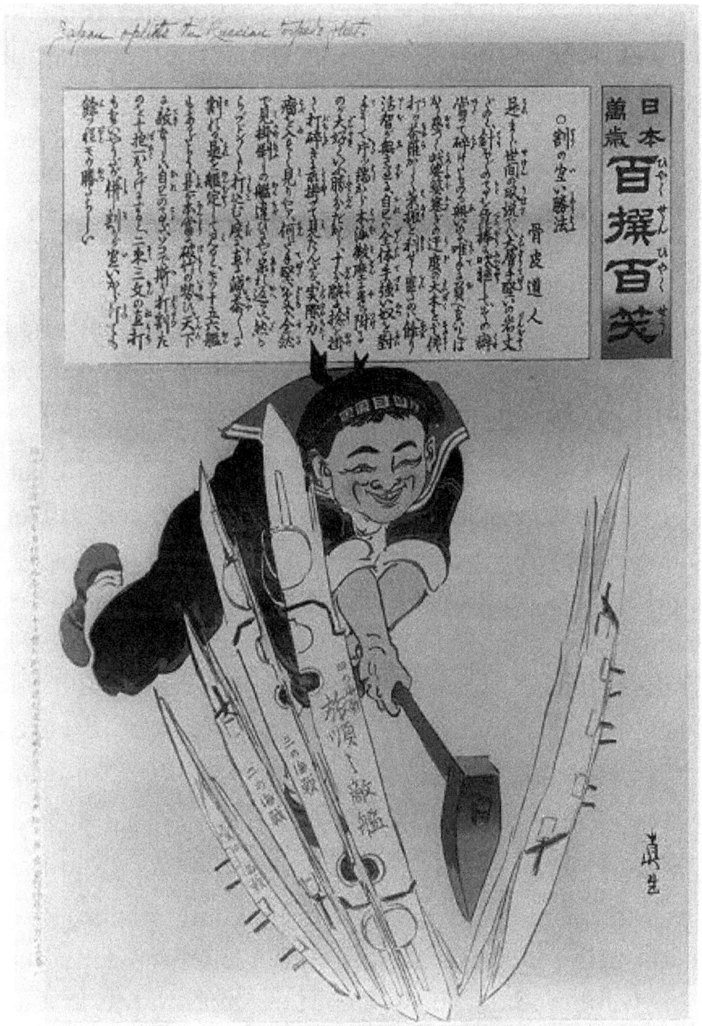

Figure 25. Russo-Japanese war art (Library of Congress)

---

[70] Evans, David C., Peattie, Mark R. *KAIGUN, Strategy, Tactics, and Technology in the Imperial Japanese Navy 1887-1941* (Naval Institute Press, 1997), 15.

The new tactics set by the *Jeune École* doctrine called for fast torpedo boats sent to break up invading fleets and cause panic and disorientation. Navies were idealized to be made up of small, fast attack craft, like light-cruisers, destroyers and ocean-going torpedo boats. With these tactics, it was the French who destroyed a Chinese cruiser during the Sino-French War of 1883-1885. Now the Japanese government, through its Diet, took notice of the lesson. They passed the first naval expansion bill in, 1882.[71] Out of this came the orders for twenty-two Motor Torpedo Boats and 26 Cruisers to be built. A full-fledged training center for torpedo boat men and officers was opened at Yokosuka in 1886. The entire training course was heavily modeled after the French Navy's schools. Two years later the first Naval Staff College would be established by Sakamoto Tosfiatsu, who was known as the "French Scholar."[72]

These efforts for modernization would make a grave difference for the Japanese in 1905 when they met the Russian Imperial Navy in the Sea of Japan. When the war opened, it was the Japanese who had the newest and most modern equipped of the combatants. This edge would mean little in the face of the weather conditions facing them.

During the first attack of the first night the crews of the torpedo boats were up against weather that was awful. As they raced full out for their targets, the crews on these boats worked on the open deck with a thin wire guard rail and a 1" toe rail on the outer edge of the boat. These were the only safety measures afforded the crew. They carried no cork life belts, jackets, or rafts. Their clothing was as protective as it could be for the era but this did not mean much in the final analysis. Soon their clothing was soaked through and the torpedoes froze in the tubes. Some men froze at their stations while others suffered frostbite.[73]

As the water broke over the open deck, crews fought to keep the torpedoes and torpedo firing pistols, from being frozen in the torpedo tubes while not being washed over the side. Visibility was hampered by the wind, freezing water and darkness. Depth perception was at a low point for the helmsman and boat commanders. In spite of the hundreds of hours in

---

[71] *Ibid.*, 15
[72] *Ibid.*, 67.
[73] Gray, 151.

intense training it would later prove the torpedoes were launched from too great a distance or too close in.[74]

A Japanese officer said in a letter to a friend, "We ought to be able to close within twenty yards of the target before she is sunk..."[75] The misplaced daring resulted in the torpedo not having enough time for the distance covered and for the safety fans to spin down and prime the warhead. Other factors that caused torpedoes not to explode were safety pins left in place and water that seeped into the warhead. In one engagement alone 150 torpedoes were fired with four hits being recorded; three struck a death blow to the Russian battleship *Sevastopol.*[76] The fourth torpedo was found on the beach. The Japanese lost two torpedo boats this night. Many boats'; crewmen were lost to Russian gun fire from their exposure on the open decks of the boats.

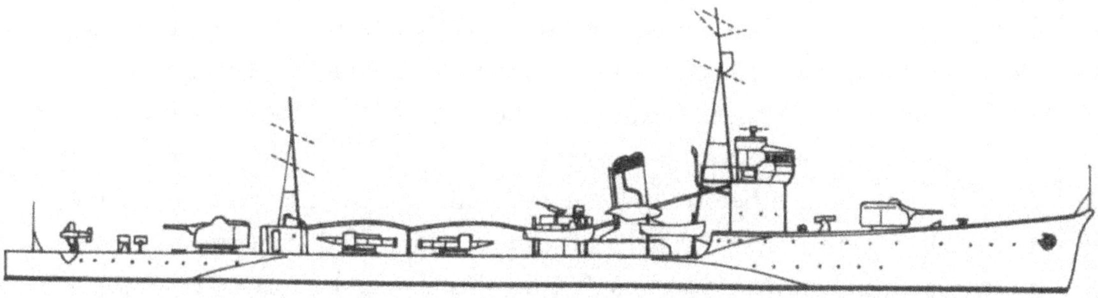

Figure 26. Japanese Torpedo Boat *Tomozuru.* (Globalsecurity.org)

The war went in the favor of the Japanese and it was the first time torpedo boats were used *en masse.* This begins to set the precedent for torpedo boat fleet tactics. Now the senior officer corps could no longer ignore the torpedo boat. The boats had earned a place in the fleet and naval strategy. Some people thought that many nations would take heed of the hard lessons learned and adopt the torpedo boat doctrines of the Japanese for themselves. This did not happen. The hard won lessons would have to be learned all over again in just a few years.

---

[74] *Ibid.,* 170
[75] *Ibid.,* 171
[76] Gray, 169.

THE TIME BETWEEN THE WARS 1914-1939

This era would see the torpedo boat go from a novelty to the beginnings of a sophisticated naval weapons system. The next phase of torpedo boat warfare saw acceptance by countries with long coastlines and the zeal to protect their interest. Only the U.S. stood out by side-stepping the torpedo boat. We had long coastlines but we had bigger seas between us and the rest of the world. The pride of being an isolationist country came back to bite us later when the Japanese attacked the Philippines. The American Navy had, like Japan, become a "big ship, big gun" oriented navy. They saw themselves as "Blue Water" navies with no interest in coastal protection capabilities.

Initially, the Washington Naval treaty seemed to be an ideal tool to reduce the possibility of a war between Japan and America. A few years later, it was evident that America and Japan shared the "big ship, big gun" doctrine and war would be inevitable and justified by either country. Their shared thinking at the time was that they would fight future naval wars with line-astern, maneuver to cross the enemy's 'T', and deliver a smashing blow to the enemy before the enemy could get to their shores. They believed in one massive battle of capital ships and a winner-take-all outcome. Both countries would suffer for this arrogant short-sidedness. By the time they played "catch-up" in World War II, it would be too late for one and costly for the other.

During the treaty era of 1923-1936 the American Navy could bring 16 battleships into combat against Japan's 10 treaty allowed battleships.[77] The Japanese favored the torpedo right behind the big gun.

The number of ships allowed along the battleship ship and cruiser limits were hard set by treaty standards. Not so for smaller auxiliary vessels. The Japanese came up with the light cruiser *Kitakami* with ten banks of three torpedo tubes each that could strike an American fleet with 200 wakeless torpedoes at a distance of 20,000 meters. The Japanese developed the type 93 torpedo (Long Lance) which was superior to the American Mark 15. The American Mark 15 could attain a speed of 45 knots

[77]Rossell, H.E., Council Member. *Types of Naval Ships* (Historical Transactions 1893-1943 Society of Naval Architects and Marine Engineers, 1945), 308.

for 15,000 yards until it reached its target, at which point it would fail to explode. The engineering flaw that permitted this was the result of years of cover ups at the highest levels of the navy. When it was time to test fire the torpedoes, it was thought to be a cost saver to load the warhead with water and not explosives. When fired, the torpedoes would hit the target but no one realized that the detonators were being crushed before the explosives could be set off. This was not discovered until early in the war when submarines and PT fired hundreds of torpedoes without a single explosion ever taking place. Many American sailors died because Japanese ships that should have gone to the bottom did not. The enemy ships just kept coming on no matter how many torpedoes were fired.

The Type 93 could make 48 knots over a distance of 40,000 yards and never leave a wake in the process.[78] The explosive charge was twice as large as an American torpedo. The Long Lance did not cripple ships, it sent ships to the bottom. Thousands of Japanese lives were saved by the effectiveness of the Long Lance torpedo.

The Germans were successful in using disguised classifications to hide the true strength of their fleet. Professor Peter Tamm perhaps summed it up best, "S-boats of the Kriegsmarine ... [were] the product of early clandestine moves by individual officers and ship yards to circumvent the Versailles Treaty. The new arm of the service, poorly prepared, now took its place in the great confrontation. As with all German units, from the onset too much was expected from too few boats. In the long run [despite] ... outstanding technical improvements....the S-boat arm was unable to hold its own against Allied defenses...sheer advantage in numbers...and continual aerial bombardment demanded too high a levy."[79]

Japan did the same by also building a huge fleet of destroyers called Torpedo Boats. Motor Torpedo Boats were never a part of the Versailles Treaty and there was no official category for a Torpedo Boat either. All navies at the time fudged on the displacement limitations with empty and fully loaded weight requirements. Today it might be called "cooking the books." By leaving off all weapons mounted and the stores of ammo for

---

[78] Evans, Peattie, *Kaigun*, 267.

[79] Frank, Hans. *German S-Boats in Action in the Second World War.* Annapolis: Naval Institute Press, 2007.

them, a ship's displacement could be within the limits of treaty. There were other tricks as well. In the final analysis, the series of naval armament treaties did nothing but fuel the fires for another war.

In 1930 the London Treaty focused on the size and number of specific vessels. This time, they confined their efforts to destroyers, cruisers, and submarines. As a result of this treaty, England needed to scrap five capital ships, America three and Japan one. At this time, Japan's navy was ever increasing in size and weapons technology.

In 1936, the navies removed quantitative restrictions. This last was an effort to allow the smaller and poorer countries an opportunity to improve their ability to protect themselves against the bigger stronger nations. "Unfortunately these were the very countries which ultimately embarked on a program of world conquest." Germany spent time and money in perfecting small fast attack craft, destroyer technology, aircraft, tanks, and artillery and infantry weapons.[80] The Germans would train their military forces, and field test weapons systems, in the Spanish Civil War. When World War II began, the S-Boats tested in Spain would find their niche in the waters of the North Sea.

During the 30's, America would sink further into the Great Depression. Until President Franklin D. Roosevelt and Congressman Carl Vinson began rebuilding the Navy in the late thirties, the country's policy of staunch "Isolationism" paralyzed the nation's military forces and crippled all efforts to stay on an even footing with other nations. The December 7th attack by Japan was fostered directly by the profound evidence of our national and military weakness. These were truly dark times for America and, as it turned out, the rest of the world.

This prewar era saw Italy, Austria, and England make more strides in the field of motor torpedo boat development than the rest of the world powers. It was an era where the scientific community began to supply support to the ships engineers and designers. As always, this era of exploration and testing spawned its share of the ridiculous and just plain dumb ideas. Nevertheless, it also gave a platform for builders and designers to reach for the hitherto unobtainable.

Starting in the 1920's the U.S. focused on scaling down to a smaller navy following the ordinances of the Washington Treaty. America did nothing to replace obsolete warships with smaller, faster, lighter, ships like light-cruiser and Torpedo Boats the size of destroyers. Instead, we illogically cried "foul play" because the smaller navies were continually building ships in complete violation the "spirit" of the treaty.[81]

The fast motor torpedo boat came into full maturity in 1935. Between 1920 and 1935, few such boats were built except for France and Italy. This kept the lead until the outbreak of World War II. England stayed with the Coastal Motor Boat (CMB) design, and rearward torpedo launching system in spite of all other countries abandoning the design.

They made no move to match the Italian and French designs until 1938. The United States did not make a significant move in its torpedo boat programs until 1939.

MOTOR TORPEDO BOAT DESIGN AND TRIALS DURING THE PREWAR ERA

Italian and French designs showed that the step hull (a sharp break in the design of the hull to allow the boat to break from the suction of the water at high speed and seemingly allow the entire craft to "step" up on the water's surface, prevalent in today's hydrofoil racing boats) had limited value and eliminated them from their designs.[82] They proved the value of V-form forward and flat bottom aft as the root of torpedo boat design. This gave the boats good freeboard forward and the bow's flare pushed the water out and kept the boats relatively dry. This feature also contributed greatly to high speed and exceptional maneuverability.

The Germans developed the round bilge design forward with a flat planning section aft. They were the first to adopt the slight step at the extreme aft section to help keep the boat level at high speeds.[83] This slight step had the effect of reducing the large size of the boat's "rooster tail." Large rooster tails coming from behind fast boats were a sign of wasted power. Underwater mufflers were another of the Germans' many construction inventions applied to small boat design.

---

[81] Rossell, 309.
[82] Rossell, 319.
[83] Fock, 110.

Figure 27. The German S-Type 100 was the pinnacle of motor torpedo boat design—the Queen of MTBs—until the British brought out the "Dog" Boats to face them down in combat.

The Germans exclusively used diesel engines. For silent running, they adopted the eight cylinder automobile engine. In many cases the Germans used MAN engines. Their experiences with many crew deaths from on-board fires for earlier motor torpedo boats had led the navy to forbid petrol engines in their S boat fleet.

Conversely, the American PT boat was a fast, a hard-hitting and an efficient fighter. Yet, it was inherently dangerous due to the use of aircraft engines that burned not just petrol, but highly volatile aviation gas. Onboard fires were not uncommon throughout WWII.

Since the invention of the torpedo boat, senior naval officers repeatedly misused the torpedo boat. Wood, and the improved science of its use, made the boats sea-worthy and solid with a long life span of heavy use. The United States considered wood flimsy and tried many steel designs. As WWII loomed, the United States finally abandoned steel and made all torpedo boats from wood. The Italians, French and later the Germans, created adhesives and methods of molded hull construction that are used to this day. The Germans created steel, aluminum and iron internal braces that gave the boats high performance and greater strength in heavy weather.

Commander John D. Bulkeley, and his experiences with his PT boat command in the Philippines, led the way for larger craft, improved performance, more seaworthy, better in rough water and added ordinance to make them more potent.

Figure 28. PT-150 displaying its early war record in the Pacific. (Frank Andruss, Sr. The Mosquito Fleet Exhibit)

The Japanese mishandled their MTB program from inception. Yet, it was the Japanese navy that first used surprise spar-torpedo boat attacks against the Chinese at Weihaiwei harbor in 1895.[84] From this point forth, the Japanese improved surprise tactics of torpedo boats until they were completely abandoned in the 1920's.[85] The IJN concern for torpedo craft went only as far as developing torpedo chasers for torpedo testing and trials.

The IJN had one brief brush with torpedo boats on August 16, 1937 when the IJN flagship, *Izumo*, an old cruiser (OAC), had been attacked by a 55' Thorneycroft built MTB-102 *Shih.* The embarrassing incident took place at the Marco Polo Bridge in the mouth of the Whangoo River, Shanghai. Once again, the Japanese learned that small fast attack craft could be dangerous to big ships. A short time later they captured to small Chinese 45' MTB's, the *Kuan Wei* and *Kuan Fu*. These effective British Power Boat Co. craft were to be used for close in attacks on the small vessels used by smugglers on the rivers. Once again, the Japanese spent hundreds of man-

---

[84] Lengerer, Hans. Motor Torpedo Boats of the Imperial Japanese Navy, Part I (*Warship International,* No.3 2007), 262.
[85] *Ibid.*, 262.

hours testing the British boats only to determine that they were only fit for torpedo chasing and not offensive tactics.

The Japanese interest in MTB warfare would raise and fall until the very last days of World War II. There never was any true make-sense focus of technological interest. While they tested numerous hull designs, they never perfected small engines for these boats. They always came back to aircraft engines that required heavy maintenance, precious aviation (and volatile) gas that could have gone to keep their air force in the air. The supreme failure of the Japanese was a complete backwardness of their technology.[86] The technical officers worked under a burden of fear and failure. Eventually the physical and mental breakdown became commonplace. There were reports of ritual suicide reported in Japanese newspapers.[87]

All Japanese M.T.B.'s were poor performers. They did not design and build their own MTB's until late 1944. Many of the few hundred boats the Japanese finally fielded could attain speeds of only 17 to 25 knots. Aside from being deathly slow, they frequently caught fire and were known to just fall apart from lousy construction materials and poor workmanship.[88] Of the 248 MTB's built, 199 were surrendered and scrapped in 1946/48. The remaining 49 were lost during the war. The Japanese wisely used most boats for harbor defense. The only serious Japanese effort in small boats was their attempt to emulate the Italian and German Motor Explosive Boats and here they met with limited small successes. They built 6,200 suicide motor boats of the naval design, "Shinyo" and 3,000 "'Maru-ni" of the army design.[89] These boats were badly built and they could never have been effective against American ships.

Figure 29. Japanese Suicide Boat "Shinyo." (U.S. Navy Archives)

---

[86] *Ibid.*, 263.
[87] *Ibid.*, 288.
[88] *Ibid.*, 288.
[89] Japanese Suicide Boats http://ptboats.or/07-0-05/museum.html.

WORLD WAR II AND THE SMALL BOATS

The motor torpedo boats of World War II saw more intense action against the enemy and nature's elements than all the battleships of the world put together. While men on a battleship might be in two or three engagements in a year, the torpedo boat men were in constant conflict on a daily basis for years. They even saw more combat than fighter pilots. At any one time between 1942 and 1945 there were over a hundred MTB's slugging it out somewhere on the waters of the globe. Torpedo boat men had their intestines scrambled, suffered broken or sprained wrist and ankles, endured constant exposure to gasoline fumes and all the while sat on top of thousands of gallons of fuel and tons of high explosives on the decks and in the holds. Torpedo boats did not gracefully sink beneath the waves when hit. They exploded into a massive fireball as bright as the sun and seldom had any survivors.

The worst part for the U.S. Navy was how ill prepared the men were who went to war on the early PT's. While our Navy ignored small boat construction, small boat training, and small boats in general, the Italians had put 20 years into their MTB's. When we finally entered the war, our men had a few weeks training while the bulk of German and Italians had had crews with five to ten year's experience. The Spanish Civil War became the weapons training ground for the German Air Force.

"U.S. PT's experienced very poor performance with torpedoes. PT crews received relatively little realistic training in torpedo attacks, little opportunity with too few torpedoes ... for training." said Norman Polmar and Samuel L. Morison, in their book,*PT Boats At War; World War II to Vietnam.*[90]

Bern Keating served for four years of World War II in the Navy as a communications officer for destroyers and as an attack officer for a special force of destroyers hunting Japanese submarines. Keating comes down hard on American journalist's propaganda efforts during WWII. He blasted such blowhards as the magazine writer who wrote that PT's carried 3"

---

[90] Polmar, Norman and Samuel Loring Morison. *PT Boats At War; World War II to Vietnam* (MBI Publishing, 1999), 36.

cannons and another writer who (many times) wrote that the PT's typically cruised at 70 miles per hour.[91]

In his book *The Mosquito Fleet,* Keating lays out the facts of the PT's real performance. He says, "Many of the PTs that fought bloody battles that follow in these pages were doing well to hit 29 or even 27 knots." PT boats that were converted to gunboats in many cases could barely maintain 25 knots and they did even better when this speed was cut. It was a hard-learned lesson that Japanese destroyers could move along all day at 35 knots.

Figure 30. The beginning of the myth of the American PT boat. (Library of Congress)

Keating debunks the myth that PT's charged full throttle to the attack. "The PT's main tactic was not the hell-roaring dash of the correspondents' romances, but a sneaky, quiet approach in darkness or fog. The PT was designed to slip slowly and quietly into an enemy formation in bad

---

[91] Keating, Bern. *The Mosquito Fleet* (Scholastic Book Services, 3rd Ed., 1968), 7- 12.

visibility, to fire torpedoes at the handiest target, and escape behind a smoke screen with which ever speed the condition of the boat permitted." He points out a few sentences later, "...it takes guts to ease along at night in an agonizingly slow approach to an enemy warship that will chew you to bloody splinters if the lookouts ever spot you." Keating quotes a PT boat sailor as saying, "Publicity has reached the point where glorified stories are not genuinely flattering. Most PT men resent the wild, fanciful tales that tend to belittle their real experience...There is actually little glamour for a PT. The excitement of battle is tempered by many dull days of inactivity, long nights of fruitless patrol, and dreary hours of foul weather at sea in a small boat."

Before they met their demise in a fiery flash, torpedo boat men had been without hot coffee, a shower, a square meal and sleep. A fourteen hour patrol was hours under the strain of constant threats. Airplanes were the worst of all enemies for the MTB's. They had no place to hide and no place to run. A firefight with an airplane was a shootout with all the odds going to the aircraft. All watercraft left a wake at night that could be seen from fifteen thousand feet for a hundred miles. All the pilot had to do was throttle back his engines, glide in behind the boat and aim his guns at the tip of the white-water V's wake and open fire. The Japanese pilots sank many American torpedo boats by this method.

After 300 years, all the all efforts of the inventors, engineers, dreamers, nut cases, and wild-eyed hysterics, the culmination of the power of the torpedo and small fast boat was finally realized. Historically, the high point of the motor torpedo boat was 1944. This was the year that saw torpedo boats attacking and sinking major naval combatants in the South China Sea, the Sea of Japan, the Philippine Sea, the Baltic Sea, the English Channel, the North Sea, the Mediterranean Sea, the South Pacific Ocean, the North Pacific Ocean, the Coral Sea, the Adriatic Sea, the Ionian Sea, the Black Sea, the Barents Sea, the Norwegian Sea and the Bering Sea. When they were not sinking big ships, they fought each other to deadly draws or fought winner-take all actions. They sowed minefields, attacked shore installations, harbor, bays and rivers. They were like fleas on the world. On occasion, they changed the course of the war.

The German *Kriegsmarine* and its S-boats almost stopped the invasion of Europe by its deadly attack on American LST's in Lyme Bay. PT boats on three different occasions drove off whole fleets of heavy Japanese cruisers, destroyers, and transport ships in the South Pacific. When they weren't sinking ships, they sank hundreds of Japanese barges loaded with thousands of troops and tons of supplies in the Philippines, Coral and Solomon Seas.

The Soviet G-5 torpedo cutters harassed Germany shipping in the Baltic, Gulf of Finland, and Barents Sea. The Italians damaged Allied shipping and disrupted one invasion effort after another in the Adriatic and Mediterranean Seas off the coast of Africa. The English MTB had fought the Germans from 1939 to 1945 in the North Sea and English Channel, while fighting the Italians in the Mediterranean. Even remote, no-where Alaska saw constant patrols of PT boats doing their best not to be pounded by heavy seas or sunk by submerged sharp rocks that were everywhere. The men of the small boats won their particular place in naval history. These men took on the biggest ships with the smallest boats and often won.

Figure 31. Soviet G-5. One of the most delicate and fastest torpedo boats in WWII. A truly fantastic MTB of WWII . Made of Duralumin, which caused untold problems, yet the boat achieved remarkable success in the cold seas of the Baltic Sea, North Sea and the Norwegian Sea. (U.S. Navy Archives)

SMALL BOATS GET NOTICED

The legacy of the MTBs is still alive today. Sixty-two years after World War II small boats saw extensive action in Vietnam and now in the Middle East where they are needed again. On January 6, 2008, five Iranian speedboats made as if they were threatening three U. S. Navy warships in the straits of Hormuz. Current navy ships are not capable of fighting off small fast attack craft. To quote Steven Noonan (employee of L-3 Communications, assigned to the Navy Program Executive Officer for Integrated Warfare Systems) from his article, "Beating the Fast Attack

Craft Threat," *Proceedings*, February, 2008, "These small boat threats, known as fast attack craft or fast inshore fast attack craft, have been a concern to the U.S. Navy for a number of years."[92] To sum up his article, the U. S. Navy has once again been caught unawares of small boat threats.

Another article appeared in *Proceedings* magazine in 2004. This article was even more direct when Lieutenant (Junior Grade) Jonathan F. Solomon, U.S. Nay said in his article, "Lethal in the Littoral:-a Smaller, Meaner LCS":[93] "... [s]mall boats pose an unconventional threat when used for harassment, attack by light standoff weapons such as rocket-propelled grenades, or the delivery of explosive devices [that can] saturate and overwhelm a force's defenses... " . Lieutenant Michael Farmer, U.S. Navy, in his article, "We Must Defeat the Small Boat Threat," says, "...[I]n areas we have overlooked ... the Navy will find itself outgunned and outmaneuvered by small, fast, and highly maneuverable opponents in the littoral environment."[94]

The Blue Water Navy had its day in the spotlight. Now the destiny of the small boat is at hand. We are here to witness this marvelous turn of event. It is this era that the small boats are going to be put to the task of protection, pursuit, and combat. It will take place in harbors, rivers, bays and shallow waters everywhere near industrialized nations.

---

[92] Noonen, Steven. Beating the Fast Attack Craft (*Proceedings*, 2008),78-79.
[93] Solomon, Jonathan F., Lt(JG) *U.S.N Lethal in the Littoral—A Smaller, Meaner LCS* (*Proceedings*, January 2004), 36.
[94] Farmer, Michael, LT U.S. Navy. "We Must Defeat the Small-Boat Threat." (*Proceedings*, January,2004), 81.

Figure 32. All U.S.Navy surface combatants have greatly increased in size since the Second World War, and the Littoral Combat Ship *USS Freedom* (LCS-1) reflects this trend. The global threat from small boats is considered sufficiently severe by the world's unchallenged naval leader that *Freedom* is designed essentially as a torpedo boat hunter. (U.S. Navy)

The torpedo designed by Syrian engineer Hassan al-Rammah in 1280 AD (Ahmad Y. al-Hassan)

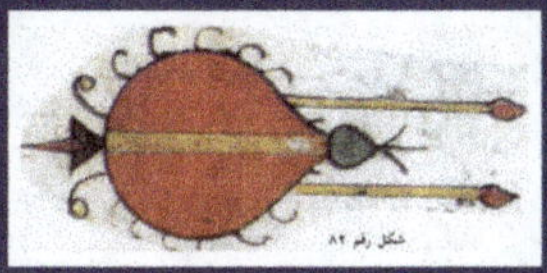

THE SHIP KILLERS: TIMELINE 1280–1899

THE SHIP KILLERS

A DEFINITIVE HISTORY OF THE TORPEDO BOAT

CSS David: the first purpose-built spar torpedo boat (illustration by author)

VOLUME 2: TIMELINE 1280–1899

BY JOE HINDS

ISBN 978-1-934840-60-3

90000

9 781934 840603

NIMBLE BOOKS LLC

8.00 x 10.00

0.281

8.00 x 10.00

CONTENTS

FIGURES

DEDICATION

Joe Judge, Curator Hampton Roads Naval Museum
Joe has spent the best part of nine years proofing my work,. He has helped me find research resources, mounted a full blown art show for my work and spending endless hours, at home, correcting my infantile spelling and grammar errors. Joe has always had words of encouragement when I hit low ebb. You, the reader, would not be reading this but for Joe Judge.

A GLOBAL TIMELINE OF THE TORPEDO BOAT

The title says it all. This is, I believe, the first truly global view of the history of the torpedo boat. This is a fascinating piece of naval history that, in my view, has never been examined in a a systematic, comprehensive, global and detailed fashion. There are hundreds of books on MTB's, PT's, MAS boast, and S boats. The cover usually carries a speeding something flashing across the cover leaving a huge wake and photos of valiant young men with death grips on machine guns. That describes about 95% of the small war-boat books out there. I have looked on these with great interest but I also put down each book with the thoughts, "Why were they important? What did they contribute to victory or defeat? How did the design evolve?" And I had hundreds of other unanswerable questions.

I set out in 1999 to produce a set of illustrations depiction the overall evolution of the torpedo boat. I had to dig in the most obscure documents to find the origin of designs, the story's of successes and failures and the most difficult part-the colors of the boats. I purchased sixty sets of torpedo boat plans and four came with color information. My illustration of the Type S-100 boat took two years to nail down the exact color mix that the i used. I spent another two years hunting down information of the CSS *David*, the first military model of a spar-torpedo boat.

In the beginning, I stored research material in 3" thick 3-ring binders and in the first year, I had filled three. I now have twenty-five binders and enough material for another ten 3-ring binders. I had to stop buying books after my first one-hundred and went to the best source in the world, the Interlibrary Loan Service from our state library. I have scanned a good part of over 300 books and I still run across information I have never seen before.

The idea for displaying the sequential history of the torpedo bat came from a man named Lieutenant G.E. Armstrong, RN, 1896, in his book *Torpedoes and Torpedo-Vessels*. In the back of Armstrong's book is a tabular listing on events in the development of the, torpedo boats, mines, and underwater explosive devices, up to his time. It was through this style of tracking information that I learned of a torpedo vessel I had never heard of before, John Ericsson's, *"Destroyer."* It turns out that there are a great many MTB enthusiasts who had also never heard of it. What became clear from Armstrong's work was the importance of these boats in naval matters of his time. The design evolution became quite clear. The shift from spar-torpedo to Whitehead's "Fish" torpedo was evident.

The book has been separated into ten volumes. As the reader progresses through the timelines in this and the following volumes, he can see the obvious differences between those navies who maximized the Motor Torpedo Boat and those who neglected it. It becomes glaringly clear that as the war advanced, the need for the motor torpedo boat grew ever more important. This fact becomes more evident in 1944 when the world wide use of the torpedo boat was on a par with air forces and armies of tanks.

This is not a book of endless lists, I was never after collecting massive amounts of data to list every boat ever built. I was not after yet another set of lists that showed only torpedo boats sunk. There are over 1,000 small stories of individual actions, fleet actions, accidents and events of stupendous valor performed by torpedo boat men. Many of these events are only eight words long and some are maybe thirty words long, yet I hope that none of them is without interest. The collective work shows the full scope of this missing piece of naval history.

This whole story loops back to relevance in our world today. In 2000, the destroyer *USS Cole* was almost sunk by a small boat manned by desperate men. In 2007, Iranian navy sent out small speedboats to taunt the British navy in the Gulf of Oman and embarrassed a world power to near tears. In 2009, navies of every nation are being embarrassed by pirates in the Gulf of Aden. Once again, men in small boats are proving the limited power of "Big" ships with "Big" guns in shallow waters. Funny how the lessons of history are constantly being taught.

TIMELINE 1280–1899

Date	Craft, Name, Event or Country	Event, Builder and Boat Specifications	City, River, Lake, Sea, Ocean
1280	al-Rammah Water-Borne Surface Torpedo	The works of Syrian military engineer Hasan Al-Rammah, called "Najm-al-Din" (Star of the Faith), have been well known for hundreds of years, but it is only recently that his efforts and achievements have been fully understood in the West. Al-Rammah explained and perfected the recipes for several types of explosive gunpowder. He was very adept at explaining the fundamentals of high-grade gunpowder by purification processes for potassium nitrate, a key ingredient in high grade gunpowder.[95] [96] Several scholars have reached the conclusion that one of the devices al-Rammah designed was intended to be a torpedo.[97] [98] The significance of this new information about al-Rammah is that it moves back the origin of torpedoes and naval explosive devices by more than three hundred years to a different continent and a different civilization. Now the record is corrected and history can be formally amended.	Syria during the Mamluk dynasty

---

[9595] Ahmad Y. Al-Hassan and Donald R. Hill, *Islamic Technology: An Illustrated History* (Cambridge: Cambridge University Press and UNESCO, 1986, paperback, 1992) and Al-Hassan, *History of Science and Technology in Islam,* http://ww.history-science-technology.com.

[96] Professor Mohammed Mansour, Foundation for Science Technology and Civilization, U.K.

[97] J.R. Partington, *History of Greek Fire and Gunpowder,* (Baltimore: The John Hopkins University Press, 1999) "... has been supposed to be a torpedo."

[98] Willy Ley, *Rockets, Missiles and Space Travel* (New York: The Viking Press, 1958): "But Hasan Al-Rammah added one unsuspected novelty: a rocket-propelled torpedo..."

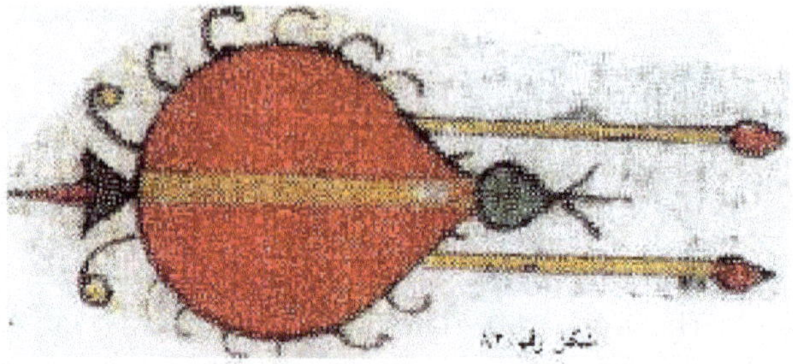

Figure 33.al-Rammah torpedo. The pear-shaped vessel , made from sheet-iron, was filled with gunpowder and incendiary materials. One or more rockets were used for propelling it towards its target. The long rods acted like rudders and

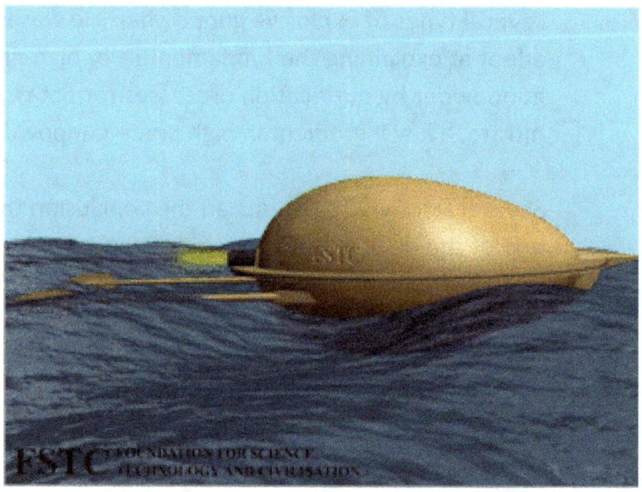

Figure 34. A conceptual model of the floating torpedo describved by Hasan al-Rammah, created by FSTC Ltd. (Foundation for Science and Technology, Ltd.)

Date	Craft, Name, Event or Country	Event, Builder and Boat Specifications	City, River, Lake, Sea, Ocean
1585[99]	Siege of Antwerp "Eighty Years War"	Small boat warfare among European navies began here, with Italian engineer Frederico Gianibelli at the siege of Antwerp. He had to remove enemy forces from a bridge over the river Scheldt. His two small skiffs, loaded with many barrels of gunpowder and covered with marble gravestones, were sent drifting down the river as with a clockwork mechanism for the fuse ticking away. At the bridge, the explosive force of the boats sent marble shrapnel flying against the 2400-foot-long bridge killed 800 soldiers of the Duke of Parma, Alessandro Farnese, and blew a hole 200 feet wide through the barricade on the bridge.	River Scheldt
1624[100]	1620-1624 Cornelius Van Drebbel Submarine Boat	Van Drebbel used designs from William Bourne, 1578, to make the earliest known European navigable submarine boat. Van Drebbel performed several successful dives at depths of 12 to 15 feet. He made three test vessels, the last capable of carrying 16 people and powered by six oars. This model was demonstrated to King James I as well as several thousand onlookers. The submarine stayed submerged for three hours and traveled between Westminster and Greenwich.	Thames River, England
1627[101]	Van Drebbel's Floating Petard	King James commissioned Van Drebbel to set up shop and manufacture "Floating Petards" for use by the British Navy to break the French blockade of La Rochelle. They were unsuccessful and were determined to be a greater danger to the British than to the French.	La Rochelle, France
1730[102]	John T. Desaguliers Underwater Rocket Torpedo	The inventor made several successful tests sinking small boats on with underwater explosive rockets.	Plymouth, England

[99] Sleeman, 330.
[100] *Ibid.*
[101] *Ibid.*
[102] *Ibid.*

Figure 35. The Eighty Years War. The Liberation of Antwerp. (Museum of Antwerp).

Figure 36. The emblem of the Dutch "Sea Beggars." The Dutch fought against Spanish occupation from 1550 to 1583. The Sea Beggars legend still lives on in Dutch folk lore as privateers, pirates, patriots and thieves. (Museum of Antwerp)

Figure 37. This scanned postcard shows a popular and mythical representation of Drebbel's diving boat.

Figure 38. BBC sponsored a team to build a reproduction of the real Drebbel craft after years of careful research. The full size operating model is on display at Heron Square, Richmond on the Thames. ("Gunner54" on Wordpress.com, used with permission.)

Date	Craft, Name, Event or Country	Event, Builder and Boat Specifications	City, River, Lake, Sea, Ocean
1774[103]	Mr. Day Diving Boat	An eager inventor who made two test runs with his "Diving Boat." On the second run, the boat and inventor stayed on the bottom.	Plymouth, England
1775[104]	Captain David Bushnell	Leaning on the history of the efforts of European inventors, David Bushnell made several successful experiments exploding charges of gunpowder underwater.	New York Harbor
1776[105]	Sgt. Ezra Lee *Turtle*	The American diving boat *Turtle,* driven by Sergeant Lee, attempted to attack and sink the British ship *HMS Eagle* in New York harbor. The unwieldy craft and lack of oxygen almost killed the sergeant.	New York Harbor
1777[106]	Capt. D. Bushnell *Turtle*	Bushnell went after the prize vessel *HMS Cerberus* with his newly created "Drifting Torpedoes." The torpedo was caught by crewmembers aboard a prize schooner astern the *Cerberus*. Here, the mine exploded killing three crewmen and destroying a small boat onboard.	New London
1778[107]	January 6 Caleb Carman 'Battle of the Kegs'	The "Drifting Mines" made by Caleb Carman, were truly effective devices. Several barrels of gunpowder were fitted with a flintlock-firing device sensitive enough to set off the mine once it touched a ship. The British fleet anchored in the middle of the Delaware River and moved during the night into Philadelphia harbor to avoid being trapped in ice. One of the mines struck a small bridge and exploded killing four sailors and wounding several more.	Philadelphia Delaware River

---

[103] *Ibid.*

[104] *Ibid.*

[105] *Ibid.*

[106] *Ibid.*

[107] Sleeman, 330, and Battle of the Kegs <http://www.carman.net/battleofkegs.htm> and Evarts and Peck, *History of Burlington County,* 463.

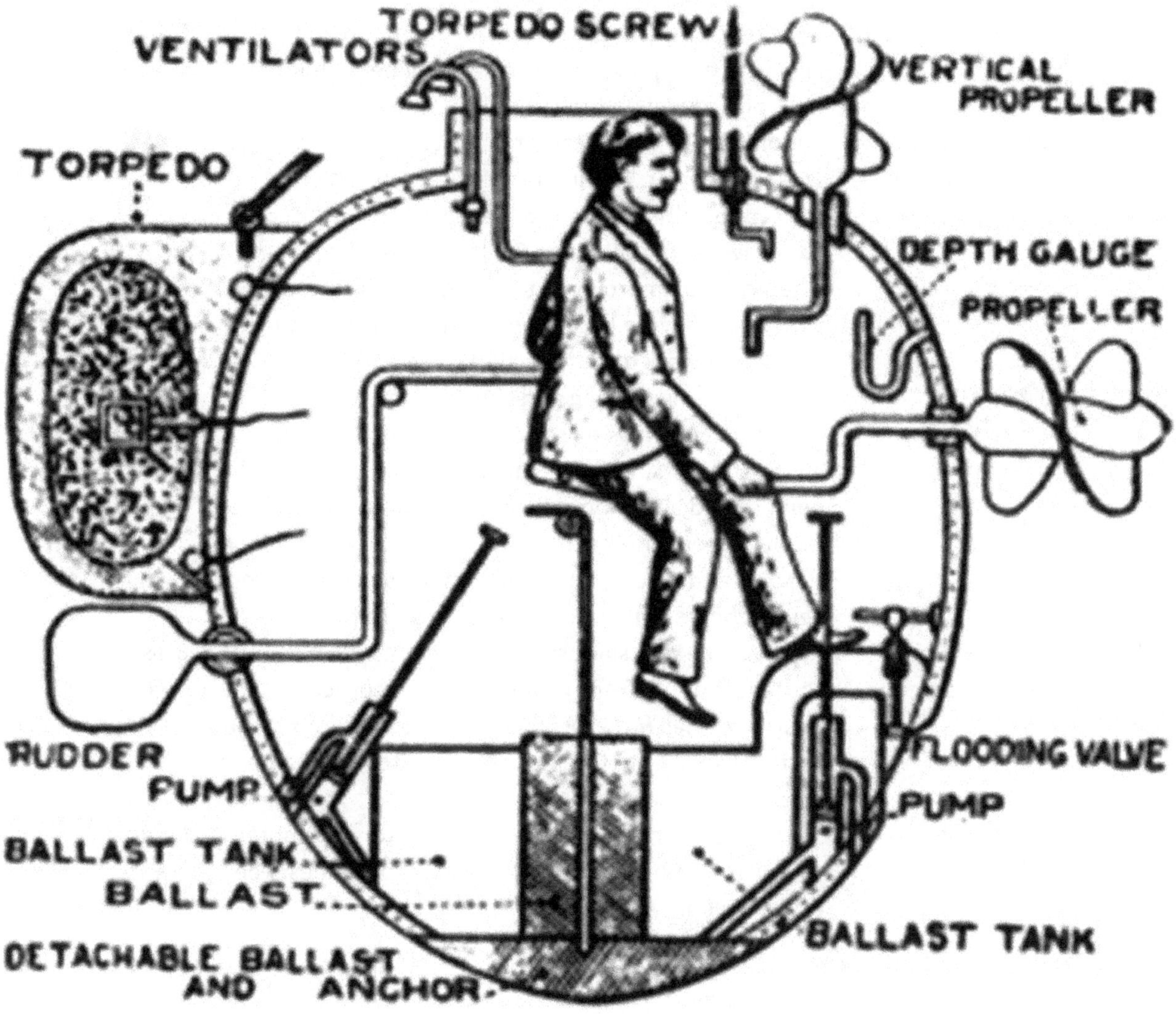

Figure 39. The best known illustration of the *Turtle*. Officially, America's first torpedo boat. (Library Of Congress Prints and Photographs Division)

Date	Craft, Name, Event or Country	Event, Builder and Boat Specifications	City, River, Lake, Sea, Ocean
1797[108]	Révéroni Saint Cyr Submarine Cannon	The French inventor proposed a catamaran hulled ship with a 48-pound carronade slung between the hulls and towed underwater.	France
1797[109]	Robert Fulton Torpedoes	Fulton turned from a career in art to science and weapons of war. His first fascination was with the concept of submarines and torpedoes. He had access to the works of both Saint Cyr and Drebbel and attempted to improve on their designs. Early torpedo experiments were largely monetary failures, but laid the groundwork for future trials.	France
1800[110]	July 24 Robert Fulton 'Nautilus'	His careful engineering and construction of the submarine *Nautilus* enabled Fulton to have many successful tests. This was the first known use of compressed air for breathing underwater. He could easily maneuver his submarine, maintained controlled descent and ascent, and could remain submerged for several hours with a "Snorkel"-type system.	Brest, Seine River, France
1801[111]	Robert Fulton Drifting Torpedo	This was the first successful experiment with a drifting torpedo. Fulton's efforts at exploding a torpedo against a small (40') vessel were successful. This is the first known event of a ship sunk by a "torpedo" from a distance (628 feet).	Brest France
1801[112]	Robert Fulton Drifting Torpedo	With the British and French at war, Fulton attempted to sink a British war ship with drifting torpedoes. The attack failed.	Boulogne France

[108] Sleeman, 330.
[109] *Ibid.*
[110] *Ibid.*
[111] Sleeman, 331.
[112] *Ibid.*

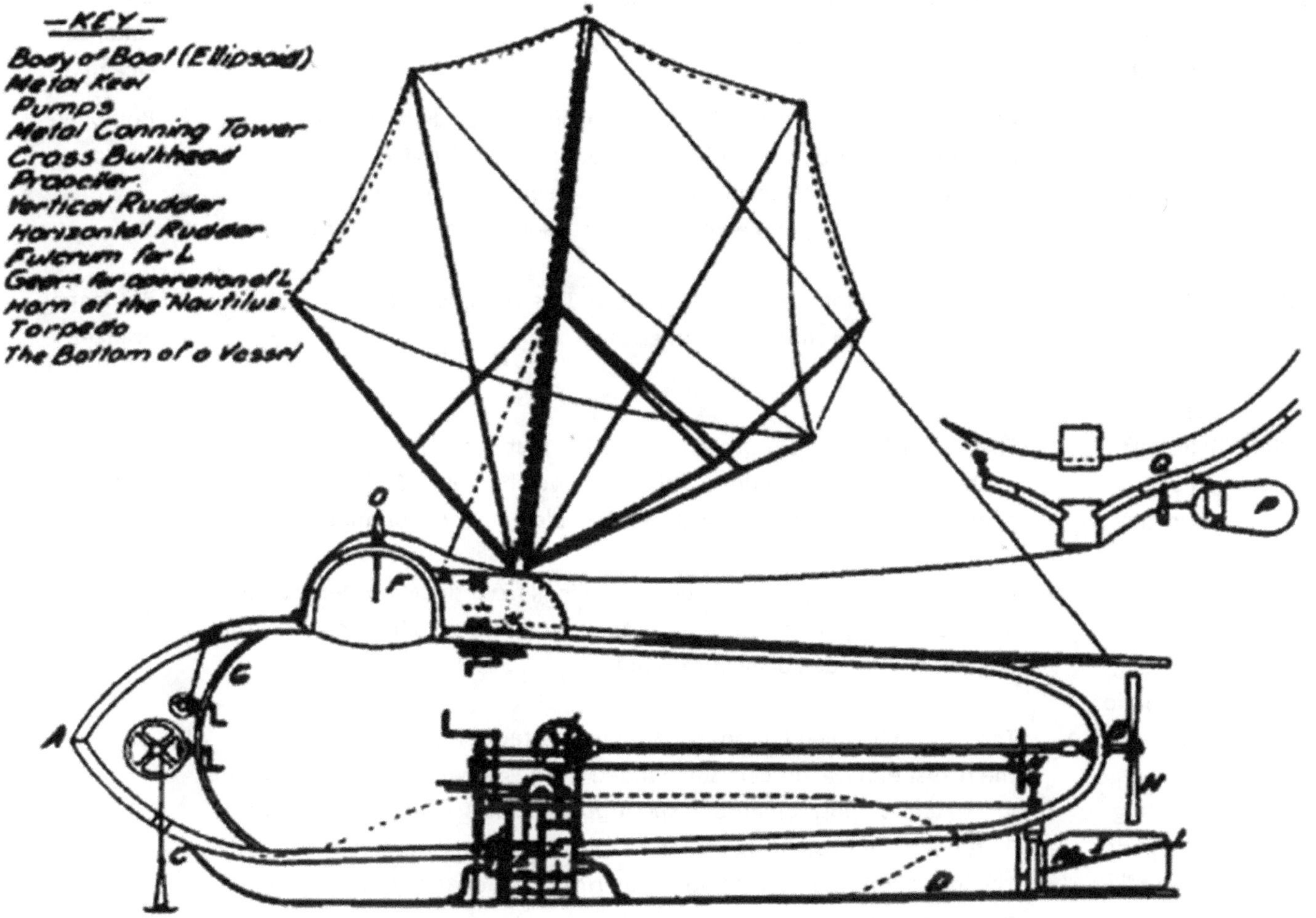

Figure 40. A fully operational submarine. Fulton carried canisters of oxygen that enabled him to stay successfully submerged for hours. (Library Of Congress Prints and Photographs Division)

Date	Craft, Name, Event or Country	Event, Builder and Boat Specifications	City, River, Lake, Sea, Ocean
1804[113]	July 20 Robert Fulton Drifting Torpedo	After a falling out with the French naval authorities, Fulton sold his plans for "Underwater Bombs and other Aquatic devices" to the British Admiralty. His attack against the French fleet failed due to faulty construction. A torpedo exploded and caused no damage to any ship.	Boulogne France
1805[114]	Robert Fulton Drifting Torpedo	A second attempt that ended in failure. This was in part due to faulty construction and shoddy workmanship.	Boulogne France
1805[115]	Robert Fulton Drifting Torpedo	The old English brig *Dorothea* became the test vessel for further Fulton experiments with torpedoes. This time, two charges of 180 pounds of gunpowder with a clockwork-timed fuse blew up most dramatically.	Dover England
1807[116]	Robert Fulton Drifting Torpedo	Back in America, Fulton secured a large hulk from the Navy Department for a test of his torpedoes. He needed to several attempts before any of the torpedoes exploded against the ship.	New York
1810[117]	Robert Fulton Drifting Torpedo	The Navy scheduled Fulton to attempt an attack on the sloop *USS Argus*. At the test site, naval officers surrounded the vessel with netting and booms in a defensive position around the *Argus.* The attack failed as expected. Fulton was furious.	New York
1812[118]	Mr. Mix Drifting Torpedo	The inventor attempted an attack on *HMS Plantagenet* with several drifting torpedoes. All six attempts failed.	Lyn Haven Bay U.S.A.
1813[119]	Mr. Mix Drifting Torpedo	Mr. Mix next made a similar attack on *the HMS Ramilles* at harbor and failed. His drifting torpedo did destroy a schooner sitting alongside the target ship.	New York Harbor U.S.A.

[113] *Ibid.*
[114] *Ibid.*
[115] *Ibid.*
[116] *Ibid.*
[117] *Ibid.*
[118] *Ibid.*

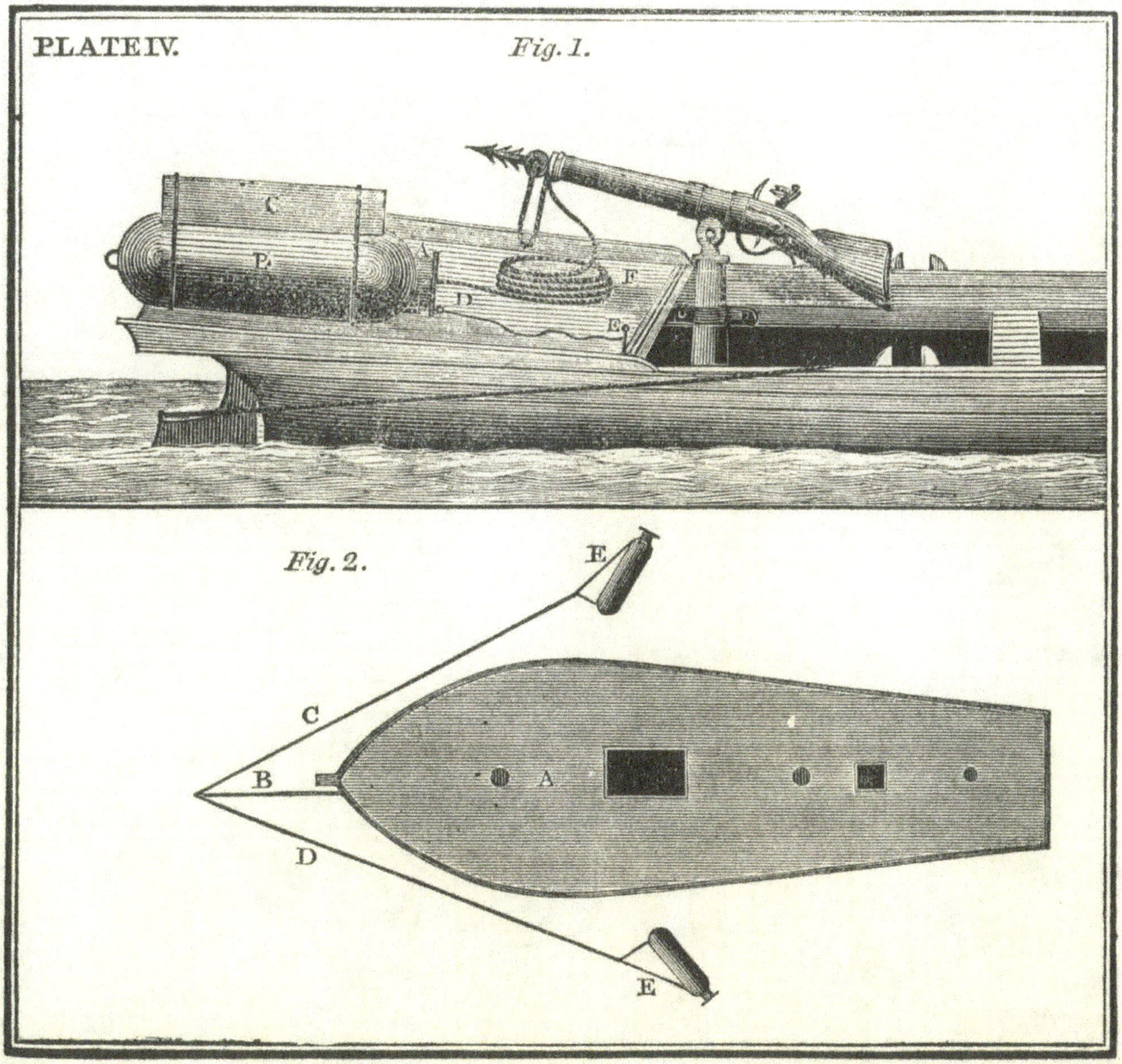

Figure 41. Fulton designs a simple and workable delivery system for his "Torpedoes." Once the harpoon struck an enemy ship the torpedo would be attached to the ships hull and the torpedo would be carried into the hull sinking the ship. (Library of Congress Prints and Photographs Division)

---

119 *Ibid.*

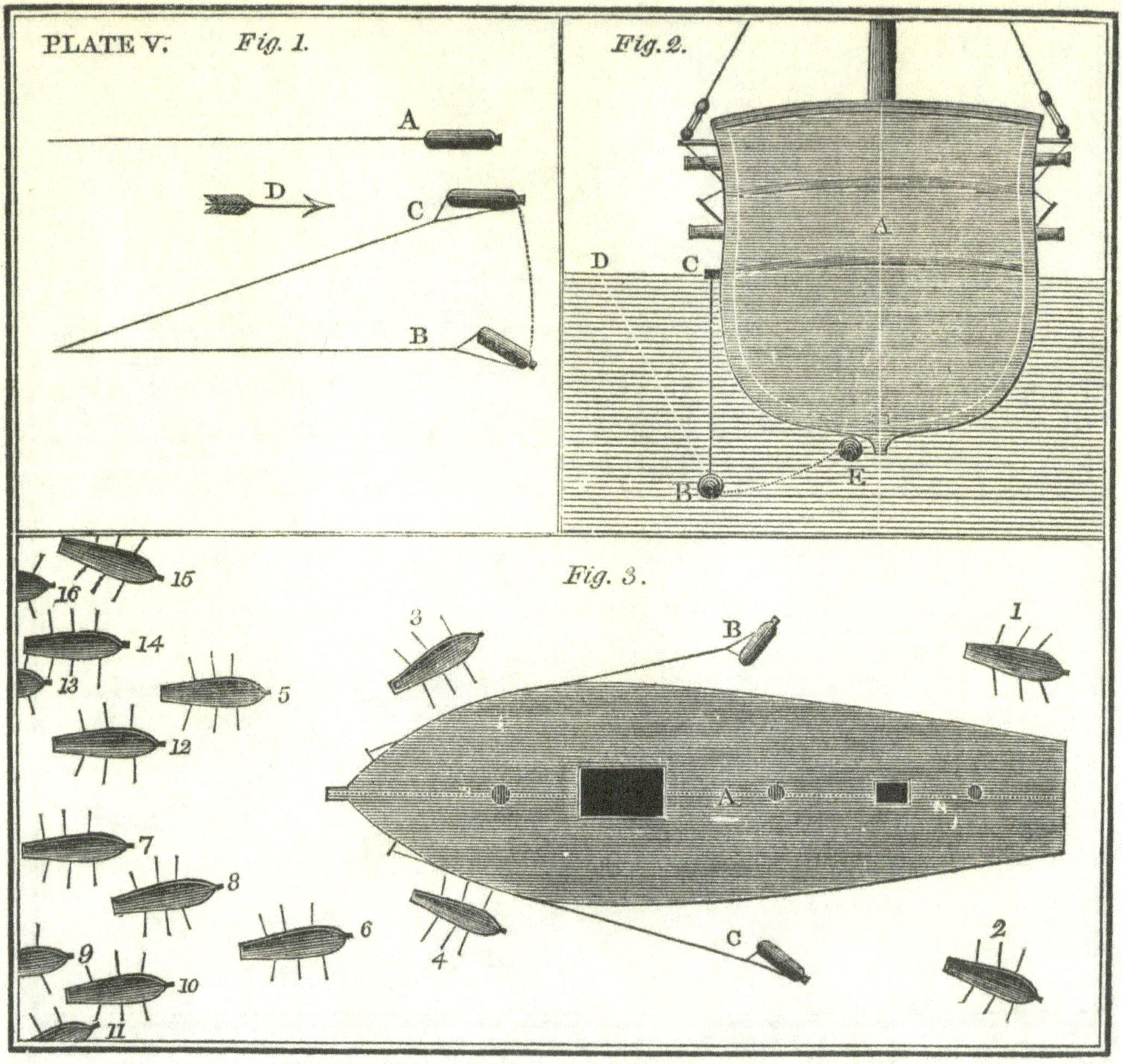

Figure 42. This shows how a fleet of small boats could attack and sink enemy ships in, rivers, harbors or estuaries. (Library of Congress Prints and Photographs Division)

Date	Craft, Name, Event or Country	Event, Builder and Boat Specifications	City, River, Lake, Sea, Ocean
1814[120]	Robert Fulton Submarine Gun	In New York harbor a replica of a ship's bulkhead attached to a raft became Fulton's target. Fulton's gun fired a shot that penetrated the test piece, which had the same thickness as a first-class ship of the line. He fired from between 12 and 15 feet of the target.	New York Harbor U.S.A.
1820[121]	Captain Johnson Submarine	Johnson's submarine carried its torpedo on its back. At the target ship, crewmembers would attach the torpedo to the bottom of the ship with screws. The trial was reported as successful but the Admiralty refused to sanction the vessel or torpedo. They declared it as too "diabolical."	Moulsford, Berks England
1826[122]	John Ericsson Sub-Aquatic Cannon	John Ericsson began to sketch ideas for a turreted war ship with his "Sub-Aquatic Cannons." He laid out the principles of a gun operated by a hydraulic and compressed air system. These ideas later became the *USS Monitor* and the torpedo ship *Destroyer.*	Sweden
1829[123]	July 4 Col. Samuel Colt Submarine Battery	Before Sam Colt was a known as a gun maker, he experimented with batteries, insulated copper wire, and underwater explosives. His torpedo (mine) destroyed a large raft on Ware Pond.	Ware Pond U.S.A.
1839[124]	Gen. Paisley, R. E. Submarine Torpedo	The general used a galvanic-firing device on his test against the HMS *Royal George.*[125] His test was successful.	Portsmouth, England

120 *Ibid.*

121 *Ibid.*

122 Conant, W. Church. "John Ericsson, The Engineer." *Scribner's Magazine,* VII (January-June 1890).

123 Sleeman, 331, and Grant, Ellsworth S., *The Colt Legacy: The Colt Armory in Hartford 1885-1980* (Providence, Rhode Island: Mowbray Company, 1982), 1-6. The latter was first published as "Gun Maker to the World," *American Heritage,* XIX (June 1968), 86-91.

124 Sleeman, 331.

125 Barnes, John Sanford. *Submarine Warfare, Offensive and Defensive: Including a Discussion of the Offensive Torpedo System, Its Effects Upon Iron-clad Ship Systems, and Influence Upon Future Naval Wars.* New York: Van Nostrand, 1969, 91.

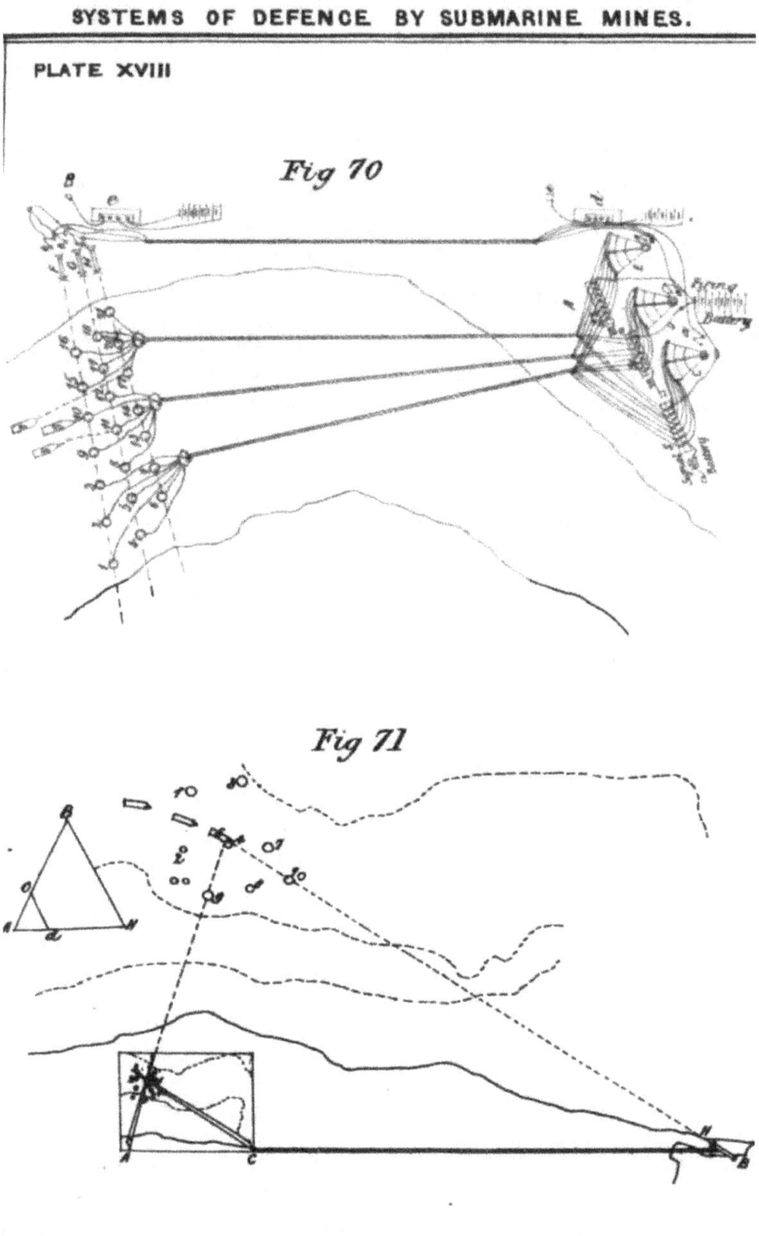

Figure 43. This was Colt's plan for a minefield with two operators who could locate the exact spot when a ship was over a "Torpedo." His mine fields were to contain 16 to 29 mines. (U.S. Government Printing Office)

Date	Craft, Name, Event or Country	Event, Builder and Boat Specifications	City, River, Lake, Sea, Ocean
1842[126]	Col. Samuel Colt Electric Submarine Torpedo	Colt had made several improvements of insulated copper wire, wet-cell batteries, and triggering mechanisms. He sank the target ship five miles away.	New York, U.S.A.
1842[127]	July 4 Col. Samuel Colt Electric Submarine Torpedo	This time his mine operator was on board another U.S. Navy man-of-war a great distance from the target ship when he sank it.	Castle Gardens, New York
1842[128]	August 20 Col. Samuel Colt Electric Submarine Torpedo	Similar experiment of the submarine electric mine. The operator was five miles away when he exploded the mine sinking the ship.	Potomac River U.S.A.
1843[129]	April 20 Col. Samuel Colt Electric Submarine Torpedo	In this test, the target vessel was a 500-ton ship moving at five knots at a distance of five miles. The crew on the trial ship left just before the time for the torpedo exploded.	New York, U.S.A.

Figure 44. Samuel Colt was a leader in the industry of "Electric-Torpedo Warfare." (Library of Congress).

---

[126] Sleeman, 332, and Grant, 1–6.
[127] Sleeman, 332, and Grant, 1–6.
[128] *Ibid.*
[129] *Ibid.*

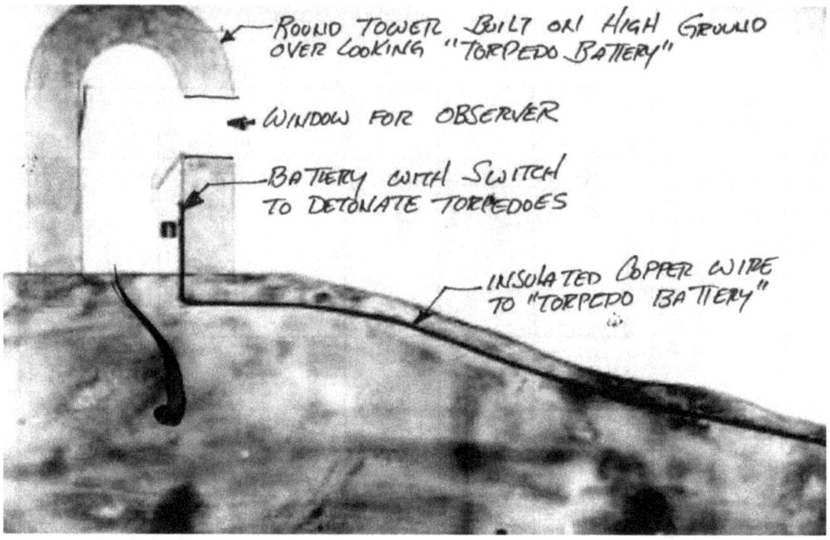

Figure 45. This "Torpedo Tower" was the heart of Colt's submarine mine field. He planned for 100' towers to be built, each with trigger mechanisms to operate the galvanic batteries. The operator was expected to sit in a cold tower for months on end, waiting foran enemy ship to park itself over the "Torpedo Barrage Field." Original art by Samuel Colt, annotations by the author. (U.S.Government Printing Office)

Figure 46.*The Last Experiment of Mr. Colt's Submarine Battery.* This painting was done by A. Gilbert. It is doubtful that Gilbert attended the demonstration. He was supplied sketches declared by Colt to be factually "correct." (U.S. Government Printing Office)

Date	Craft, Name, Event or Country	Event, Builder and Boat Specifications	City, River, Lake, Sea, Ocean
1844[130]	Capt. Warner "Invisible Shell"	On July 20, Captain Warner destroyed the ship *John o'Gaunt* with an underwater-device. He marketed his invention as an "invisible shell," but a subsequent Parliamentary investigation revealed that "the destruction of the *John o'Gaunt* was a trick of the same class as the blowing up of the punt on the fish-pond, which consisted merely of shells sunk and anchored under the water, and a long rope attached to the punt, which at a signal given was drawn by a team of horses, and which on striking the composition blew up the vessel. The destruction of the *John o'Gaunt* was just the same, except that a steamer was employed to drag the vessel to unavoidable destruction instead of a team of horses."[131][132]	Brighton, England
1845[133]	January 1 Col. Samuel Colt Electric Submarine Torpedo	The last and best testing of the "Electric Submarine Torpedo" against a target ship 40 miles distance from the operator.	New York, U.S.A.
1846[134]	Professor Schonbein Gun-Cotton	The professor was the inventor of an explosive that would have a profound effect on the world. By 1863, Professor Abel had brought this form of explosive up to standards for military use.	New York, U.S.A.
1846[135]	Sobrcro Nitro-Glycerin	This discovery was intended for blasting purposes. It would become a deadly agent in warfare in the hands of M. Alfred Nobel in 1863.	New York, U.S.A.

---

130 Sleeman, 332.

131 *HC Deb 25 June 1847 vol 93 cc921-46* http://hansard.millbanksystems.com/commons /1847/jun/25/captain-warners-inventions#S3V0093P0-01185

132 See also Burke, Edmund (ed.) *Annual Register, or a view of the History and Politics of the Year 1844* (London: Longwood and Company, 1845) for a vivid description of the mysterious explosion.

133 Sleeman, 332, and Grant, 1–6.

134 Sleeman, 332.

135 *Ibid.*

Date	Craft, Name, Event or Country	Event, Builder and Boat Specifications	City, River, Lake, Sea, Ocean
1854[136]	Russian Navy Stationary Submarine Mine	This is the first military use of stationary submarine "mines" (torpedoes). The Russians were able to place several mines near the English men-of-war *HMS Merlin* and *HMS Firefly.* The only effect noticed after detonation was the crew was wet.	Kronsdtadt Harbor St. Petersburg
1856[137]	John Ericcson Sub-Aquatic Torpedo Gun	A young Ericcson spent years working out the theories for the armored battle ship and underwater artillery. He approached the Emperor Napoleon III with designs for an armored, turreted battleship and his sub-aquatic "Torpedo Gun". Ericcson and his concepts impressed the Emperor, but the Emperor never contacted Ericcson again.	France
1860[138]	U.S.Navy Brutus DeVilleroi *USS Alligator*	DeVilleroi began building working submarines in the 1830's.His success in France prompted him to move to America. He built a test model, 35' long, for demonstration to the U.S. Navy. The design was accepted and he immediately began work on a full size submarine.	Delaware River Philadelphia

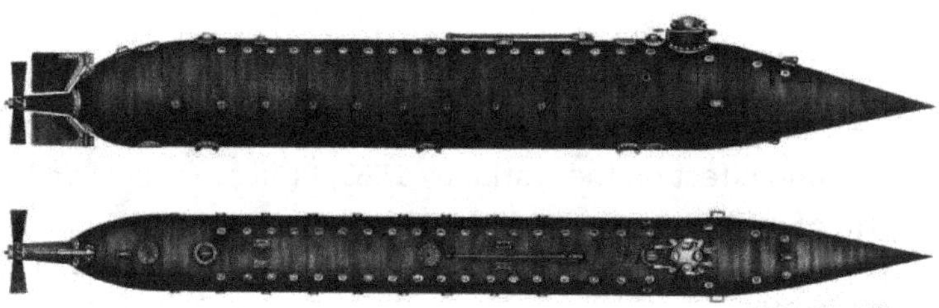

Figure 47. The *Alligator* was never in the category of a "Diving Boat" or "Semi-Submersible." It was a true submarine in that the crew could breathe and work with clean oxygen generated by chemicals carried on board the vessel. (Illustration by author).

---

[136] *Ibid.*

[137] Banard, Charles. "Ericsson's *Destroyer* and Her New Gun." *Scribner's Monthly Magazine,* 21 (March 1881, No. 5), 689-693.

[138] Maloney, Janet M. "Portrait of Possibility: The Submarine Alligator." *Civil War Times,* XLIV (No. 5, December, 2005), 34-40 and Bolander, L.H. The *Alligator, First Federal Submarine of the Civil War* (Annapolis: Naval Institute Press, June 1938).

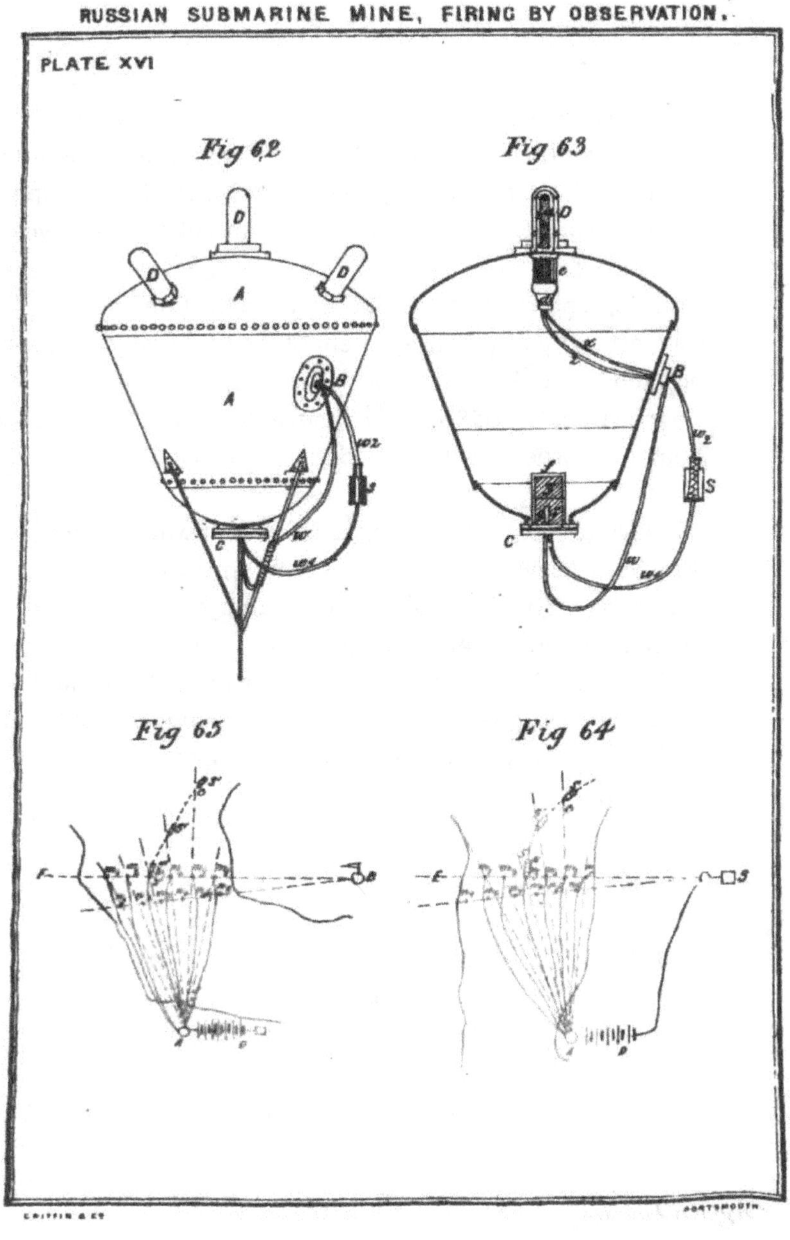

Figure 48. Russian stationary submarine mines. Russian was the first country in the world to embrace torpedo and mine warfare as a part of a country's naval doctrine. (Sleeman)

Date	Craft, Name, Event or Country	Event, Builder and Boat Specifications	City, River, Lake, Sea, Ocean
1861[139]	April *USS Pawnee*	The *USS Pawnee* saved the *USS Cumberland* from capture by Confederate forces in Norfolk, Virginia. The *Pawnee* received the surrender of Alexandria, Virginia on May 24, 1865, by a ship's boarding party. Confederate sources claim sinking the *Pawnee.* Claim disputed!	Atlantic Ocean
1861[140]	July 31 *USS Richmond*	This ship, captured by the U.S. Navy from the Confederates, fought in almost every major engagement of the Civil War. Confederates reported her sunk by torpedoes. Claim disputed!	Atlantic Ocean
1862[141]	February 14 *USS Susquehanna*	A launch from the *USS Susquehanna*, a side-wheel frigate, struck a submarine torpedo and received slight damage.	Wright River Georgia
1862[142]	February 18 Confederates Submarine Torpedoes	U.S. Navy gunboats attempting to force their way up the Savannah River were delayed by stationary submarine torpedoes laid by the Confederates. The fleet sustained no damage but naval warfare had turned a corner. The American navy was now on the defensive.	Savannah River, Georgia
1862[143]	December 12 *USS Cairo*	The ironclad *USS Cairo* ran over two large Confederate stationary submarine torpedoes. The explosions shattered most of the ship and it sank in 12 minutes with a great loss of life, off Haines Bluff, Mississippi.	Yazoo River Expedition

---

139 Dictionary of American Naval Fighting Ships, V, 239.

140 DANFS, VI, 102-103.

141 Infernal-Machines.com. "Civil War Torpedoes, Ships Sunk or Damaged by Civil War Torpedoes," http://www.infernal-machines.com.

142 Sleeman, 333.

143 Sleeman, 333, and DANFS, II, 9.

Figure 49. The *Pawnee* was in almost every single action of importance during the times of the protracted war. She was claimed sunk several times by the Confederates and obviously, they were wrong about this also. (Library of Congress Prints and Photographs Division).

Figure 50. The capture of the *Richmond* so early in the war became an embarrassment to the citizens of Richmond, Virginia, who considered their city to be "sacred." (Library Of Congress Prints and Photographs Division)

Figure 51. This sketch shows the fanciful flights of fancy that people took in regards to the new technology. The artist shows the *Winona* completely submerged which was impossible. The David Class torpedo boat was a semi-submersible but not a submarine. Notice that on the bow the artist drew in the dive planes that were used on later models of the *David*. (Library of Congress Prints and Photographs Division)

Figure 52.The *USS Cairo* was the first craft of the Civil War to be sunk by a submarine torpedo. (Library of Congress)

Date	Craft, Name, Event or Country	Event, Builder and Boat Specifications	City, River, Lake, Sea, Ocean
1863[144]	January 1 *USS Kinsman* (ex-*Gray Cloud*)	Built as the side-wheel steamships *SS Grey Cloud,* the ship was seized in New Orleans and fitted out as a U. S. Army gunboat. The ship became damaged in a firefight with the *CSS Cotton* (ironclad gunboat) and ran over a contact submarine torpedo mear Franklin, Louisiana, in the Bayou Teche. The explosion unshipped the rudder. The ship did sink in Berwick Bay, Brasher City, Louisiana, on February 23, 1863. Six men died in the explosion.	Franklin, Alabama, Bayou Teche
1863[145]	January 8 *Julia Hamilton*	The English sloop was working as a blockade runner when she was captured by the U.S. Navy. The ship was sold to the American navy as a tender for the Northern Blockading Squadron. The Confederates reported her sunk by torpedoes. The claim cannot be confirmed.	Atlantic Ocean
1863[146]	February 28 *USS Montauk*	The *USS Montauk,* monitor type, destroyed the blockade-runner *CSN Rattlesnake* in the Ogeechee River. The boat then made contact with a submarine stationary torpedo and sustained damage. By April 1, 1863, the *Montauk* was part of the fleet that attacked Charleston.	Ogeechee River, Georgia
1863[147]	April 7 *USS Weehawken*	The single turreted monitor ship put up quite a fight in Charleston Harbor. The ship received 53 hits from Confederate shore batteries. As the ship withdrew to calmer waters, it struck a stationary submarine torpedo. The torpedo exploded directly under the ships heavy keel and caused no damage.	Charleston Harbor Charleston, S.C.
1863[148]	July 13 *USS Baron de Kalb* (ex-*Saint Louis*)	The USN stern-wheel ironclad gunboat, *USS Baron de Kalb* (originally *USS Saint Louis*) sank after running over two stationary mines near Yazoo City, Mississippi.	Yazoo River

---

[144] Infernal-Machines.com. "Civil War Torpedoes, Ships Sunk or Damaged by Civil War Torpedoes" www.infernal-machines.com . U.S. Naval Historical Center. "USS Ships-USS Kinsman (1863-1863)." http://www.history.navy.mil/photos/sh-usn-usnsh-k/kinsman. DANFS, III, 655-656.

[145] DANFS, 573.

[146] Sleeman, 333.

[147]Sleeman, 333, and DANFS VIII, 190-191.

[148] Sleeman, 333, and DANFS I, 98.

Figure 53. The only damage done to the *Kinsman* from striking a submarine torpedo was that the ships rudder was unshipped. The repairs were effected in a few hours and she was back in battle form. (Library of Congress Prints and Photographs Division).

Figure 54. The *USS Weehawken* is pushing a torpedo clearing raft on its bow. This "anti-torpedo device" did work amazingly well. It carried several drag chins suspended around the edges of the raft to detonate the submerged "Torpedoes." This was another invention of John Ericsson. (Library Of Congress Prints and Photographs Division)

Date	Craft, Name, Event or Country	Event, Builder and Boat Specifications	City, River, Lake, Sea, Ocean
1863[149]	August 8 *USS Commodore Barney*	The *USS Commodore Barney*, gunboat, sank after hitting a stationary submarine torpedo. The Confederates had placed a submarine torpedo with 1,750 pounds of black-powder fired by an on-shore electrical battery. The ship became damaged and 20 members of the crew died. The engine no longer functioned and the ship was towed for the rest of its tour in the area. The ship was fully repaired and returned to service, April 13, 1864.	James River Virginia
1863[150]	August 21 *CSS Torch*	The *Torch* was a unique design. It was a screw-propelled steamer with a triple-spar arrangement. The boat carried six spar charges, each weighing 75 pounds. Her attack on the *USS New Ironsides* failed for an unknown reason.	Charleston Harbor
1863[151]	September *USN John Farrow*	The U.S. transport *John Farrow* ran over a submarine torpedo and sustained serious damage.	Charleston Harbor
1863[152]	October 5 Confederate Navy Spar-Torpedo Boat I	This was the first use of a Spar-Torpedo boat built and used by any military force. The *CSS David* was built to sink the giant ironclad, *USS New Ironsides,* in Charleston Harbor. The attack was successful. The sixty-pound charge at the tip of the spar caused enough damage to keep the *New Ironsides* out of action the rest of the war.	Charleston Harbor
1863[153]	Confederate Navy *CSS Marion, Ettiwa*	The Confederate Navy lost two ships, *CSS Marion* and *CSS Ettiwa,* to their own submarine torpedoes while laying mines intended for American navy ships.	Charleston Harbor
1863[154]	Confederate Navy *Squib*-class Torpedo Boat *CSS Shultz*	The Confederate used the Squib class spar-torpedo boat as a courier boat referred to as a "Flag of Truce Boat." The *CSS Shultz,* when returning to Richmond after a prisoner of war exchange, blew up in an explosion from a Confederate torpedo.	James River Virginia

---

149 Sleeman, 333 and DANFS II, 153-154.

150 DANFS II, Confederate Naval Vessels, 364.

151 Sleeman, 333.

152 Sleeman, 333 and DANFS, Confederate Forces, II, 513-514.

153 Sleeman, 333 and DANFS, Confederate Forces, VII, 518, 547

Figure 55. Illustration of the *USS Commodore Barney* striking a Confederate torpedo. (Library Of Congress Prints and Photographs Division)

---

[154] Smart, Larry R. "Evolution of the Torpedo Boat," *Military Affairs*, 23 (No. 2, Summer, 1959) 97-101. DANFS, Confederate Forces, II, 565.,

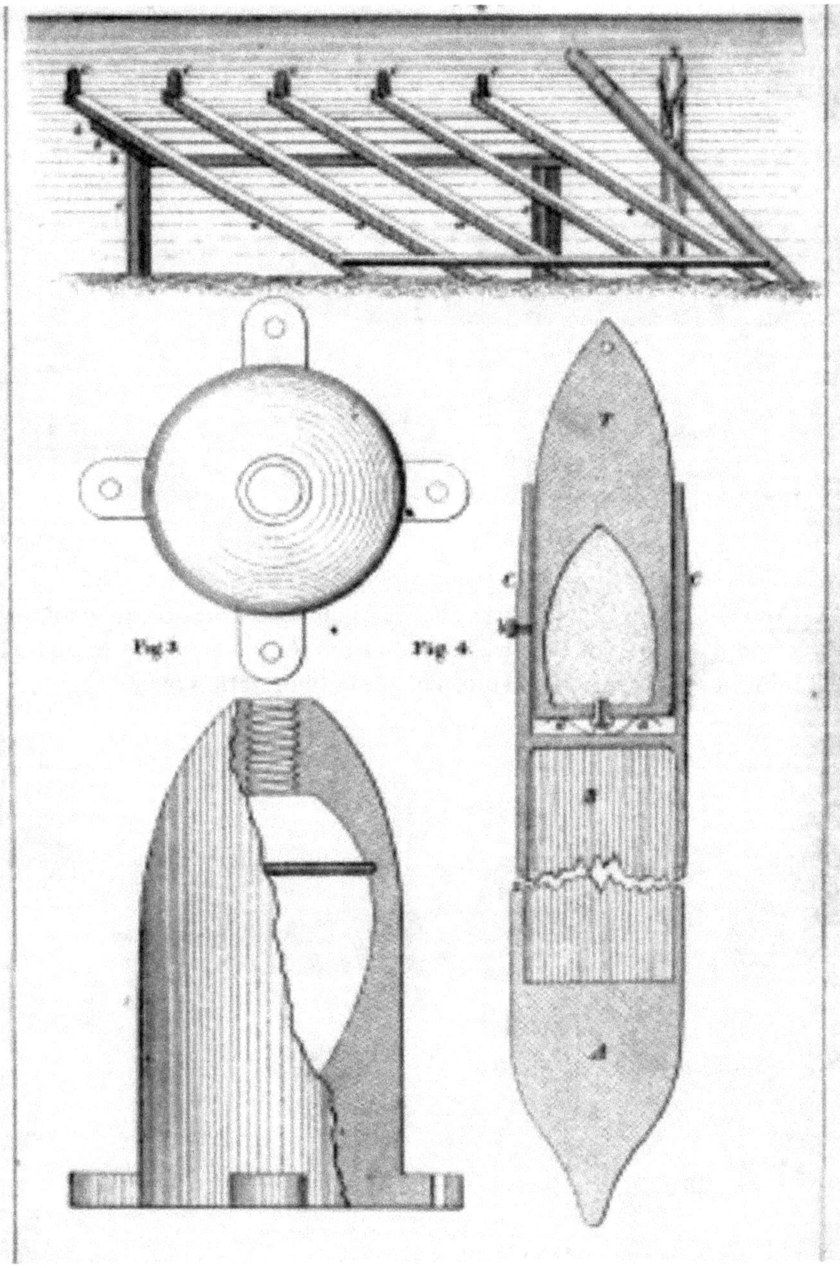

Figure 56. The Confederate Torpedo Department perfected the frame torpedo and they proved to be deadly. (Sleeman, 40)

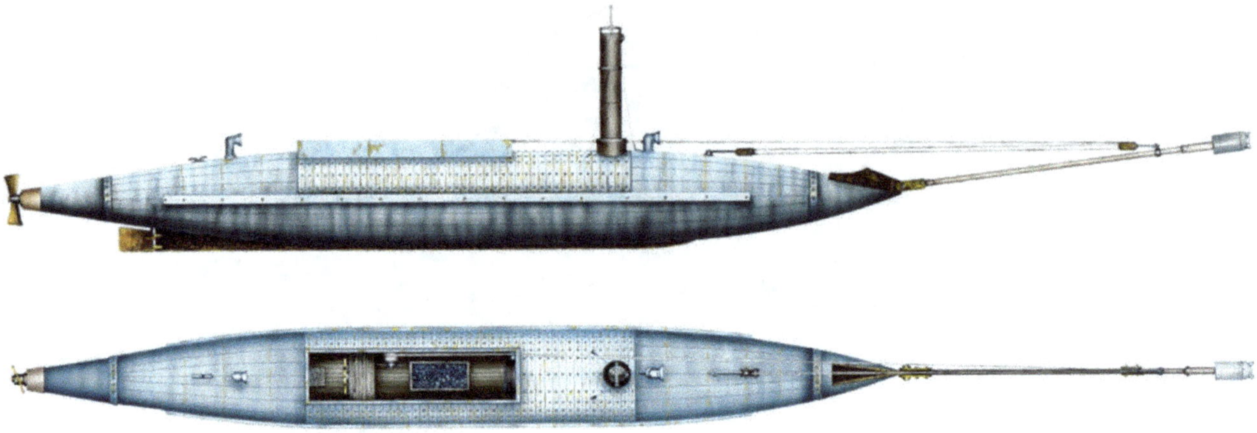

Figure 57. The *CSS David* was the first purpose-built spar-torpedo boat built by any Government. It was a direct design taken from the only spar-torpedo boat design in the world of Robert Fulton of 1830. (Illustration by author).

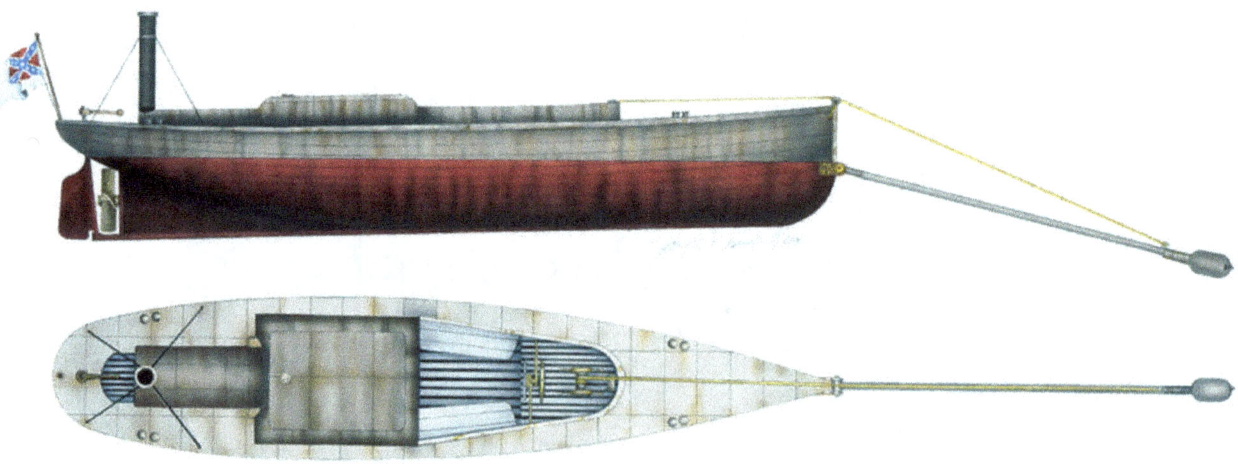

Figure 58. The craft was well built but manned by any one of any real competence. The boat became know and used for "Flag of Truce" boat or a "Surrender Boat". None of this design ever sank an enemy vessel. (Illustration by author)

Figure 59. ***USS Housatonic*** **under attack. (Library of Congress Prints and Photographs Division)**

Figure 60. The ***USS Maple Leaf*** **and** ***General Hunter*** **fell victim to two or more large 'Submarine Torpedoes' planted by members of the Confederate Torpedo Department. (Library Of Congress Prints and Photographs Division)**

Date	Craft, Name, Event or Country	Event, Builder and Boat Specifications	City, River, Lake, Sea, Ocean
1864[155]	February 8 *USS Rose (ex-Ai Fitch)*	The screw tug was operating in New York Harbor when obtained by the U. S. Navy. They fitted her with one 20-pounder Parrot rifle and one 12-pounder gun. The Confederates reported her sunk. No confirmation.	Atlantic Ocean
1864[156]	February 17 *USS Housatonic*	The *H. L. Hunley* sank the *USS Housatonic,* and herself, after striking the ship with a spar torpedo. This would be the first spar-torpedo vessel to sink a capital ship. The *Hunley* sank due to faulty design and construction.	Charleston Harbor
1864[157]	February Spar-Torpedo Boat *CSS Midge*	The *Midge* was made along the lines of the *CSS David* in Charleston, South Carolina. US Army forces captured the Midge in February and moved the boat to New York Navy Yard for display.	Charleston, South Carolina
1864[158]	March 6 *USS Memphis*	The *Memphis,* built in Scotland for the Confederate Navy, and captured by the *USS Magnolia* and the *U. S. Navy* purchased her from the Prize Court and placed into service, September 4, 1862. A *Squib* class spar torpedo boat attempted to sink *the USS Memphis* in the North Edisto River. The spar mounted torpedoes malfunctioned and the *Memphis* suffered no damage.	North Edisto River South Carolina
1864[159]	April 1 *Maple Leaf*	The Confederates, using floating torpedoes, sank the *Maple Leaf,* a transport ship.	St. John's River Florida
1864[160]	April 15 *USS Eastport*	During the early stages of the Red River Expedition the ship was damaged by a Confederate torpedo and run aground. The damage was beyond repair and destroyed to prevent capture.	Grand Ecore

---

155 DANFS, II, 518

156 Sleeman, 333, and DANFS, III, 370-371f

157 Confederate Ships-Midge (1864-1865) http://history.navy.mil/photos/sh-us-cs/csa-sh.csash-mr/midge.htm; Ships Sunk or Damaged by Civil War Torpedoes <http://www.infernal –machines.com/_sgg/f10245.htm> and DANFS, II, 548-549.

158 Sleeman, 333, and DANFS, IV, 317-318.

159 Sleeman, 333, and "Ships Sunk or Damaged by Civil War Torpedoes" <http://www.infernal-machines.com/_sgg/f10245>.

160 DANFS, II, 32.

Date	Craft, Name, Event or Country	Event, Builder and Boat Specifications	City, River, Lake, Sea, Ocean
1864[161]	April 9 *USS Minnesota*	The *Squib* made a night attack and damaged the *USS Minnesota* without sinking her. Most *Squib* pilots lacked training to cause real damage to U.S. Navy ships.	Newport News, Hampton Roads
1864[162]	April 18 *USS Wabash*	A *Squib* class spar torpedo boat made an attack on the *USS Wabash*, a naval frigate, and the attempt failed. The pilot showed lack of training by the Confederate Navy Spar-Torpedo Boat Department.	Charleston Harbor
1864[163]	May 6 *USS H.A. Weed*	The U.S. transport, *USS H.A. Wee*d, sank after detonating a submarine mine.	St. Johns River
1864[164]	May *CSS St. Patrick*	The small *David* class spar-torpedo boat, built by John P. Hailigan, Mobile Alabama, failed to inflict any damage on the USS *Octorara* in Mobile Bay. The mission was aborted when her boiler burst, killing one man and causing serious injury to another.	Mobile Bay,AL
1864[165]	June 19 *USS Alice Wood*	The U.S. transport, *USS Alice Wood,* is destroyed by a submarine mine.	St. Johns River
1864[166]	August 5 *USS Tecumseh*	The monitor class *USS Tecumseh sank* from a submarine torpedo while attacking the defenses of Mobile Bay. The explosion was large enough to cause the ship to immediately role and "turn turtle'. It sank in 25 seconds with 93 men including its very courageous captain, Commander T.A.M. Craven.	Mobile Bay, AL

---

161 DANFS, Confederate Forces, II, 567.

162 Sleeman, 334, and DANFS, VIII, 5-7.

163 Sleeman, 334.

164 Smart, Larry R. "Evolution of the Torpedo Boat" *Military Affairs,* 23, No. 2, Summer, 1959, 97-101; DANFS, Confederate Forces, II, 567.

165 Sleeman, 334.

166 Sleeman, 334, and DANFS, VII, 78.

Figure 61. The fear of Torpedoes became a major concern for the ships commanders. Morale dropped to new lows after the introduction of the stationary and floating torpedo. (U.S. Government Printing Office)

Figure 62. While Matthew Fontaine Maury was an officer in the U.S. Navy, he met with Samuel Colt on several occasions. He became interested in Colt's "Submarine Batter" and knew first hand of all of Colt's many accomplishments. It was Maury who established the Confederate Torpedo Department early in the war. Maury is also famous as a pioneer in oceanography. (U.S. Government Printing Office)

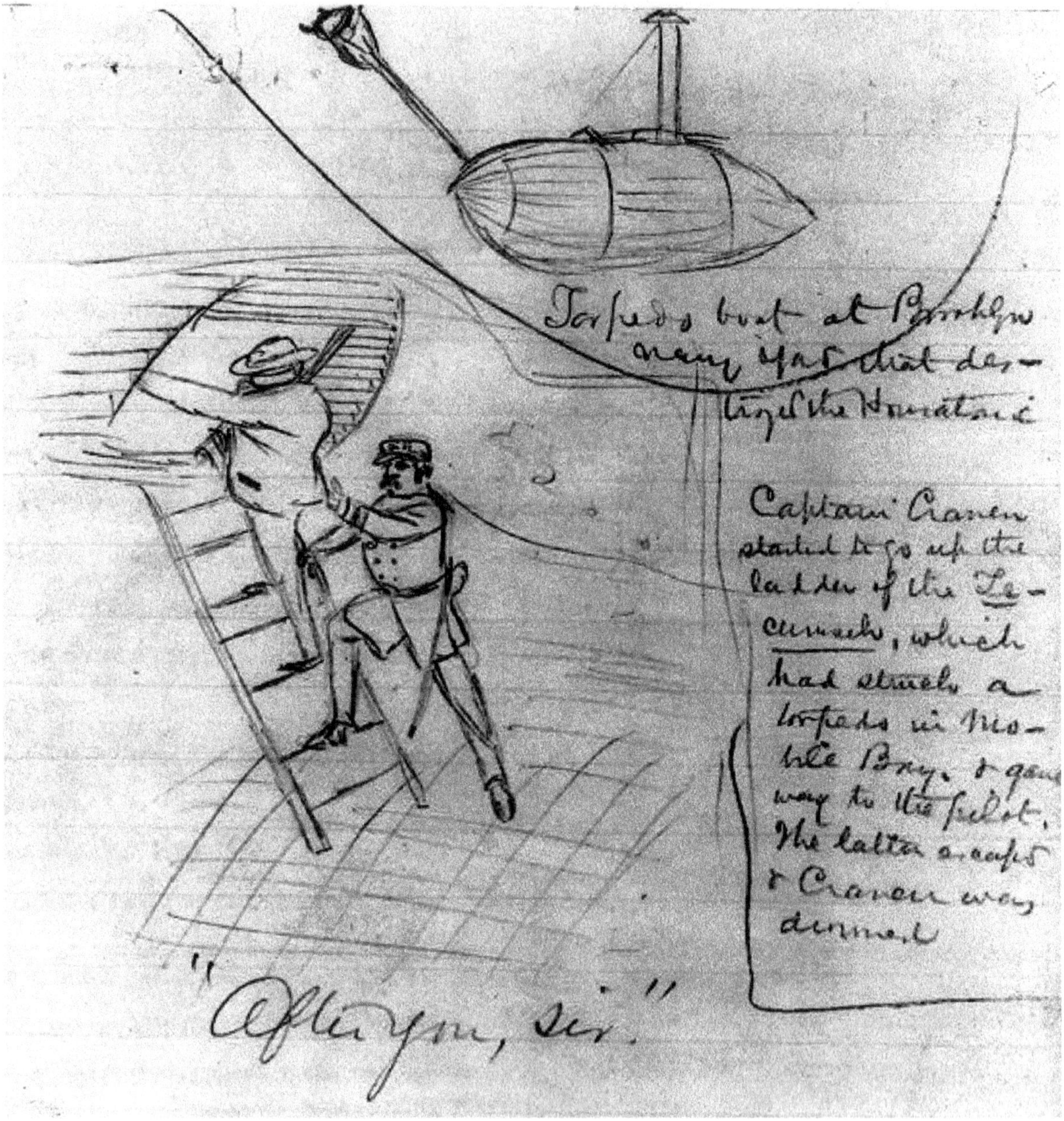

Figure 63. The *USS Craven* had been attacked by a *David* Class spar-torpedo boat. While the vessel was sinking Captain T.A.M. Craven died while assisting his men out of the ship. (Library of Congress Prints and Photographs Division)

Date	Craft, Name, Event or Country	Event, Builder and Boat Specifications	City, River, Lake, Sea, Ocean
1864[167]	September 9 *CSS Squib*	Built in Richmond, Virginia, at Tredegar Ironworks, the *Squib* quickly became a class of spar torpedo boat. Small and agile, it is the best spar torpedo boat of the Civil War. The Squib moved in to Hampton Roads to attack the frigate, *USS Minnesota* with a single spar-torpedo with a 53 pound charge. The attack failed causing next to no damage to the ship. It is believed that she was sunk or captured in the final days of the Civil War in the Battle for Trent's Reach, James River.	Hampton Roads, VA
1864[168]	October 27 USN Picket Boat No. 1	This is only successful attack by a spar-torpedo boat of the Civil War. The converted Picket Boat came armed with a Wood Lay torpedo of a greatly advanced design. Lt. Cushing and his small crew accepted the assignment to attack the *CSS Albemarle*, the South's newest and most powerful ironclad. The Albemarle's destruction was complete.	Roanoke River Plymouth, N.C.
1864[169]	November 27 *Greyhound*	The *USS Greyhound,* a transport ship, suffered its loss by the act of a Confederate saboteur. Confederates were supplied with "Coal Torpedoes". These explosive looked like lumps of coal and were easily placed in the coal bunkers on U.S.Navy ships.	James River. VA
1864[170]	December 12 *USS Narcissus* (ex-Mary Cook)	Like many other gun boats the *Narcissus* performed as a minesweeper in Mobile Bay. Caught in a severe storm off Mobile the ship struck a floating torpedo and sank in 15 minutes. There was no loss of life. The ship was raised and put back into action	Mobile Bay, AL
1864[171]	December 9 *USS Otsego, USS Bazeby*	The American steamers *USS Otsego* and *USS Bazeby* were destroyed with a great loss of life by Confederate submarine torpedoes. The Otsego struck first and the Bazeby, while coming to her aid, struck the other mine.	Roanoke River

[167] DANFS II, 567.

[168] *Official Records of the Union and Confederate Navies in the War of the Rebellion*. "Torpedo used by Lieutenant Cushing in the destruction of the Confederate States Ram Albemarle,"Series I, Volume 10, 1900, 623.

[169] Sleeman, 334.

[170] Sleeman, 334, and DANFS V, 11-12.

[171] Civil War Torpedoes, Ships Sunk or Damaged by Civil War Torpedoes <http://www.infernal-machines>, Accessed by author November 24,2007; Sleeman, 334

LIEUT. CUSHING'S TORPEDO BOAT SINKING THE ALBEMARLE ON ROANOKE RIVER, N. C.

Figure 64. This is the only spar-torpedo boat of the Civil War to actually and fully sink an enemy ship. Lt. John Cushing took the well designed, boat and torpedo, by Engineer William W. Wood, and sank the newest and most heavily armed Ironclad of the South. (U.S. Navy Historical Department)

Figure 65. Lieutenant William B. Cushing planned and executed a daring and daunting mission to attack and sink the newest, most powerful ironclad, in the Confederate fleet. He lost a good number of men captured or killed in the attack. (U.S. Navy Historical Department)

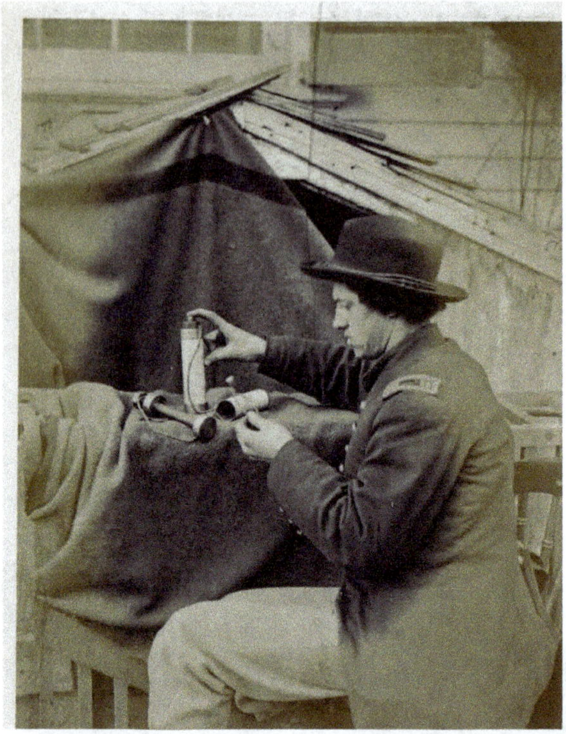

Figure 66. This U.S. Army officer is setting up the Haupt's torpedo for destroying a bridge. After the many successes of the Confederate Torpedo Department, the U.S. services followed suit. (Library Of Congress Prints and Photographs Division)

Figure 67. U.S. Navy "Boat Flag" 1862. (U.S. Government Printing Office)

Date	Craft, Name, Event or Country	Event, Builder and Boat Specifications	City, River, Lake, Sea, Ocean
1864[172]	M.A. Nobel "Dynamite"	Nobel was able to make the raw Nitroglycerine into a safer yet powerful explosive form, Dynamite.	U.S.A.
1864[173]	Captain Lupuis & Mr. Whitehead	The Captain asked Mr. Whitehead for assistance in developing his "Automobile Torpedo Boat." Whitehead's interest in the Captain's problem led to the full operational modern torpedo.	Fiume, Austria
1865[174]	January 15 *USS Patapsco*	While covering picket boats dragging for submarine torpedoes the monitor ironclad *USS Patapsco* stuck and detonated a submarine torpedo, deep under her keel The ship sank in less than one minute. Sixty-two men went down with the ship.	Charleston Harbor, S. C.
1865[175]	January 20 *USS Osceola*	The *USS Osceola,* a side-wheel, double-ended gunboat, became crippled after striking a Confederate drifting torpedo. Not verified.	Cape Fear River, N. C.
1865[176]	January 25 US Navy *Spuyten Duyvil*	The USN Torpedo Boat No.5 *Spuyten Duyvil* successfully launched two Wood Lay torpedoes in a test off Mystic, Connecticut. Chief Engineer William W.W. Wood and 1st Assistant Engineer, John L. Lay designed this and five other craft.	New Haven, Conn. Atlantic Ocean
1865[177]	January 23 *Spuyten Duyvil*	For the first time since her arrival, the spar-torpedo boat was ordered into position to support the twin-turreted monitor class warship, *U.S.S. Onondaga.* This was a serious misuse of the craft. The *Onondaga* would have been better served if the *Spuyten Duyvil* had attacked the Confederate Navy's James River Fleet.	Trent's Reach James River

---

[172] Sleeman, 334.

[173] Gray, Edwyn. *The Devil's Device: Robert Whitehead and the History of the Torpedo* at 34. London: Seeley, Service & Co., 1991.

[174] Sleeman, 335, and DANFS, V, 224-225.

[175] Sleeman, 335, and DANFS, V, 180.

[176] *Official Records of the Union and Confederate Navies in the War of the Rebellion*. Report of Assistant Engineer, John L. Lay, U.S.N, Commander, USN Torpedo Boat No.5, *Spuyten Duyvil.*

[177] Official Records of the Union and Confederate Navies in the War of the Rebellion: Report of Assistant Engineer, John L. Lay, U.S.N, Commander, USN Torpedo Boat No.5, *Spuyten Duyvil.*

Figure 68. U.S. Navy Commodore's Broad Pennant. (U.S. Navy Historical Department)

Figure 69. Hand drawn illustrations such as this one of *USS Patapsco* were the early equivalent of camera. Unfortunately, most bear no signature or date.

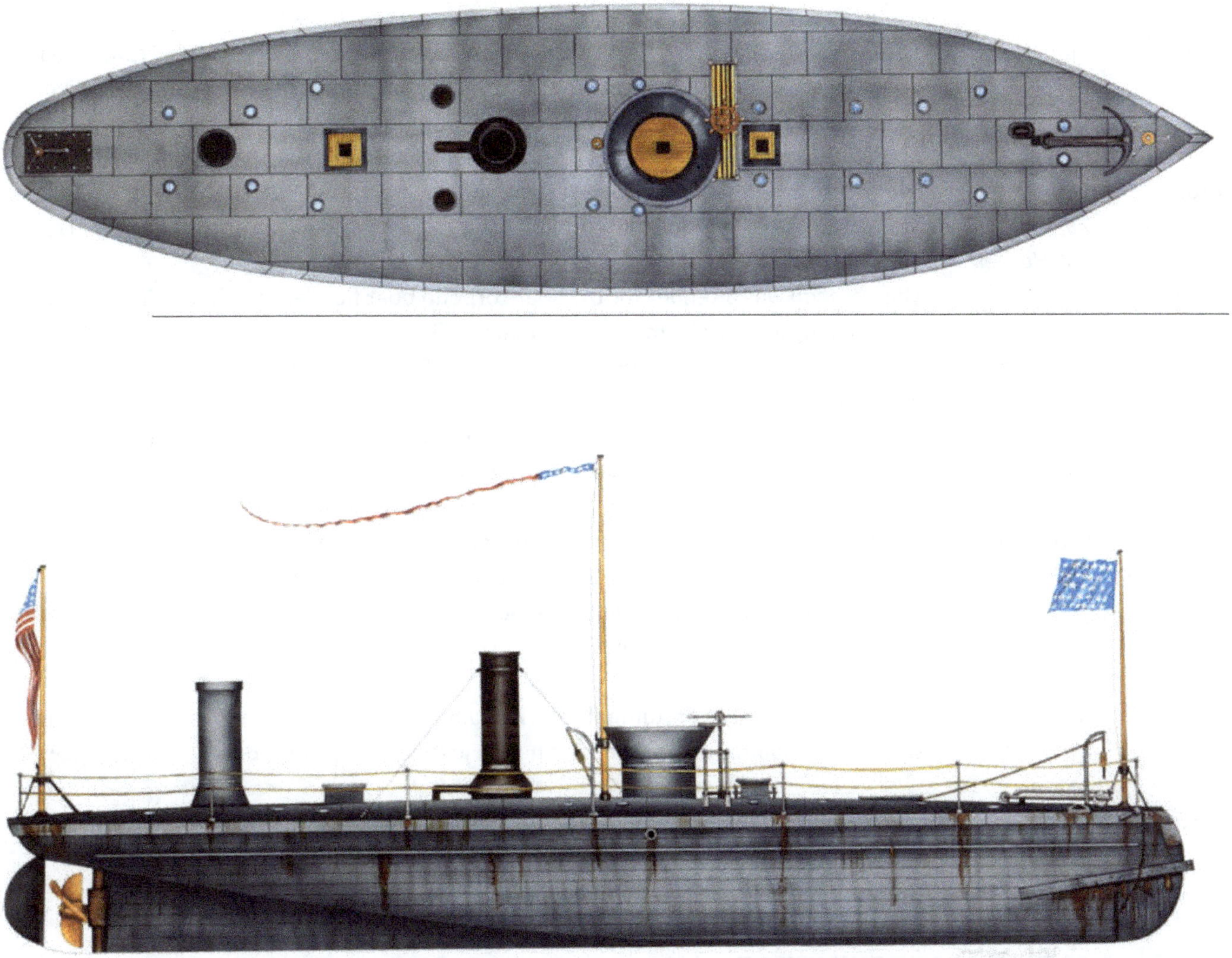

Figure 70. The U.S. Navy *Torpedo Boat No. 5, Spuyten Duyvil* was a most successful feat of engineering since the USS *Monitor*. It came out of the yards working perfectly. (Illustrations by author)

Date	Craft, Name, Event or Country	Event, Builder and Boat Specifications	City, River, Lake, Sea, Ocean
1865[178]	January 23/24 *CSS Wasp*	The *Wasp* was one of four spar torpedo boats attached to the Confederate James River Fleet. The boat's crew attempted to remove obstructions laid by earlier Confederate forces That had made the break-out of the fleet an impossibility.	Trent's Reach James River
1865[179]	January 23/24 *CSS Hornet*	During the attempted breakout of the Confederate James River Fleet, the *Hornet* collided with the much larger ship, *C.S. Allison* in Trent's Reach.	Trent's Reach James River
1865[180]	January 23/24 *CSS Torpedo*	The *Torpedo* was a tender for the spar-torpedo boat flotilla. The ship was 70' long and at 150 tons substantial in a fight with her twp 20-pounder guns. The ship failed to make it out of the Trent's Reach entanglements.	Trent's Reach James River
1865[181]	January 28 *CSS Saint Patrick*	The *St. Patrick* attacked the *USS Octorara* in Mobile Bay of January 28. The spar torpedo misfired and the *Octorara* sustained no damage.	Mobile Bay Mobile, AL
1865[182]	February *CSS Midge*	The *Midge* was similar in design to the *David*. It was built in Charleston, South Carolina, in 1864. The *Midg*e was captured by U. A. Army forces in February 1865. The boat was taken to the New York Navy Yard and put on exhibit until 1877. The boat's fate is then unknown.	Charleston Harbor Charleston,S.C.
1865[183]	March 1 *USS Harvest Moon*	This was the flagship of Admiral John A. Dalhgren; *USS Harvest Moon* sank from a submarine torpedo in an area just swept for torpedoes.	Winyan Bay Georgetown
1865[184]	March 4 *USS Thorne*	The U.S.Navy transport *USS Thorne* hit a stationary submarine torpedo and sank in the Cape Fear River.	Cape Fear River

[178] DANFS, Confederate Naval Vessels, II, 367.

[179] *Ibid.*

[180] *Ibid.*

[181] Torpedo Boats < http://www.infernal-machine.com/ sgg/f10235.htm> Accessed by Author November, 2005; DANFS, VII, 503

[182] Confederate Ships-Midge (1864-1865) <http://www.history.navy.mil/photos/sh-us-cs/csa-sh/csah-mr/midge.htm> Accessed By Author, 2006

[183] Sleeman, 334 and DANFS III, 259

[184] Sleeman, 334.

Figure 71. This is a portrait of the Confederate submarine sent out to sink the *USS Minnesota,* a U.S. Navy ship of the line. Like most Confederate efforts to use "Diving-Boats" to destroy U.S. Navy ships, the mission was a dismal failure. (Library of Congress Prints and Photographs Division)

Figure 72. This illustration, by an unknown solider or sailor, paints the grimy picture that "Torpedoes" were to the men who manned the ships on the river fleets, completely terrifying. These barrel torpedoes could seldom be seen in time and when in groups of three or more could be quite deadly. (Library of Congress Prints and Photographs)

Date	Craft, Name, Event or Country	Event, Builder and Boat Specifications	City, River, Lake, Sea, Ocean
1865[185]	March 12 *USS Althea*	The screw-tug *Althea* (originally *Alfred A. Wotkyns*) had been assigned to the Northern Blockading Squadron and then ship later assigned to the James River Squadron. Here the boat was used as a tug, tender and spar-torpedo boat. A stationary submarine torpedo sank the ship on March 12th. The boat was raised later and stayed in service until 1866.	Blakely River
1865[186]	March 12 *USS Jonquil*	The *USS Jonquil* (originally *J.K. Kirkman*) joined in with others clearing submarine torpedoes in Charleston Harbor. Crewmembers were in the process of de-fusing the torpedo when it exploded. Three men were wounded and nine knocked in to water. The minor repairs were completed by the next morning and the ship was back to work, clearing torpedoes.	Charleston Harbor Charleston, S.C.
1865[187]	March 28 *USS Milwaukee*	The U.S. Navy monitor class *USS Milwaukee* hit a stationary submarine torpedo and sank.	Blakely River
1865[188]	March 29 *USS Osage*	The U.S. Navy monitor class, USS Osage struck a drifting torpedo and sank.	Blakely River
1865[189]	April 1 *USS Rodolph* Gun-Boat No. 48	The side-wheeler, gun-boat, tin clad, *USS Rodolph*, No.48, had been one of the most combative boats in the US Navy. The boat served its last year as a "Mine Sweeper.". While towing a barge with salvage equipment, it struck a submarine torpedo and sank. 4 crewmen died and 11 were wounded.	Blakely River

185 Sleeman, 335, and DANFS I, 38.
186 Sleeman, 335, and DANFS VII, 559.
187 Sleeman, 335, and DANFS, IV, 362.
188 Sleeman, 335, and DANFS, V, 565.
189 Sleeman, 335, and DANFS , VI, 147.

Figure 73. The Confederate States of America went through several versions of a national standard. It turns out there was no unity in ever selecting a design. Every Confederate state wanted their own flag. Up into the closing days of the war, Southern politicians still argued about the flag. (U.S. Government Printing Office)

Figure 74. The South used a large number of "Flag of Truce Vessels." (Library of Congress)

Date	Craft, Name, Event or Country	Event, Builder and Boat Specifications	City, River, Lake, Sea, Ocean
1865[190]	April 3 *CSS Wasp, Hornet, Scorpion*	The *Squib* class *CSS Wasp* helped to refloat the spar-torpedo boat *CSS Hornet* in the James River. Later the boat tried to tow the *CSS Scorpion* off shore where she had run aground and the *Wasp* became grounded in the attempt. The *Scorpion* was abandoned and captured by U.S. Navy forces.	Trent's Reach James River
1865[191]	April 13 *USS Ida* Gun Boat	The U.S.Navy gunboat *USS Ida* struck a stationary submarine torpedo. The boilers burst, timbers on the starboard side were crushed in, the deck turned into a mass of splinters. Three crewmembers wounded and two died.	Mobile Bay Mobile, AL
1865[192]	April 14 USS *Scotia* Gun Boat	The "90-Day" gunboat *Scotia* built by Jacob Birley, J.P. Morris & Company, Philadelphia, 1862. Struck a submarine torpedo off Mobile and sank with a great loss of life.	Mobile Bay Mobile, Al
1865[193]	May 12 *USS R.B. Hamilton*	The U.S.Navy transport *R. B. Hamilton* hit a stationary submarine torpedo and sank.	Mobile Bay Mobile, AL
1865[194]	June 6 *USS Jonquil* Gun Boat	The crew of the gunboat *USS Jonquil* attempted to raise a set of 'Frame' torpedoes when the torpedoes accidentally detonated. The gunboat was destroyed with a great loss of life.	Ashley River
1866[195]	September 2 Paraguayan Navy	The Paraguayan Navy took notice of the success of "Torpedo Warfare" in the United States and adopted the technology. Their first use of stationary submarine torpedoes netted them the Brazilian war steamer *Rio Janeiro,* during a bombardment of the harbor and town.	Currupaity, Paraguay

[190] DANFS, Confederate Forces, II, 565.

[191] DANFS, III, 414-415.

[192] Sleeman, 335, and DANFS VI, 382-383.

[193] Sleeman, 335, and National Park Service, "Full Speed Ahead" <http://www.nps.gov/archive/vick/vistr/sitebltn/damntorp.htm>.

[194] Sleeman, 335, and DANFS III, 559.

[195] Sleeman, 335.

Figure 75. This rendition caused unmanageable problems for Confederate troops. When this largely white flag first appeared, it was commonly mistaken for a flag of surrender. Quickly, another flag was created with a broad red panel on its right side. (U.S. Government Printing Office)

Figure 76. The conspirators in the Lincoln assassination were kept handcuffed aboard the new ironclads. (U.S. Navy Historical Department)

Date	Craft, Name, Event or Country	Event, Builder and Boat Specifications	City, River, Lake, Sea, Ocean
1868[196]	Robert Whitehead "Fish Torpedo"	His first trials were highly successful and resulted in Austria being the first European nation to adopt the "Torpedo" doctrine.	Fiume Austria
1869[197]	Captain W. Harvey	Harvey favored perfecting the "towed" torpedo concept. His trials were never successful and his torpedo dropped from use.	Russia
1870[198]	John Ericsson Compressed Air Torpedo	His torpedo is all iron construction and steered by compressed air running through a flexible tube on a reel. From a shore installation or a ship, the device could be set up for use. Ericsson's torpedo tested by Commodore W.N. Jeffers, USN, received rave reviews for trouble-free performance and speed.	U.S.A.
1870[199]	German Navy	The Germans developed torpedo fleets, *Freiwillige Seewehr* (Voluntary Sea Force) using existing small boats fitted with torpedo equipment. They were to guard the Elbe and Weser estuaries.	Germany
1871[200]	England Rev. C. Ramus	Through his own inventiveness and dedicated research, the Reverend developed the concept for a "stepped" ship. This went with his patent for a planing boat without a "step." Models were made and tested of his concept boats that weighed in at 2,500 tons. His were the first ship models to be tested in indoor tanks.	England
1871[201]	British Navy *Miranda*	The *Miranda* was a lightly built steel boat with a speed of 16.4 knots. This is considered to be the precursor of the modern motor torpedo boat.	Great Britain

---

[196] Sleeman, 335, and Gray, Edwyn. *The Devil's Device: Robert Whitehead and the History of the Torpedo Service* (Seeley & Co. Ltd, 1991), 58-59.

[197] Sleeman, 335.

[198] Brown, Allen D. "Torpedoes and Torpedo Boats," *Harper's New Monthly Magazine,* LXV (June to November, 1882), 36-47

[199] Groner, Erich. *Die deutschen Kriegssniffe, 1815-1945* (J.F. Lehmanns Verlag Munchen, 1968).

[200] Fock, *Fast Fighting Boats*, 13.

[201] Fock, *Fast Fighting Boats*, 11.

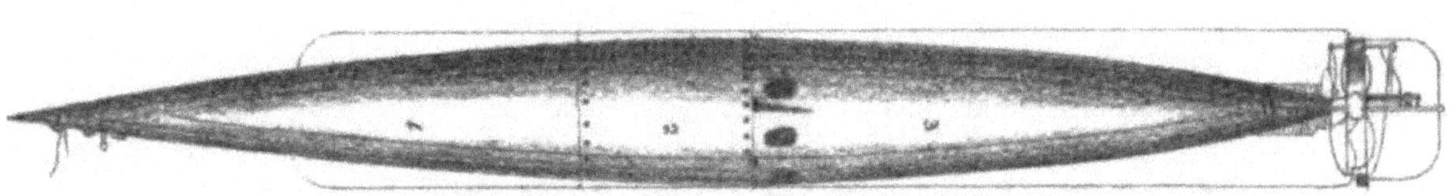

Figure 77. Robert Whitehead's first "Fish Torpedoes" were slow and erratic but they did lead to the ultimate stable design. (Sleeman, 226)

Figure 78. Ericsson created a clumsy but effective "Automobile Torpedo" which extended his reach into advanced technology. Later Brennan would create his wire-driven model based on this experiment. (U.S. Navy History Department)

Date	Craft, Name, Event or Country	Event, Builder and Boat Specifications	City, River, Lake, Sea, Ocean
1871[202]	German Navy *Devrient*	The boats, I-III, were modest performers. They functioned as defensive vessels in Wilhelmshaven Harbor, July 8, 1881. **Speed:** 8.0 knots; **Displacement:** 24 tons; **Armament:** 1-Bow spar w/ 17kg. charge, w/impact fuse; **Length:** 20.3m; **Beam**:3.30m; **Draft:** 1.9m;**Engines:** 1-250hp steam; Single Screw'	Devrient, Danzig Germany
1871[203]	German Navy *Waltjen*	Again, three boats built for spar experiments and later changed to minelayers for Wilhelmshaven Harbor. **Builder:** Waltjen & Co. **Speed:** 7.7 knots, **Displacement:** 24 tons, **Armament**: 1-Bow spar w/ 17kg charge, w/impact fuse; **Length**: 14.6m; **Beam:** 3.33m; **Draft:** 1.8m; **Engine:** 1-60hp steam; Single Screw.	Bremen, Germany
1873[204]	U.S. Navy 1st Asst Engineer John L. Lay	Lay introduced the Locomotive Controllable Torpedo. This was the beginning of "Wire-Guided" torpedo design for this era and the concept still used by the world's military arms manufacturers today.	U.S.A.
1873[205]	Royal Norwegian Navy *Wasp*	The Thorneycroft Company built its first spar-torpedo boat for Norway. **Speed:** 15 knots, Length: 57'	England
1874[206]	British Navy *Vesuvius*	An extremely strange design, but within keeping to the Victorian era of fantasy engineering and construction. The ship had no traditional funnel. Fumes exited via a long row of exhaust ducts along both sides of the rear deck. The boat burned only high-grade coke to prevent smoke. These features prefigured modern "stealth" designs.	England

---

[202] *Die deutschen Kriegssniffe*, 1815-194, I, 32.

[203] *Die deutschen Kriegssniffe*, 1815-1945, I, 32.

[204] Hughes, W. S. "Modern Aggressive Torpedo Boats." *Scribner's Magazine*, I (Issue 4, 1887), 427-437.

[205] Gray, 141. and *Historical Transaction 1893-1943*, 356.

[206] Personal communication with Gordon Smith, Naval-History.com, May 5 2008 and Gray, 142.

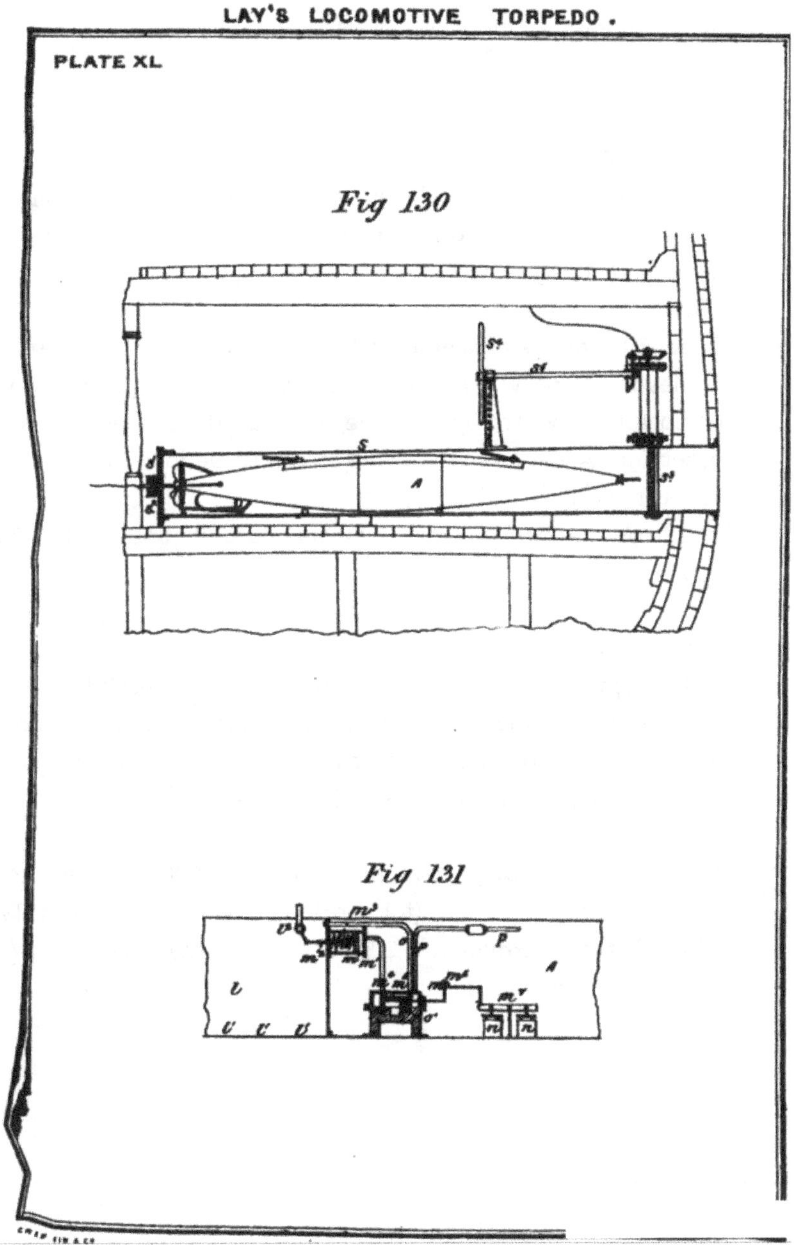

Figure 79. Lay's Locomotive Torpedo. (Sleeman, 232)

Date	Craft, Name, Event or Country	Event, Builder and Boat Specifications	City, River, Lake, Sea, Ocean
1874[207]	British Navy "Brennan Torpedo"	Irish born, Australian national, Louis P. Brennan, patents the device on February 1, 1878. Brennan invented the 'Wire-Driven Dirigible Torpedo". It used a steam engine to drive the winding engine with 18-gauge wire to control direction and speed. 1882. **Speed:** 20 knots; **Displacement:** 3.5 tons; **Length:** 24', **Diameter:** 24"; **Warhead:** 200 lbs. wet-guncotton; **Range:** 1.5 miles (2,640 yards)	England
1874[208]	U.S. Navy *Intrepid*	The Boston Navy Yard built a large, iron hulled, multiple spar torpedo boat backed up by heavy guns. Called a Steam Torpedo Ram, it was not a successful design.	U.S.A.
1875[209]	Royal Dutch Navy	The Yarrow company builds its first spar-torpedo boat for the Royal Dutch Navy. **Speed:** 17 knots	England
1876[210]	U.S. Navy *Lightning*	The Herreshoff brothers build their first highly successful spar-torpedo boat, Hull No.20. It was capable of going backward at the same speed as going forward by reversing engine and not using rudder. The boat could stop in its own length by reversing its engine. **Speed:** 17.5 knots; **Displacement :** Hull only, 2,800 lb; full weight, 6,900 lb; **Length:** 58'; B: 6'-3", **Draft:** (forward)-14", (Aft): 22"; **Engines:** Twin 5" Cyl. dia. x 10" stroke; **Armament:** 2-Steel spars 22' w/torpedoes w/ 301 lbs. dynamite; **Crew:** 2	U.S.A.

207 Gray, 130-134.

208 *Historical Transactions* 1893-1943, 356, and U.S. Navy Historical Center. "Intrepid." http://www.history.navy.mil/danfs/i2/intrepid-ii.htm

209 *Historical Transactions 1893-1943*, 356

210 Simpson, Richard V. *Building The Mosquito Fleet: The U. S. Navy's First Torpedo Boats*, (Arcadia Publishing, 2001). *Historical Transaction* 1893-1943, 356.

Figure 80. Australian Louis Brennan. carried forward Ericsson's idea of remote control of a torpedo through an early for of 'wire-guided' technology. Brennan Torpedo Stations were built in Europe and Asia. (Hong Kong Coastal Defense Museum)

MR. LOUIS F. BRENNAN, THE INVENTOR OF THE TORPEDO.

Figure 81. Louis Brennan.

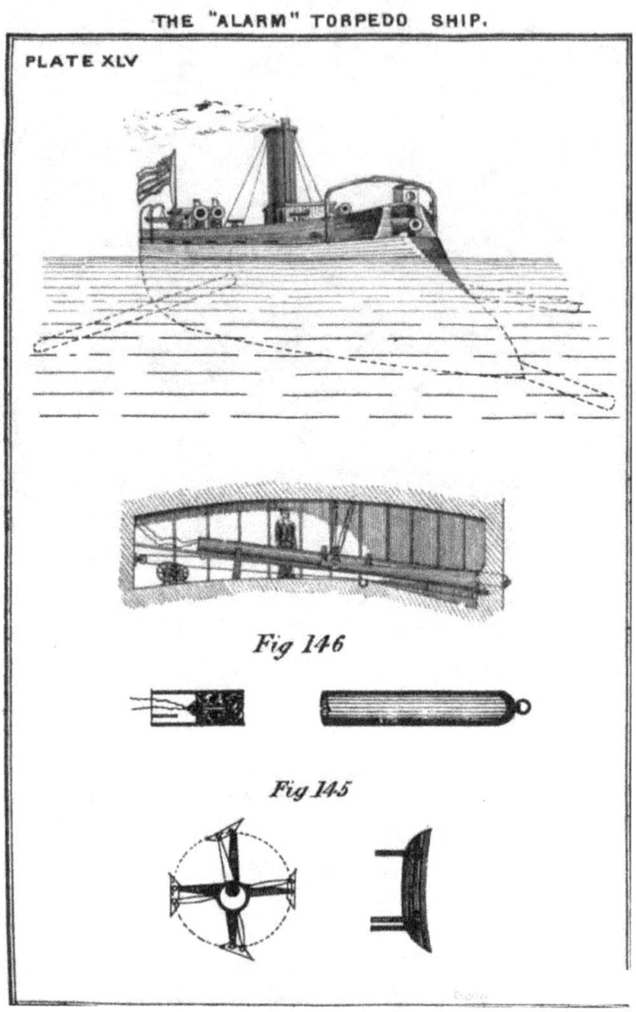

Figure 82. The *Alarm* was a very unsuccessful design. The "Old hands" in the U.S. Navy wanted to desperately hang on to the ram bow. This made the new torpedo boat virtually useless. (Sleeman, 274)

Figure 83.*USS Lightning* (Sleeman)

Date	Craft, Name, Event or Country	Event, Builder and Boat Specifications	City, River, Lake, Sea, Ocean
1876[211]	German Navy *Ziienten*	This is Germany's first purpose built torpedo boat. Designed for Whitehead torpedoes and no spars. The Thames Ironworks of England built it. The boat carried two underwater torpedo tubes, one aft, and one in the bow. The boat later carried 6-50mm (1.95") guns. **Speed:** 16 knots; **Displacement:** 1,152 tons; **Length:** 260'-6";B:28'; D:15'-2"; **Engines:** 2-Horizontal Compound steam; 2- screws, 2-Mast, 4-sails; **Armament:** 2-15" torpedo tubes, 6-50mm guns; **Crew:** 95	Germany
1877[212]	May 12 Russian Navy	The Russian launched a torpedo boat attack against several Turkish ships. The Russian were using towed torpedoes and a torpedo hit one Turkish ship, it failed to explode.	Batoum, Black Sea
1877[213]	May 27 Russian Navy *Shutka* and *Mina.*	Russian spar torpedo boats successfully attacked the Turkish ships *Fettu Islam, Duba Saife*, and *Kilidj Ali.* The monitor *Duba Saife* was sunk. Two of the Russian torpedo boats were made of iron, the *Shutka* and *Mina.*	Matchines River, Danube
1877[214]	May 29 *HMS Shah*	During the Battle of Pacocha, for the first time a warship fired a Whitehead 'Fish' torpedo against an enemy ship, the Peruvian Ironclad Huascar. The distance was too great and the torpedo ran out of fuel in the attempt.	Pacific Ocean

[211] Jackson, Robert. *Destroyers, Frigates, and Corvettes* (Barnes & Noble, Inc., 2000), 278.
[212] Greger, Rene "The Origins of Russian and Soviet MTBs" (*Warship,* IX, Issue 33), 38; Sleeman, 336.
[213] Sleeman, 336, and Greger, 38.
[214] Sleeman, 335; Memorials and Monuments in Plymouth City Centre (*HMS Shah*) www.memorials.inportsmouth.co.uk/city-centre/shah.htm ; The Huascar, http:/ members.lvcos.co.uk/Juan39.THE_HUASCAR.nm.

Figure 84. Russian Spar Torpedo Boats attacking a Turkish Man-of-War. (Richard Worth Collection)

Date	Craft, Name, Event or Country	Event, Builder and Boat Specifications	City, River, Lake, Sea, Ocean
1877[215]	June 9 Russian Navy *Torpedo Boat No.1 (SK-1)*	The Russian torpedo boat commanders were reckless in pressing their attacks against the Turkish fleet. An attack of six wooden boats against the four heavily armed ironclads led to the lead boat, *Torpedo Boat No. 1* with her commander Lieutenant Poutschin and his crew were all taken prisoner. The other boats scattered.	Sulina, mouth of the Danube Black Sea
1877[216]	June 20 Russian Navy Spar Torpedo Boat *Choutka*	This action helped prove the foolhardiness of daytime attacks by spar torpedo boats. The single boat went after a Turkish monitor and the small boats commander became severely wounded from the larger ships gunfire. The wire to trigger the spar charge was cut by enemy fire and the attack failed.	Rutschuk, on the Danube Black Sea
1877[217]	June 23 Russian Navy Spar Torpedo Boats	Once again, the daylight tactics of the Russian boat commanders meant a doomed mission. Two boats set out to attack a Turkish monitor. The spirited defense of the Turks proved to be too much and the boats pulled back.	Mouth of the Aluta, Danube, Black Sea
1877[218]	August 22 Russian Navy Spar Torpedo Boats	The Turkish commander of the *Assari Shefket* adjusted to the Russian tactics by placing small guard boats out from his ship. When the four Russian boats moved to attack, they were driven off by heavy and relentless small arms fire.	Soukoum Kaleh, Black Sea
1877[219]	October 10 Russian Navy Electric Contact Mine	The Russians placed electric contact mines near the Turkish defenses of Sulina and sank a gunboat *Suna.* About fifteen officers and men were killed in the explosion	Sulina, Black Sea
1877[220]	December 27 Russian Navy Torpedo boats	The Russians, with four torpedo boats, attacked the Turkish Squadron near Batoum. Two of the boats carried Whitehead "Fish" torpedoes. Both torpedoes ran out of fuel and were picked up by the Turks.	Batoum, Black Sea

---

[215] Sleeman, 336; Greger, "The Origins of Russian and Soviet MTB's", 39-40.

[216] Sleeman, 336

[217] Sleeman, 336.

[218] Sleeman, 336.

[219] Sleeman, 336.

[220] Sleeman, 336.

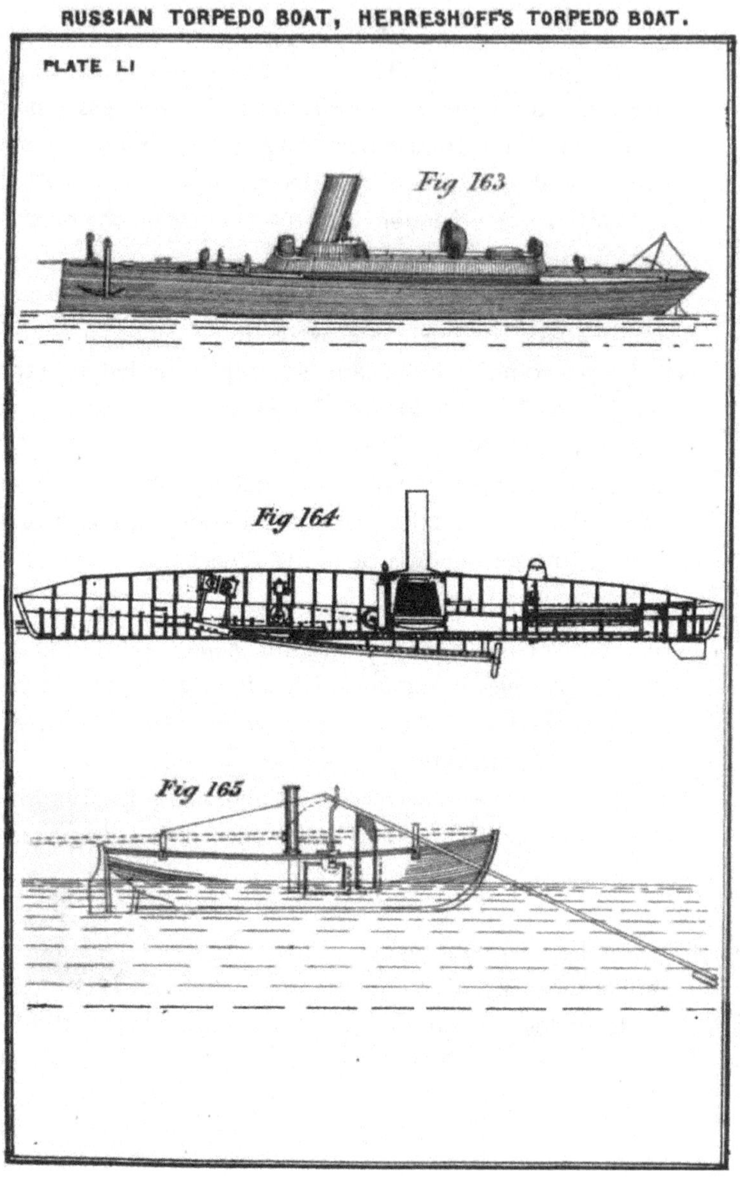

Figure 85. Three types of Russian Spar-Torpedo Boats (Sleeman, 304)

RUSSIAN SUBMARINE MINE, FIRING BY OBSERVATION.

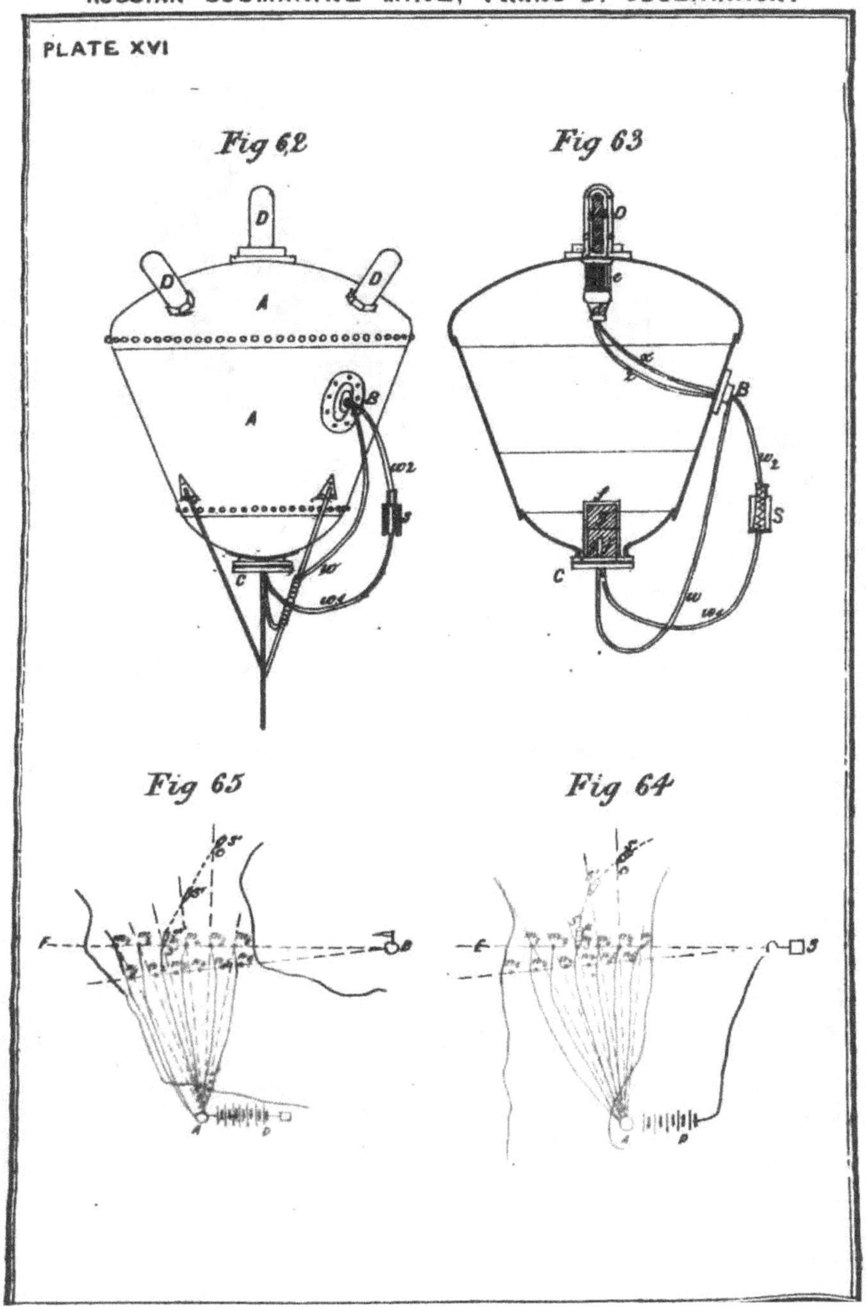

Figure 86. Russian submarine mine fired by contact or observation. (Sleeman, 120)

Figure 87. USN *Wooden Torpedo Boat No. 1,* or *S.S. Stiletto.* The Herreshoff torpedo boat was a great breakthrough in the marriage of technology and craftsmanship. This is the third boat to be a USN No. 1 craft. (U.S. Navy Historical Department)

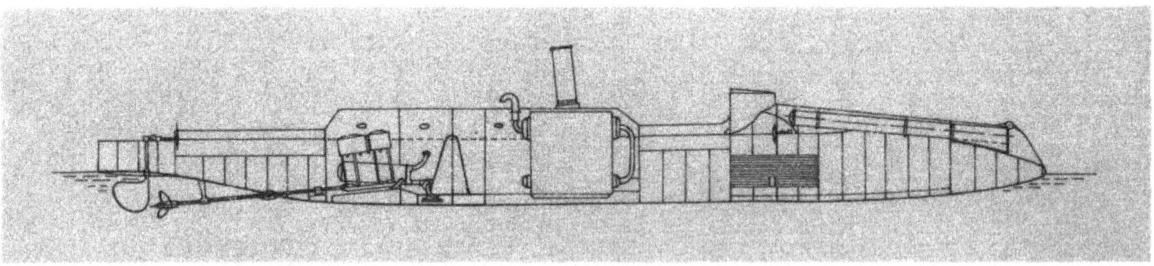

Figure 88. Italian Navy *Nibbio.* (Sleeman)

Date	Craft, Name, Event or Country	Event, Builder and Boat Specifications	City, River, Lake, Sea, Ocean
1877[221]	U.S.Navy *Stiletto*	The U.S. Navy bought a fast steam yacht built by the Herreshoff brothers. *The S.S. Stiletto* is designated as *Wooden Torpedo Boat No.1.* **Speed:** 18.2 knots: **Length:** 95'-0"; **Beam:** 11'-6"; **Draft**: 3'-0", **Displacement:** 31 tons; **Crew:** 15; **Range :** Unknown	Bristol, Rhode Island
1878[222]	January 25 Russian Navy *Chesma*	Two Russian torpedo boats set out to attack Turkish ships and sent two Whitehead "Fish" torpedoes in a Turkish revenue steamer. The Russian Navy had adopted very small MTB designs. They were known as "Minonosk" (torpedo cutters). The boats sank the Turkish guard ship *Initibah.* This was the first successful use of the Whitehead torpedoes. It was the final attack in the Russo-Turkish war (1877-78)	Batoum, Black Sea
1878[223]	Italian Navy *Nibbio*	Italy's first single purpose torpedo boat the *Nibbio* is built by Thorneycroft. The Italians were among the first nations to realize the usefulness of shallow draft, fast, boats to guard their shores.	Italy, Adriatic Sea Tyrrhenian Sea
1878[224]	British Navy *Hull No. 44*	The Herreshoff brothers build their first all steel spar torpedo boats. A separate 2-1/2 hp engine drove the Sturtevant blower for the engine. The first crew called the boat "The Coffin." Mr. G.R. Dunnell of London purchased the boat and gave a demonstration to the Royal Navy on the Thames. The Royal Navy purchased *Hull No.44* and designated the boat as: *H.M. 2nd Class Torpedo Boat No. 63* **Speed:** 16 knots; **Displacement:** 7.5 tons; **Length:** 59' ; **Beam:** 7'-6"; **Draft:** 2'-9"; **Engines:** 1-Compound steam engine, single screw; Crew:4	America England

---

[221] Simpson, Richard V, *Building The Mosquito Fleet: The U. S. Navy's First Torpedo Boats*, (Arcadia Publishing, 2001); *Historical Transaction* 1893-1943, 356

[222] Sleeman, 336; Grant, I.A. "The Herreshoff Spar Torpedo Boats of 1878-1880" *Warship International,* 34, No. 4, 1987, 350-355

[223] Fock, Harald. , *Fast Fighting Boats 1870-1945: Their Design, Construction and Use* (Naval Institute Press, 1978), 11

[224] Grant, I.A.. "Herreshoff Spar Torpedo Boats of 1878-80" (*Warship International* No.3, 1977), 259.

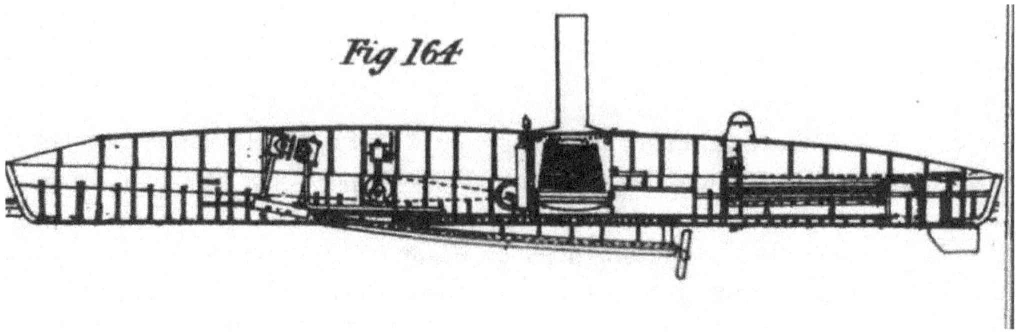

Figure 89. The British Navy purchased *Hull No.44* From the Herreshoff brothers and the British sailors dubbed the new boat "The Coffin." (Sleeman)

Figure 90. Feeding the era's coal-fired engines was hard, dirty work. (Library of Congress Prints and Photographs Division)

Date	Craft, Name, Event or Country	Event, Builder and Boat Specifications	City, River, Lake, Sea, Ocean
1879[225]	Italian Navy *Avvoltoio*	This was a breathtaking design launched into the modern era of naval warfare. She was rated as a 2nd Class Torpedo Boat, fast and lightly armed. This boat was among the first of this class to carry a "Bow Rudder." **Speed:** 21.3 knots; **Displacement:** 25 T; L:86'0"; **Beam:** 11'-0"; **Draft:** 2'-9"; **Engines:** Compound condensing direct-acting, 2-cyl, 100hp, 1 screw; **Armament:** 2-14" torpedo tubes, one 1-pounder. revolving cannon; **Crew:** 20	Italy
1879[226]	December 11 Herreshoff vs. Thorneycroft Race	The Herreshoff boat was up against the Thorneycroft 2nd Class Torpedo Boat No. 56. The course was 2⅔ land miles. The boats had to start from the tug *H.M. Manly* and from a cold-water start run to Spit Buoy. The finished official resultes were "Thorneycroft: total time from cold, 22min. 21 sec., Herreshoff: 24 min.30 sec." The British boats won the race but the Admiralty liked the rapid starting capacity of the Herrshoff boilers. Herreshoff supplied Thorneycroft with rapid-fire boilers until 1882.	England
1880[227]	April 10 Chilean Navy *Guacolda*	The torpedo *Guacold*a attacked the Peruvian corvette *Union* with two armed, and charged spar torpedoes. One was broken off in a collision with a fishing boat, the other spar charge went off when in came in contact with the boom logs surrounding the ship.	Bay of Callao, Peru
1880[228]	Herreshoff *Hull No. 64*	The Herreshoff brothers made this boat for the Russian Navy. It would influence Russian Torpedo boat design for decades. This would be the last spar torpedo boat the company made. **Speed:** 23 knots; **Length:** 59'-0"; B: 6'-6"; **Draft:** 4'-7", **Engines:** Compound condensing 2-cyl; **Armament:** 4-spar torpedoes.brothers	Bristol, Rhode Island

[225] Jackson, 41.
[226] Grant, I.A. 253.
[227] Sleeman, 336.
[228] Grant, I.A. 253-261

Figure 91. The attack on the *Cochran.*

Figure 92. Chilean naval ensign.

Date	Craft, Name, Event or Country	Event, Builder and Boat Specifications	City, River, Lake, Sea, Ocean
1880[229]	April 23 Chilean Navy *Janequeo, Guacolda*	The two spar torpedo boats *Janequeo* (Yarrow) and *Guacolda* moved in to attack Peruvian steam launch. The electrically fired torpedo exploded prematurely causing no damage other than a severe shaking.	Bay of Callao, Peru
1880[230]	May Peruvian Navy	The Peruvians used drifting torpedoes in an attempt to destroy the Chilean blockading ships at the mouth of the Boqueron Channel. The Chilean guard ship discovered the McEvoy drifting torpedoes and destroyed them.	Boqueron Channel, Callao, Peru,
1880[231]	June Chilean Navy *Janequeo, Guacolda*	The *Janequeo* and *Guacolda* make an attack on three large Peruvian gunboats in the Bay of Callo. A mutual explosion of the torpedo sinks one Peruvian gunboat and the *Janequeo.*	Bay of Callo, Callo, Peru
1880[232]	July 3 Peruvian Navy	The Peruvian Navy stepped up the destruction of ships by loading a supply ship with provisions and 300 pounds of dynamite. It was a trap rigged to explode only when the Chileans started to unload the prize provisions. The ship was set adrift and when the Chileans on the *Loa* started the unloading process, it blew, sinking the *Loa* and killing 100 men of the ship.	Bay of Callo, Callo, Peru
1880[233]	October 13 Peruvian Navy	The Peruvians adopted the design of a large framework with 70 barrels of gunpowder from the Confederate Torpedo Department. A clockwork firing mechanism was mounted to a small lighter. The lighter was moored to the anchorage of the Chilean ironclad *Cochrane.* During the night, the lighter drifted out of position and when it exploded with a huge eruption, the ironclad was not scratched.	San Lorenzo Anchorage

---

229 Sleeman, 336.
230 Sleeman, 336.
231 Sleeman, 337.
232 Sleeman, 337.
233 Sleeman, 337.

Date	Craft, Name, Event or Country	Event, Builder and Boat Specifications	City, River, Lake, Sea, Ocean
1880[234]	August John Ericsson *Destroyer*	The first public viewing of the *Destroyer* piqued interest around the world. Here at the construction site his ship, Ericsson said, " Ironsides are doomed... we can sink an enemy without ram, steam launch or spar torpedo... " Ericsson's design was about to make a huge impact on naval architecture and naval tactics. Destroyer- Torpedo Gun **Length:** 25'-6"; **Diameter:** 16"; **Total weight:** 1,500 pounds; **Charge:** 300 pounds guncotton; **Range:** 400'-700'; **Speed of charge:** 300' in 3 seconds	Sandy Hook
1880[235]	December 6 Chilean Navy	Three torpedo boats, *Frezia, Guacolda,* and an English Thorneycroft attacked the Peruvian gunboat *Arno.* The ironclad put up a fight and sent a round through the hull of the *Frezia,* which sank her. The other two boats speedily withdrew and the Arno escaped.	Bay of Callo, Peru,
1881[236]	January 3 Peruvian Navy	This was the first use of the newly purchased Herreshoff Torpedo Boat. The boat set out to attack a Chilean ironclad. The ironclad could not be found, and the two towed-torpedoes proved nothing but a hindrance.	Ancona
1882[237]	Imperial German Navy *Schutz* (Defense)	The German designers laid down six 1st class torpedo boats of transverse steel construction, galvanized, with six watertight compartments. These boats were the *Schutz* (Defense), *Flink* (Agile), *Scharf* (Sharp), *Tapfer* (Brave), *Kunh* (Bold), *Vorwarts,* (Onward), *Scicher.* **Speed:** 17.9 knots, **Armament:** 1-3.7cm MG, 2-35cm Bow tube w/4-reloads.	Germany

---

[234] *John Ericsson, A Register of His Papers in the Library of Congress,* Manuscripts Division, Writings File, 1883–1880, Box 5, Reel(s) 5–6, papers 8, 9, 19, 40, 71, 72, 134, 135, 236, 241, 242, 250. *General Correspondence* 1821–1880, Box 4, Reel(s) 4–5. Van Der Weyder, P.H, M.D, Editor. "Captain Ericsson's New Torpedo Boat, *The Manufacturer and Builder* X (No. 10, October, 1878). White, Ruth. *Yankee From Sweden, the Dream and the Reality in the Days of John Ericsson* (Henry Holt and Company, 1960). Barnard, Charles "Ericsson's *Destroyer* and Her New Gun," *Scribner's Monthly Magazine,* XXI, No. 5, (March 1881) 689-693.

[235] Sleeman, 337.

[236] Sleeman, 338.

[237] Fock, 12.

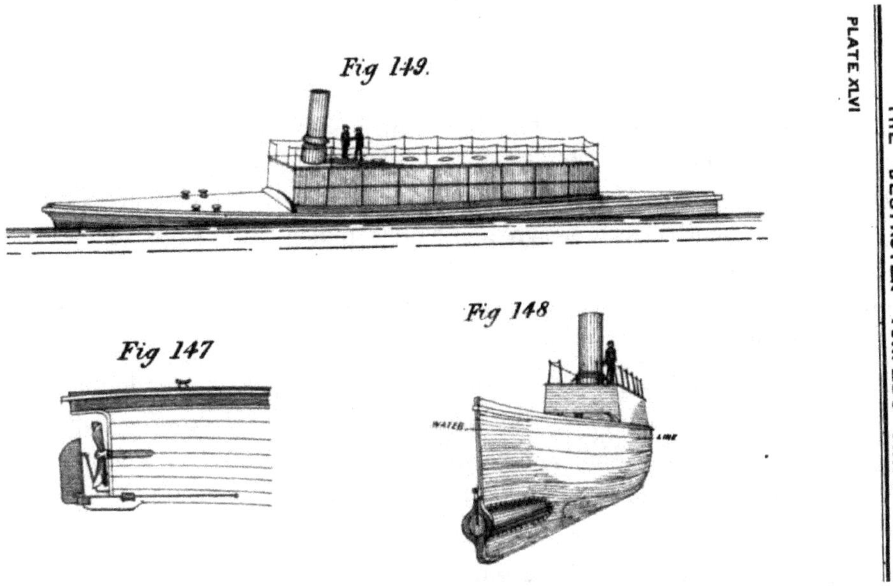

Figure 93. Ericcson's *Destroyer.* (Sleeman, 278)

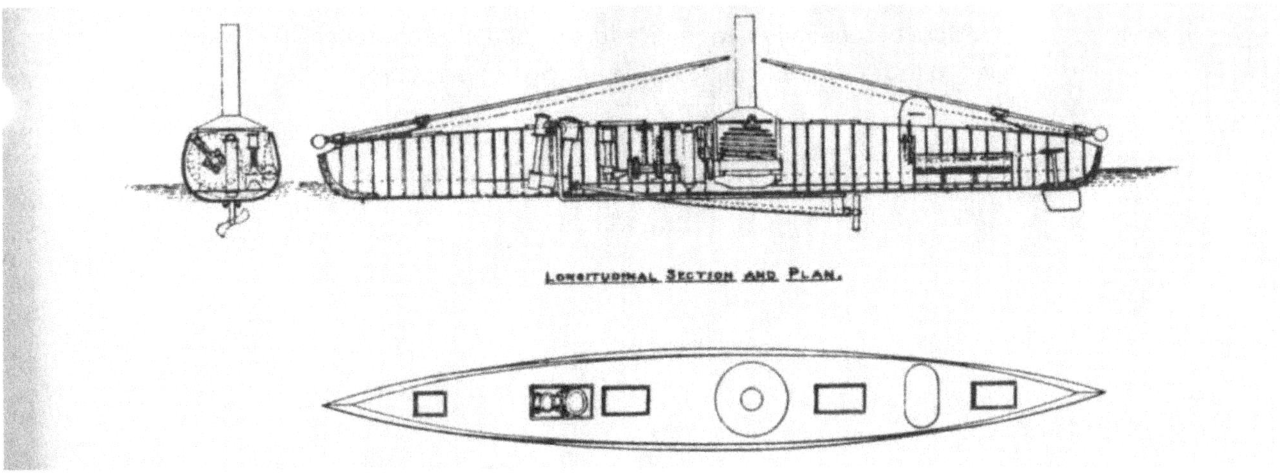

Figure 94. Herreshoff Torpedo Boat made for the Peruvian Navy. (Sleeman, 304)

Date	Craft, Name, Event or Country	Event, Builder and Boat Specifications	City, River, Lake, Sea, Ocean
1882[238]	Japanese Navy *Jeune École*	The Japanese adopted the philosophy of the French *Jeune École* ("young school") and their Diet called for the construction of 48 French designed torpedo boats. This philosophy became the master architect for the Japanese Navy as it designed and built smaller faster ships and deadly torpedo boats.	Japan
1883[239]	British Royal Navy *MTB-85*	The Thorneycroft Yards developed the best-designed torpedo boat in the 1800's. It was a "day boat" which required a "mother ship" for extended operations. The mother ship allowed the crew adequate rest; while they slept, mess, mechanics could work on the boat's engines and rearm and top with coal for the next day's operations. **Speed:** 17.3 knots; **Displacement:** 12.89 tons; **Length:** 19.20m; B: 2.25m; **Draft:** 0.46m; **Armament:** 2-Bow Tubes w/ 4-35.6cm torpedoes.	England
1883[240]	Dr. Gustave Laval	The doctor invented the concept of blowing air through slits in the bottom of speedboats to reduce friction and increase speed. This work led to today's modern air cushion and hydroplane craft.	Germany

[238] Evans, David C. and Peattie, Mark R. *Kaigun Strategy, Tactics, and Technology in the Imperial Japanese Navy 1887-1941* (Annapolis: Naval Institute Press, 1997), 15.
[239] Fock, 12-13.
[240] Fock, 16.

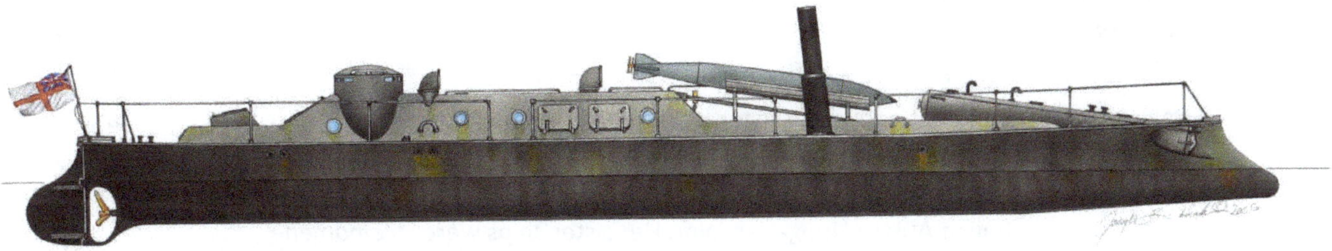

Figure 95. HMS *2nd Class TB 85.* (Illustration by author)

Figure 96. U.S. Navy sailors getting clothes prior to a prisoner exchange during the Civil War.(Library of Congress Prints and Photographs Division)

Date	Craft, Name, Event or Country	Event, Builder and Boat Specifications	City, River, Lake, Sea, Ocean
1883[241]	Italian Navy *Euterpe* *TB-No.37*	This boat and seven sister boats were built by Thorneycroft for coastal defense. With high speed, shallow draft, these boats could hunt down larger ships in the Adriatic. They were decisive weapons platforms in the Italian-Austro-Hungarian War. Her sister ships were; *Melpomene, Tersicore, Polimnia, Urania* and *Calliop.* **Speed:** 19 knots; **Displacement:** 13.3 tons; **Length:** 63'-0"; **Beam:** 7'-6"; **Draft:** 3'-9"1-Verticle triple expansion engine, 1 screw; **Armament:** 1-25mm MG, 2-14" TT; **Crew:** 9-15,	Italy
1883[242]	Italian Navy *Erato*	The Thorneycroft Yards produced a second design for lightweight class of torpedo boat. This last evolution of designs would greatly influence Italian small craft design for operations in the Adriatic and Mediterranean Seas.	Italy
1884[243]	August 23 Chinese Navy *Young-Woo*	The French introduce its newest design in torpedo boats in the French-Chino wars of 1883-85. The French torpedo boat sank the Chinese frigate *Young-Woo* by a spar-torpedo off shore of Foochow.	Foochow, China
1885[244]	February 15 Chinese Navy *Yu-Yuen*	The Chinese lost the frigate *Yu-Yuen* to two French spar-torpedo boats. The frigate sank in the harbor guarded by heavy gunned shore batteries. This action would move the Japanese to adopt France's naval philosophy and spur close relations in ship construction for decades.	Sheipu, China
1885[245]	July Chinese Navy *Wei-Yuen*	The ship was built as an armed sloop. She later disarmed and designated as a training ship. The Japanese *Torpedo Boat No. 23* torpedoed and sank her at Wei Hai Wei.	Battle of Wei Hai Wei China

---

[241] Jackson, 139.

[242] Jackson, 127.

[243] Sleeman, 338.

[244] Sleeman, 338.

[245] Evans and Peattie, *Kaigun.*

Figure 97. The Chinese flagship *Yangwu* and the corvette *Fuxing* under attack by French torpedo boats *No. 46* and No. *45*. *Combat naval de Fou-Tchéou* ('The naval battle at Foochow'), by Charles Kuwasseg, 1885. (Wikipedia Commons)

Figure 98. Chinese national ensign.

Date	Craft, Name, Event or Country	Event, Builder and Boat Specifications	City, River, Lake, Sea, Ocean
1885[246]	German Navy 1st Class TB	The Germans brought their plans for their new model torpedo boat to J.S. White & Co., Isle of Wight. These boats were designed to be carried aboard larger ships and were direct predecessors of the fast motor torpedo boats. **Speed:** 15 knots; **Length:** 37.72m; **Draft:** 2.6m; **Armament:** 1-35cm Bow, above water, 1-3.7cm MG, **Engines:** 1-verticle expansion;	
1885[247]	Imperial German Navy *S-1*	The line of boats *S-1 to S-6* was built by the yards of F. Schichau, Elbing, Germany. At this time, these boats gained a reputation for excellent handling. **Speed:** 19.3 knots; **Displacement:** 84 tons; **Length:** 37.72m; **Beam:** 4.92, **Draft:** 1.07m, fwd, and 2.3m aft; **Crew:** 15; **Armament:** 1-35cm TT; **Engines:** 3-cyl. triple expansion; One Screw; **Range:** 1,400nm @ 10k, 250nm @18k.	Elbing, Germany
1885[248]	French Navy *Bombe*	The vessels of this class carried three masts and had a ram shaped bow. Much of the vessel was covered in 0.5" armour plate. Speed: 18-19 knots; **Displacement:** 369 tons; **Armament:** 2-14" torpedo tubes, 4-3 pounder guns, 3 1-pounder revolver guns, **Crew:** 70; **Engines:** 2-vertical expansion; 2 screw	France
1885[249]	*Chinese Navy Fu Lung*	The boat was the most modern of the Chinese fleet. It was a steel-hulled, first-class ocean going torpedo boat. The torpedoes were built into the bow and reloading could take place in armour-covered spaces. At the fall of Wei-Hai-Wei the boat was captured by the Japanese and renamed *Fukuryu.* **Speed:** 18 knots; **Displacement:** 128 T: **Length:** 144'-4"; **Beam:** 16'-5"; **Draft:** 7/-6"; **Engines:** 1-triple expansion, 1 screw: **Armament:** 2-14"TT with reloads, 3-small machine guns.	

---

246 Fock, 12.
247 Fock, 14.
248 Jackson, 56.
249 Jackson, 169.

Date	Craft, Name, Event or Country	Event, Builder and Boat Specifications	City, River, Lake, Sea, Ocean
1885[250]	November 24 Spanish Navy *Orion*	Germania-Werf in Kiel, Germany built what is considered the best torpedo boat the Spanish navy ever had until they bought German S boats in 1942. **Speed:** 20 knots; Disp:88 T; **Length:** 37.70m;B: 4.78m; **Draft:** 2.04m; **Engines:** 1,000 hp, 1- Screw; Range: 2,000 nm; **Crew:** 18; **Armament:** 2-356cm torpedo tubes, 2-25mm/42 cal R$V MG.	Spain
1886[251]	July 22 John Ericsson *Destroyer*	The British Admiralty held trials for Ericsson's boat and Torpedo Gun. The *Destroyer* ran a straight course and blew up its target at 300' in 2 seconds. His gun proved viable and the warhead deadly.	England
1886[252]	Spanish Navy *Habana*	The Spanish were very aggressive in keeping a modern fleet of 13 1st Class torpedo boats for its large contingent of ironclads and capital ships. This was the fastest torpedo boat of the era. Speed: 24.5 knots; **Displacement:** 68 tons; **Length:** 127'-7"; **Beam:** 12'-7"; **Draft:** 5'-0" **Armament:** 2-14"TT Bow mounted; 1-Heavy machine gun; **Engines:** 1 screw, 1-triple expansion eng. **Crew:** 24	Spain
			Spain

[250] The Spanish American War Centennial Website. "Spanish Torpedo Boats, First Class." http://spanamwar.com/spantorpedoboats.htm#JULIAN%20...

[251] Barnard, Charles. "Ericsson's *Destroyer* and her New Gun." *Scribner's Monthly* Magazine XXI, No. 5 (March, 1881), 689-693.

[252] Jackson, 214.

Figure 99. John Ericcson's *Destroyer.* (Illustration by author).

Figure 100. The *Destroyer,* profile. (Illustration by author)

Date	Craft, Name, Event or Country	Event, Builder and Boat Specifications	City, River, Lake, Sea, Ocean
1886[253]	Spanish Navy *Ariete, Rayo,* 1st Class Torpedo Boats	The *Ariete (Ram)* and *Rayo (Ray) were* built by Thorneycroft. The boats set speed records for their day. The *Ariete* made 27 knots in record time. None of the Spanish torpedo boats were not ordered to Cuba and the Spanish-American War. **Speed:** 26 knots; **Length:** 45m; **Beam:** 4,43m; **Draft:** 1.4m; **Armament:** 2-14" Whitehead torpedo tubes, w/2/reloads, 4-42mm Nordenfelt mg; **Engines:** 2 screws; **Crew:** 25	
1886[254]	Spanish Navy *Destructor*	This was one of the first vessels to be fitted with triple-expansion engines. **Speed:** unknown; **Displacement:** 458 tons; **Length:** 192'-6"; **Beam:** 25'; **Draft:** 7'-0"; **Armament:** 5-15" torpedo tubes, 1-89mm cannon	Spain
1886[255]	British Royal Navy *Rattlesnake*	These ships were built to counteract the quickly expanding fleet of torpedo boats being built by many countries. Russia and its large torpedo boat fleet were deemed an immediate threat with the threat of war over Afghanistan. The ship had 19mm (0.75 in) steel protective decks and carried a 102mm gun forward. **Speed:** 19.2 knots; **Displacement:** 550T;**Length:** 200' ; **Beam:** 23' ; **Draft** :10'-4; **Engine:** 2 triple expansion, 2 screws; **Armament** 4-14"torpedo tubes, 1-102mm gun; **Crew:** 66	Britain

[253] The Spanish American War Centennial Website, *ibid.* http://www.spanamwar.com/spantorpodboats1stclass.htm#Azor.
[254] The Spanish American War Centennial Website, *ibid.* http://www.spanamwar.com/spantorpodboats1stclass.htm#Azor.
[255] Jackson, 274.

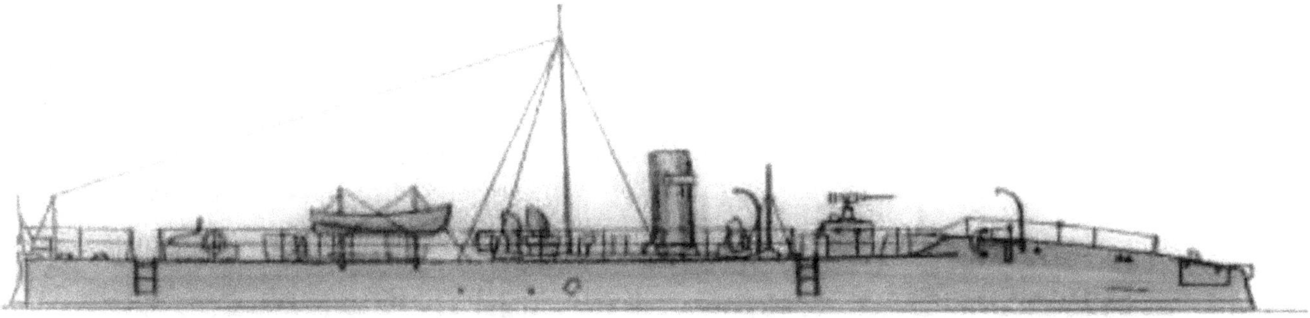

Figure 101. Spanish Torpedo Boat *Habana.* (U.S. Navy Historical Department)

Figure 102. Spanish Navy *Destructor.* (U.S. Navy Historical Department)

Date	Craft, Name, Event or Country	Event, Builder and Boat Specifications	City, River, Lake, Sea, Ocean
1886[256]	France *Balny*	The French Normand shipyards built 10 boats modeled on their highly successful torpedo boat, *TB POTI,* for the Imperial Russian Navy. This version had one major flaw: it rolled badly in high seas. **Speed:** 19 knots**; Displacement**: 66 T; **Length:** 134'; B: 11'; **Draft:** 3'-3"; **Armament:** 2-14' TT in bow, two 1-pounder Guns; **Engines:** 1-triple compound, 1-screw **Crew:** 22	France
1886[257]	Russian Navy *Viborg*	This model showed the strong hold that "Old School" navy still had on the ship building industry, worldwide. Even though no capital ship carried sails, the top brass still wanted to put sails on something. This boat was the largest torpedo boat of the period. The Thompson's Yard at Clyde Bank, Scotland, became specialists in this field. The forward bow area had heavy armor plate to the water line. **Speed:** 20 knots; **Displacement:** 169 T; **Length:** 142'-6"; **Beam:** 17'; **Draft:** 7' **Armament:** three 15" torpedo tubes, 2-37mm revolving Hotchkiss cannons, **Engines:** 2-Vertical compound, 2 screw, Three masts w/ sails; **Crew:** 21;	Russia
1888 [258]	Japanese Navy Torpedo Boat *Kotaka*	Japans first torpedo boat is built in England to their specifications. This model would become the forerunner of the torpedo boat destroyer. The hull was built in sections and then shipped to Japan for assembly. This design, while slower then other torpedo boats of the time, did prove that larger torpedo boats could play a major role when they accompanied major warships of the high seas fleets. **Speed:** 19 knots; **Displacement**: 203 tons; **Length:** 164'; **Armament:** 4 1-pounder quick firing gun, 4-14 inch Schwartzkopf torpedos, 2 tubes bow, 2tubes stern	Japan

---

[256] Jackson, 45.
[257] Jackson, 299.
[258] Evans and Peattie, 17.

Date	Craft, Name, Event or Country	Event, Builder and Boat Specifications	City, River, Lake, Sea, Ocean
1889[259]	Italian Navy *Avvoltoio*	The second ship of this name had evolved into a true ocean-going vessel. The ship had the same rigging as a three-masted schooner and could keep up with the fastest of larger ships. They possessed a good radius of action and saw a great deal of service in the Dardanelles against the Turkish fleets. **Speed:** 26.6 knots; **Displacement:** 130 tons; **Length:** 152'; **Beam:** 17'-2"; **Engines:** 2-vertical triple expansion, 2 screws; **Draft:** 7'-9" ; **Armament:** 3-14" torpedo tubes, 2 3-pounder guns; **Crew:** 65	Adriatic, Black Sea
1889[260]	Russian Navy *Kapitan Saken*	This vessel was by herself considered a threat to her other European neighbors. She had five torpedo tubes (17") in fixed positions; two in the Bows, one in the stern, and two tubes on swivel mounts midships. There were four large caliber guns, two each side on sponsons. Extra loads of torpedoes were carried below the armoured deck and moved into position on small rails.	Baltic Sea
1889[261]	Austria *Flamingo*	This was one of ten small torpedo boats built to offset the advances of the Italian navy in torpedo boat design. They had been based on an earlier successful design of the Schichau boats. She was sunk by enemy gunfire on August 23, 1914, off Pola. **Speed:** 19 knots; **Displacement:** 91.4 T; **Length:** 130'; B: 15'-9"; D 6'-3"; **Engines:** 1- triple expansion, 1 screw; **Armament:** 2-13.8" torpedo tubes, 2-MG	Austria, Adriatic, Black Sea, Dardanelles

[259] Jackson, 42.
[260] Jackson, 237.
[261] Jackson, 153.

Figure 103. Imperial Japanese Navy Torpedo Boat *Kotaka.* (Wikipedia)

Figure 104. Austro-Hungarian Navy Torpedo Boat 26 *Flamingo.*

Date	Craft, Name, Event or Country	Event, Builder and Boat Specifications	City, River, Lake, Sea, Ocean
1890[262]	British Navy *Gossamer*	The British built these larger and better-armed boats as a counter to the new threat of the improved French designs. The13 ships of this class came with rapid fire 120mm quick-firing guns mounted fore and aft. The boats were originally known as the *Sharpshooter* class. **Speed:** 19 knots; **Displacement:** 735 tons; **Length:** 230'; **Beam:** 27'; **Draft:** 10'-6"; **Engines:** 2 triple expansion, 2 screws; **Crew:** 91	English Channel, North Sea, Sea of Ireland
1890[263]	U.S.Navy *Cushing TB-1*	The Herreshoff company built their first modern torpedo boat for the U.S. Navy. The *Cushing* was all steel and the design came from the Newport Naval Torpedo Station, Rhode Island. She carried the familiar two bow torpedoes and one on a swivel mount midships. Torpedo reloads came from below decks. Its original cost: $82,750 **Speed:** 23 knots; **Displacement:** 116 Tons; **Length:** 140'; **Beam:** 15'-1"; **Draft:** 4'-10"; **Engines:** 1,720 HP; 2-vertical quadruple-expansion engines; **Armament:** 3-18" torpedo tubes, 3 6-pounderguns; **Crew:** 22	Atlantic Ocean
1892[264]	French Navy *Grondeur*	The boat was a test bed for the newly made Thorneycroft Water Tube boiler. Like the Herreshoff lightweight boiler, it was much lighter and improved speed performance as well as cut down the time between cold boilers to ready to run time. **Speed:** 23.5 knots; **Displacement:** 131 tons; **Length:** 147'-10"; **Beam:** 14'-6"; **Draft:** 4'-6"; **Engines:** 2 triple expansion, 2 screw: **Armament:** 2-14" torpedo tubes, 2-74mm guns; **Crew:** 75; Range: Unknown	France

[262] Jackson, 196.

[263] Simpson, *ibid.;* DANFS.

[264] Jackson, 206.

Figure 105. The USN's *TB-1 Cushing* was named for U.S Navy Lieutenant John Cushing and his successful attack on the CSS *Albemarle*. (U.S. Navy Historical Department)

DECK TUBE AND PROJECTILE OF A TORPEDO BOAT

Figure 106. Torpedo boats world wide had the disadvantage of always working in the open spaces in even the worst of gun battles at sea. The death toll among these crewmen was always high. (U.S. Navy Historical Department)

Date	Craft, Name, Event or Country	Event, Builder and Boat Specifications	City, River, Lake, Sea, Ocean
1893[265]	February 3-4 Japanese Navy *Torpedo Boat No.6*	Tem Japanese torpedo boats made the first close-in attacks in the East Asian waters on February 3-4 at Wei-Hai-Wei Harbor, China. Swartzkopff torpedoes were used. The boats made several night attacks in freezing weather with the torpedo men suffering from severe frostbite. They hit and sank the Chinese battleship *Tung-Yuan* and the *We Yuen, Sai Yuen, Chen Yuan.*	Wei-Hai-Wei Harbor, China
1895[266]	France Torpedo Boat *Forban*	The Normand Yards built the boat that quickly became known for outstanding performance, open ocean sea-keeping qualities, and the most powerful engines of any torpedo boat in the world. The boat achieved the speed of 30 knots for the first time ever for a torpedo boat. This boat drove the British to step up design and production of Torpedo Boat Destroyers. **Speed:** 30+ knots; **Displacement:** 152 T; **Length:** 144'-4"; B: 15'-3"; **Draft:** 4'-5"; **Engines:** 2 triple expansion, 2 screws; **Armament:** 2-14" torpedo tubes; **Crew:** 57; **Range:** 1,000 nm.	Bay of Biscay, English Channel, Mediterranean
1898[267]	Russia Torpedo Boat *Nibord*	This was the first seagoing torpedo boat to run on fuel oil. Later in WWII, the Russians themselves would revert to highly flammable petrol engines for their motor torpedo boats.	Black Sea, Baltic Sea
1898[268]	July 29 British Navy *1st Class TB TB.28*	The Boat was stationed of the waters of South Africa as a training vessel for stockers. The boat ran aground in Kalk Bay, Valsbaai Bay, Cape Town. She was considered a "Construction Total Loss," towed to sea and used as a target ship. Builder: Chiswick, 1886 **Displacement:** 60 tons, **Length:** 129', B: 12'; **Armament:** 4-torpedoes	Cape Town, South Africa

---

[265] Evans and Peattie, 41, 49; Gray, *ibid.*, 150-151.

[266] Jackson, 160.

[267] Greger, *ibid.; Historical Transaction*, 357

[268] Hepper, David. *British Warship Losses in the Ironclad Era, 1860-1919*, (Chatham: 2007), 15.

Date	Craft, Name, Event or Country	Event, Builder and Boat Specifications	City, River, Lake, Sea, Ocean
1898[269]	September 25 U.S Navy *T.A.M. Craven* *Torpedo Boat* *No.10*	The boat was named for the Civil War U.S. naval hero Captain T.A.M. Craven, commander of the USS *Tecumseh.* Plans for the boat were provided by the famous French firm, Normand of Le Harve. The boat performed training and experimental work at Naval Torpedo Station, Newport, Rhode island. In 1913, off the coast of Georgia, a boiler exploded, injuring several men and came close to foundering. She was rescued by a tugboat the Estelle and towed through heavy seas to Fort Screven. The old boat was finally decommissioned on November 13, 1913 and then sunk as a target on November 15, 1913.	Atlantic Ocean
1899[270]	Italian Navy Torpedo Boat *Pellicano*	The boat had inherent good sea-keeping abilities but her machinery was never up to the task designed. This was the last heavy torpedo boat that Italy would ever build. She did prove that high speed and maneuverability were essential and a smaller size vessel was imperative. **Speed:** 21 knots; **Displacement:** 183 T; **Length:** 159'-10"; **Beam:** 18'-10"; **Draft:** 5'-0"; **Engines:** Unknown; **Armament:** 2-14"torpedo tubes, 2-37mm guns; **Crew:** 25	1899[271]
1899[272]	U.S. Navy *Dahlgren* *Torpedo Boat* *No. 9*	The *Dahlgren* became a class of torpedo boat in the U.S. Navy. Her machinery was much improved over that of the *Craven.* The boat stayed in service as training vessels and during World War I. She was commissioned April 1, 1917, as *Coast Torpedo Boat No. 4,* an escort and harbor entrance patrol boat. She was decommissioned on March 11, 1919. **Speed:** 30 knots; **Displacement:** 146 T:**Length:** 161'-4"; **Beam**: 16'-4"; **Draft:** 4'-8'; **Engines**; 2 triple expansion, 2 screw; **Armament:** 2-18 torpedo tubes, 4-1-pounderGuns; Crew:28	Atlantic Ocean

[269] DANFS, I, 3.

[272] U.S. Navy Historical Center. "Dahlgren." http://www.history.navy.mil/DAFS/Di/Dahlgren-i.htm.

[272] Jackson, 270.

DAHLGREN

THE SHIP KILLERS

THE DEFINITIVE ILLUSTRATED HISTORY OF THE TORPEDO BOAT

CMB-4 was one of the British boats that made a successful raid on the gigantic Russian fortress, Kronstadt, in August 1919. This small boat and a few others almost destroyed an entire Russian fleet while in its harbor. After this event, British naval doctrine included the torpedo boat in its overall naval strategy. (Illustration by author)

VOLUME III: TIMELINE 1900 - 1939

BY JOE HINDS

NIMBLE BOOKS LLC

CONTENTS (INCLUDING FIGURES)

DEDICATION

Hans Frank, retired admiral of the German navy, has spent months assisting me with a deeper understanding on the the activities of Kriegsmarine S-boats. He has patiently explained nomenclature, relevance to events and corrected some pretty dumb statements I have made. He very kindly pointed out the places that I just plain got the facts wrong. For his efforts, I have a better book.

PREFACE TO THIS VOLUME

In the last days of 1898 some hard to accept facts came to light for U.S. battle fleets. The big guns of both Spanish and American fleets had little to do with victory or failure. The Spanish squadron was totally destroyed and in the last battle, the U.S. did not lose a single sailor. Of the 130 strikes on Spanish warships only two came from capital ships, the rest from cruisers and destroyers. Now you have to keep in mind that the U.S. squadron expended 6,000 shells of all sizes in this action. In the ensuing years this figure would only change a few percentages; in the twenty years from Santiago to Jutland the shooting by big ships was not improved.

These deficiencies in the faith of battle fleets opened the door to a new naval doctrine called the *Jeune École* (Young School). The new naval doctrine was a French invention from the 1880's and was worked out by some daring men not opposed to new ways of thinking. The founder of this discipline was Admiral Hyacinthe-Laurent-Theophile Aube. He joined the French navy when he was seventeen and was assigned to many stations in Africa, the Far East and West Indies. From his travels and experiences he reasoned that the capital ship was doomed. He saw for himself in 1883 that battleships were useless against the torpedo attacks. From making a study of after action reports from the U.S. Navy senior officers of the various blockading fleets in 1898, their single biggest fear was the torpedo and torpedo boat.

In 1905 the Japanese Navy shocked the world by defeating the mighty Russian battle fleet not with heavy gunfire, but torpedoes launched from destroyers and torpedo boats. This pattern accelerated during World War I.

According to Rodrigo Garcia y Robertson [273],"World War I revealed just how much the world had turned. A quarter million tons worth of capital ships were lost to enemy torpedo vessels during the conflict...capital ships sank 17,000 tons worth of torpedo vessels...25,000 tons of friendly warships accidentally run over and sunk by their own capital ships". He added, "Mines

[273] *Failure of the Heavy Gun at Sea,* Technology and Culture, 1987, The Johns Hopkins University Press.

and torpedoes accounted for 70 percent of combat losses among major warships."

Once World War I ended, Italy, France, Austro-Hungary, Russia, and a host of other smaller nations went into high gear to develop torpedo boat fleets. Germany came on board in the 1920'2 with England not far behind. Torpedo boats were not regulated by the interwar naval treaties, which added to their appeal, and engine technology was advancing rapidly. The race was on for superiority in small fast and deadly boats. After decades of procrastination, caused partly by geography (wide oceans separating our ports from possible enemies) and partly by classic senior officer bull-headedness, the U.S. Navy would not join the fray until after war commenced in September 1939.

TIMELINE 1900 - 1939

Date	Craft, Name, Event or Country	Event, Builder and Boat Specifications	City, River, Lake, Sea, Ocean
1904[274]	French Navy Count Recope	The Count is given credit for the first "authentic" torpedo boat. His integral bow-firing torpedo tube is an offshoot of the same device first used by the *Spuyten Duyvil*, 1864 and Ericsson's *Destroyer,* 1871. Reloading the tube for the crew meant they would be in relative safety bellow deck. This design feature was not adopted by torpedo boat designers until the German LM torpedo boat of 1917 was built. **Speed:** 16.3 knots; **Armament:** One integral bow-torpedo tube (TT); **Length:** 52'-5"; **Beam:** 8'-9"; **Draft:** 3'-0", **Crew:** 2; **Displacement:** 8 tons; **Engines:** one screw, one Cazes type motor, 170 hp.	France, English Channel
1905[275]	Italian Royal Navy *Pegaso*	The boat was a highly successful design and the first of 27 other craft of her class. *Pegaso* was strongly built, had good sea-keeping qualities and could maintain maximum speed for 349 nm. **Speed:** 23 knots; **Armament:** 2-17" torpedo tubes, 2-57mm Guns; **Length:** 165'-2": **Beam:** 17'-5"; **Draft:** 5'-8"; **Crew:** 50; **Displacement:** 207 tons; **Range:** 1,200 nm; **Engines**: two screws, two vertical triple expansion (later converted to oil in 1908).	Adriatic Sea and Ionian Sea
1905[276]	December 16 Japanese Navy *IJN Lt. Cdr. Seki, 14th Flotilla*	Battle of Tsushima; Russo-Japanese War The *Sevastopol* was repeatedly attacked by 25 Japanese torpedo boats, which launched 60 torpedoes, and received one hit. Her torpedo net defensive measure worked well up until a torpedo boat of the 14th Flotilla sank her and damaged the destroyer *Storoshevoi*; *Sevastopol* had to be beached.	Korea Strait and Sea of Japan

[274] Fock, Harald. *Fast Fighting Boats 1870-1945* (Naval Institute Press, 1978), 18.
[275] Jackson, Robert. *Destroyers, Frigates, and Corvettes* (Barnes & Noble, Inc., 2000), 269.
[276] Gray, *The Devil's Device*, 170.

Figure 108. Japanese soldiers and sailors getting haircuts. It is not often that the human side of a conflict comes through. In this case the message is quite clear. (The Russo-Japanese War. Japan Center for Asian Historical Records, 2004)

Our Torpedo-Boat Destroyers making repeated Attacks upon the Russian Squadron on the Night of the 27th May.

Figure 109. Japanese torpedo boats attacking at Tsushima (National Archives).

Figure 110. Damage to Russian cruiser *Pallada* at the Battle of Tsushima 1905 shows the devastating effects of a Japanese torpedo. Navies world wide were still not convinced of the role of importance that the torpedo and the torpedo boat would play in sea battles to come. (The Russo-Japanese War. Japan Center for Asian Historical Records, 2004)

Figure 111. A Japanese torpedo boat getting creamed by a Russian battleship at Tsushima. The battle was not all one-sided. (National Archives)

Date	Craft, Name, Event or Country	Event, Builder and Boat Specifications	City, River, Lake, Sea, Ocean
1905[277]	December 17 Japanese Navy *IJN Cdr. Fujimoto*	Battle of Tsushima; Russo-Japanese War Japanese torpedo boats sank the Russian flagship, *Suvorov*. In subsequent night action Japanese torpedo boats sank the following Russian ships: *Admiral Nakhimoff*, cruisers *Navarin*, *Demitri Donskoi*, *Pallada* and battleship *Sisoi Veliki*.	Korea Strait and Sea of Japan
1906[278]	September 5 Russian Imperial Navy Spar torpedo boat	Battle of Tsushima; Russo-Japanese War On September 5, 1905, a Russian ship launched one of its spar-torpedo boats to attack the many Japanese torpedo boats swarming the fleet. The Russian sank one Japanese torpedo boat sank. The Russians used boats made by Lewis-Nixon, New Jersey, U.S.A. **Speed:** 25 knots, **Length:** 88.85'	Korea Strait and Sea of Japan
1905[279]	German Navy *S-125*	The S-125 was the first German torpedo boat to be fitted with a Parsons steam turbine built by Schichau, Elbing. There were 22 boats from G-110 to G-113 and S-114 to S-131 in the massive German building program. **Speed:** 26.7 knots; **Armament:** 3 x 450 mm torpedo tubes, 3 x 4-pounder quick firing guns, two machine guns, **Length:** 205', **Beam:** 23'; **Displacement:** 440 tons; **Engines**: two Parsons turbines @ 6,000 to 6,500 hp; **Range**: 1,400 nm @ 15 knots.	Baltic Sea
1905[280]	Italian Royal Navy *Fiat-Muggiano*	One of the few all steel torpedo boats ever built. She was built for testing purposes alone. Developed in 1906 by Fiat-Muggiano at La Spezia for liaison, landing and ship borne torpedo-attack. **Speed:** 16 knots; **Length:** 40.19'; **Beam:** 8.86'; **Draft:** 2.17'; **Displacement:** 6.6 tons tons; **Armament:** 1 x 47mm gun; 2 x 35.56 cm TT´s; two Maxim MGs: **Engines:** two 80 hp Fiat, petrol.	Adriatic Sea

277 Pleshakov, *The Tsar's Last Armada*, 309-310.
278 Pleshakov, *The Tsar's Last Armada*, 309-310.
279 Professional Notes *(Proceedings* XXXIII, No.4 December 1904), 1517-1518.
280 Fock, *Fast Fighting Boats*, 38.

Figure 112. The Japanese used torpedo boats on a mass scale for the first time in history. This was the first lesson in the true power of the "small boats" over Big-Ships and Big-Guns. (The Russo-Japanese War. Japan Center for Asian Historical Records, 2004)

Figure 113. As this Japanese illustration shows, Russian sailors were treated with utmost respect The Japanese spared no expense in insuring the comfort of these men was of the highest standards. Sadly, this would not be the case in World War II. The illustration shows the sailors in clean white hospital attire relaxing and watching there pet birds in a cage.

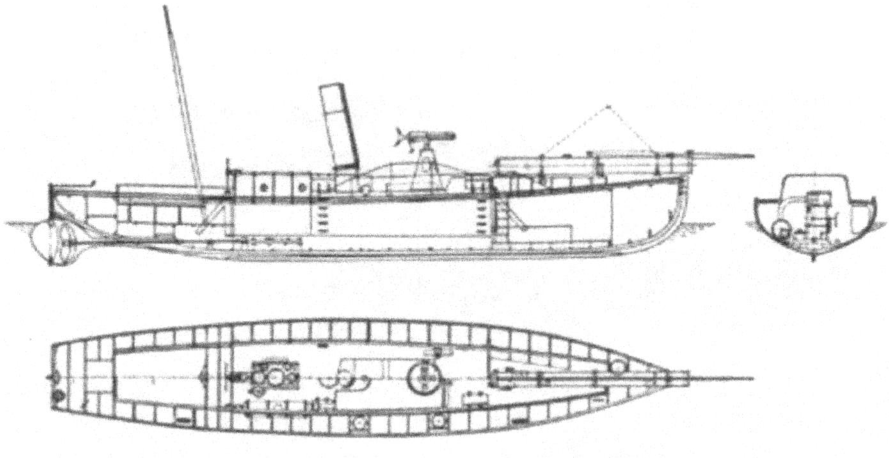

Figure 114. German Steam Launch Torpedo Boat. The Germans spent great effort in developing small steam launches for river and harbor duties. (The Friedrich Files) [281]

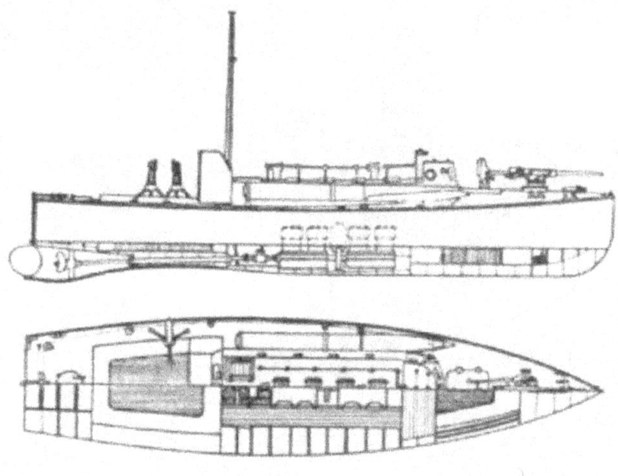

Figure 115. Fiat-Muggiano Torpedo Boat. The Italian Navy would pursue torpedo boat naval doctrines from now until 1945. This created for the Italian boat building industry a status in the world second to none. (The Friedrich Files)[282]

[281] http://picpage7.tripod.com/motorlaunch/identify.html.
[282] *Ibid.*

Date	Craft, Name, Event or Country	Event, Builder and Boat Specifications	City, River, Lake, Sea, Ocean
1906[283]	April 18 British Royal Navy TB.84 1st Class TB	During night exercises off Malta, the 84 boat was cut in half by the destroyer, HMS *Ardent*. The boat sank with no loss of life. **Builder:** Poplar, 1889 **Displacement:** 85 tons; **Length:** 130'; **Beam:** 13'-6"; **Armament:** 3 x 3-pounder guns, 3 torpedoes	Mediterranean
1906[284]	May 17 British Royal Navy *TB. 56* 1st Class TB	The boat was ordered to return to Malta from patrol duties on the Suez Canal. To conserve coal, the 56 boat was towed by the cruiser HMS *Arrogant.* Off Damietta, the tow part for the last time. The boat capsized and 7 men died.	off Egypt Mediterranean
1905[285]	British Navy *Eden*	One of 34 vessels made in 1903-1905 and classified as Sea-Going Torpedo-Boat. She has mistakenly been listed a destroyer. **Speed:** 25.5 knots; **Length:** 220', **Beam:** 23'-6"; **Draft:** 10'; **Armament:** 2x17.7 inch torpedo tubes, 1x12 pounder, 5x6 pounder; **Displacement:** 550 tons; **Engines:** Parson turbines (7,000 hp) ; six screws ; **Crew:** 70.	English Channel
1906[286]	Italian Royal Navy *Calliope*	Launched in 1906 as one of eight vessels based on the successful Thorneycroft design. Converted to oil burning for increased endurance and reduced displacement The ship gave good service and served through World War I and was not stricken from the records until 1924. **Speed:** 26.5 knots; **Length:** 53'; **Beam:** 5 m; **Draft:** 2 m; **Displacement:** 220 tons; **Armament:** 3 x 1.85-inch guns, 3 x 17.7 inch torpedo tubes two screws; two vertical triple expansion engines; **Range:** 800 nm.	Adriatic Sea

[283] Hepper, *British Warship Losses in the Ironclad Era,* 1860-1919, 19.
[284] *Ibid.*
[285] Professional Notes, (*Proceedings* XXXIII, No. 4, December 1904), 1520-1521.
[286] Jackson, *Robert Destroyers, Frigates and Corvettes* (Barnes & Noble, 2000), 50.

Date	Craft, Name, Event or Country	Event, Builder and Boat Specifications	City, River, Lake, Sea, Ocean
1906[287]	French Navy	The French launched a massive torpedo boat building program (72 boats) as a part of adapting a new naval policy. 15 were built at Harve, 12 at Rouen, 18 at Nantes, 10 at Bordeaux, 9 at Creusot Works at Chalon-sur-Saône. **Builder:** Schneider et Cie **Speed:** 26 knots: **Armament:** 3 x 17.75" torpedo tubes (one bow, two midships) with three reloads; **Length:** 124'-8"; **Beam**: 13'-11"; **Draft:** 8'-7"; **Displacement:** 97.5 tons; **Range:** 10 knots-1,800 nm; 26 knots-200 nm; **Engines:** two boilers @ 1,000 hp each, one screw.	France, English Channel
1906	Bulgarian Navy[288]	The French built six more boats for the Bulgarian Navy. *Drazki ("Intrepid,"*, *Smeli* ("Brave"), *Hrabri* ("Valiant"), *Shumni* ("Noisy"), *Letyashti* ("Flying") and *Strogi* ("Rigid").	Black Sea
1906[289]	Russian Imperial Navy Baltic Fleet *Minonski No.1 - 10*	The Russian bought 10 motor torpedo boats from Lewis-Nixon, Perth, New Jersey. The U.S. Navy had refused to evaluate these same boats. The 10 boats arrived in Sevastopol, Russia and were then transported to Kronstadt fortress for assembly. The original petrol engines were replaced by diesel engines. Russia would be the first and only navy to use diesel engines until the German *Schnellbootes* of WWII. **Speed:** 25 knots; Length: 88.58'; **Armament:** one rotating torpedo launcher on deck, 1 x 47 mm Hotchkiss QF gun, 2 light MG's; **Engines:** 2 x 300-hp petrol.	Sevastopol, Kronstadt
1907[290]	America U.S. Navy *Motor Boat Project*	This craft had been designed by A.T. Chester to be hoisted aboard large ships. It proved to be too heavy for available ship-board machinery. **Speed:** 20 knots; **Armament:** 1-stern torpedo tubes; **Length:** 18.3 m; **Beam:** 2.56 m; **Draft:** 0.68 m; **Displacement:** 12.3 tons; **Engines:** 2-100 hp gasoline, 2-screws.	Atlantic Ocean

[287] "Professional Notes" (*Proceedings* XXXIII No.4), 1576-1577.
[288] http://en.wikipedia.org/wiki/Drazki_torpedo_boat.
289 Greger, Rene, Ed. Andrew Lambert "The Origins of Russian and Soviet MTB's" *Warship IX*, 1985, 38-41, 39.
[290] Fock, *Fast Fighting Boats*, 18-20.

Figure 116. The Bulgarian *Drazki* is preserved at the naval museum in Varna, Bulgaria. She was in service from 1907 to 1944 and fought for her country in three wars. (Nikola Gruev, Wikipedia).

Date	Craft, Name, Event or Country	Event, Builder and Boat Specifications	City, River, Lake, Sea, Ocean
1907[291]	Italian Royal Navy *Astore*	The *Astore* was one of the first boats made for the open ocean in the Atlantic. She and 18 others were well designed and all gave good service in WWI. **Speed:** 25.8 knots; **Armament:** 3- 17.7" torpedo tubes, 3- 1.8 inch guns; **Length:** 160'-2"; **Beam:** 17'-5"; **Draft:** 5'-9"; **Displacement:** 12.3 tons; two triple expansion engines, two screws; **Crew:** 150	Atlantic Ocean, Mediterranean
1907[292]	Germany Navy *G-137* *Torpedo Boat*	Considered to be the fastest naval vessel of its time. Germany was chided by the British for not classifying these boats as destroyers. She ran the naval trials at Neukrug and ran the Admiralty Mile at 34 knots. **Speed:** 34 max, 33.1 mean; knots; **Armament:** 1 x15 pounder., 3 x 5 pounder., 2 MG, 3-450 mm torpedo tubes; **Length:** 235; **Beam:** 24'-10"; **Displacement:** 570 tons (all other of this class were 530 tons); four Schultz boilers, four Parson turbines @ 10,000 hp	Neukrug, Germany
1908[293]	America U.S. Navy MTB *Satterlee*	The Assistant Secretary of the Navy called on the Board of Construction to Develop plans for a 150' torpedo boat. The U.S. Navy General Board killed the idea saying it contradicted the navy strategy of sending a fleet to meet the enemy in distant waters. **Speed:** 21 knots; Disp. 29.5 tons; **Length:** 150'; **Beam:** 13'-3/4"; **Armament:** 1 x 18" torpedo tubes, 2 x 6-pounder guns; one screw; 4000 gal. petrol.	Never Tested

[291] Jackson, Robert. *Destroyers, Frigates and Corvettes* (Barnes & Noble, 2000), 35.
[292] Professional Notes (Proceedings Vol. XXXII, No.4, December 1907), 1517-1518.
[293] Friedman, Norman. *U.S. Small Combatants* (Naval Institute Press, 1987), 98.

Date	Craft, Name, Event or Country	Event, Builder and Boat Specifications	City, River, Lake, Sea, Ocean
1910[294]	British Royal Navy *Miranda IV*	Sir John I. Thorneycroft, became inspired by the writings of Reverend C.M. Ramus who had in his published a proposal for planning boats. Thorneycroft made a few changes to the ideas of Ramus and built the boat *Miranda IV.* **Speed:** 35 knots; **Length:** 25'-9"; **Beam:** 5'-9"; **Displacement:** 1.27 tons; **Crew:** 2; **Engines:** 1 -x 120hp, petrol, one 1-screw.	England
1912[295]	October 31 Greek Navy *TB No.11*	The Greek *TB. No. 11* sank the Turkish Battery ship, *Feth-I-Bulend* at Salonika, Dardanelles. She was scuttled on November 23rd.	Aegean Sea

Figure 117. Turko-Italian War in the Aegean, 1912. (Wikipedia Commons)

[294] Fock, *Fast Fighting Boats*, 23.
[295] Fock, *Fast Fighting Boats.*

Date	Craft, Name, Event or Country	Event, Builder and Boat Specifications	City, River, Lake, Sea, Ocean
1914 [296]	Germany Navy *S-33 Torpedo Boat*	The builder F. Schichau built a line of torpedo boats that married reliability, speed and firepower in an elegant design. The boat was launched on April 4, 1914 and commissioned October 4, 1914. The *S-33* was sunk by a torpedo from the British submarine *L-10* on October 3, 1918, in the North Sea with 5 dead. **Speed:** 34.2 knots; Arm: 3- x 8.8 cm/45 calibre torpedo guns; **Displacement:** 971 tons ; **Length:** 261.15'; **Beam:** 27.33'; **Draft:** 9.19'; **Engines:** 23,516 shp; **Range**: 1,100 nm @ 20 knots.	North Sea
1915[297]	March 21, 2008 British Royal Navy *TB-064* *1st Class TB*	While bound for Mudros (Moudhros) from Port Said, while in rough weather, the boat ran out of coal and took refuge in a small bay on the east side of the island of Lemnos. The crew got off before the boat was totally destroyed. **Builder:** Poplar, 1886. **Displacement:** 87 tons; **Length:** 125' **Beam:** 13'; **Armament:** 2 x 3-pounder. Guns, five torpedoes.	Aegean Sea
1915[298]	April German Navy *Muller-Thomamuhl*	The boat showed much promise, yet the high command could never give it a slot for service. **Speed:** 32.6 knots; **Displacement:** 7.8 tons; **Length:** 13 m; **Beam:** 4.0 m; **Draft:** 0.366 m; **Engines:** four Aero (two linked by chain-drive to each shaft), 65 hp with Sirocco fan; **Armament:** 2 x 45cm torpedoes (aft) , five depth charges.	Germany

[296] Groner, Erich. *Diedeutschen Kriegsschiffe 1815-1945* (J.F. Lehmanns Verlag Munchen, 1958), 178, 179.
[297] Hepper, *British Warship Losses in the Ironclad Era*, 34.
[298] Fock, *Fast Fighting Boats*, 50.

Figure 118. Experimental Italian hydroplane designed by Gaetano Arturo Crocco, 1907, reached 38 knots powered by a 80 h.p. Clement-Bayard engine. (From the collection of Stefano Dentice)[299]

Figure 119. The *S-33* had the distinction of being the only torpedo boat to be sunk by a submarine, the British *L-10* on October 3, 1918, in the North Sea. (Illustration by author)

[299] Gaetano Arturo Crocco, http://www.foils.org/gallery/crocco.htm.

Figure 120. Navy Recruiting Poster WWI. (Naval Historical Center)

Figure 121. This poster had high appeal to the British citizens of WWI. (Library of Congress Prints and Photographic Division)

Date	Craft, Name, Event or Country	Event, Builder and Boat Specifications	City, River, Lake, Sea, Ocean
1915[300]	June 10 British Royal Navy *TB.12 1st Class TB* *TB 10 1st Class TB*	As part of the Nore Defense Flotilla, the 12 boat sailed from Harwich to the Thames estuary. Near the Sunken Light vessel she hit a mine or was torpedoed. The 10 boat came to her aid and saved the crew. The boat was a total loss. **Builder:** Poplar, 1907. **Displacement:** 25 tons; **Length:** 175'; **Beam:** 17'-6"; **Armament** 2 x 12-pounder guns, three torpedoes.	North Sea
1915[301]	Austro-Hungarian Navy *Hydroplane Torpedo Boat*	Lieutenant Dogbert VonMuller-Thomamuhl presented plans for a Hydroplane Torpedo Boat. This would be the first air-cushion boat ever built. He was currently serving on the torpedo boat *60T*. For a test model, this craft was well engineered. The craft never saw service but was important in hydroplane technology. **Speed:** 32.6 knots; **Displacement:** 6.5 tons; **Armament:** 2 x 45 cm torpedo tubes, one MG; 3 x 6.5 kg depth charges; **Crew:** 5; **Engines:** four 120-hp petrol aero-engines (two per drive shaft), 1 x 65-hp Sirocco/forward fan; **Range:** 120 nm @ 30 knots.	Austria
1915[302]	November 1 British Royal Navy *TB.96 1st Class TB*	The 96 boat was on patrol, east of Gibraltar, when she was run over by the troop ship *Tringa*. Eleven men died. **Builder:** Cowes, 1894 **Displacement:** 130 tons; **Length:** 142'-6"; **Beam:** 15'-3"; **Armament:** 3 x 3-pounder guns, three torpedo tubes.	Strait of Gibraltar

300 Hepper, *British Warship Losses in the Ironclad Era,* 37.
301 Fock, *Fast Fighting Boats*, 50.
302 Hepper, *British Warship Losses in the Ironclad Era,* 1860-1919,46

Date	Craft, Name, Event or Country	Event, Builder and Boat Specifications	City, River, Lake, Sea, Ocean
1915[303]	December 12 British Royal Navy *TB. 046* *1st Class TB*	The Mercantile Fleet Auxiliary ship *Carrigan Head* left Malta on December 18th with the *TB.046* boat in tow, headed for Mudros, Lemnos Island. Rough weather caught the boats on the open Mediterranean Sea and the small boat started to take on water. The engine gave out and the crew was taken off. Two crewmen were injured in the transfer and the boat was cut loose and sunk by gunfire. **Builder:** Chiswick, 1886. **Displacement:** 60 tons; **Length:** 127'-6"; **Beam:** 12'-6"; 2 x 3-pounder. guns, four torpedo tubes.	England
1916[304]	January 26 British Royal Navy *TB.131st Class TB*	The boat was on patrol in the Firth of Forth in, foggy, misty weather about 5 miles east of May Island. The sloop *Gladiolus* was also in the area on the track from Greenock to Tyne. At 3AM the sloop struck the *TB.131* boat and split her in half. 9 men died in the accident. **Builder:** Cowes, 1907. **Length:** 185'; **Beam:** 18'; **Armament:** 2 x 12-pounder.guns, three torpedoes.	
1916[305]	British Royal Navy *CMB-1* (Exp.40') Spiral Propeller	Thorneycroft invented the spiral propeller to prevent entanglements with defensive anti-torpedo boat netting and barricades. Speed and efficiency remained with the new design. **Builder:** Thorneycroft, 40' Class**; Speed:** 33.5 knots; **Length:** 40'-0"; **Beam:** 8.53'; **Armament:** 1-45.7 cm torpedo tubes, one twin. 303-cal. MG; **Displacement:** 4.34 tons; **Engine**: one 150/275 hp V-12 cyl. Thorneycroft, petrol; **Range:** 160 nm; **Crew:** 3.	North Sea
1916[306]	September Italian Royal Navy *MAS 20, MAS 21*	The first Italian torpedo boats to be fitted with electric engines for silent running in the confines of coves and bays in the Adriatic Sea and particularly the Fasana Channel. The boats were fitted with two 5-hp electric motors that gave the boats a range of 20 miles for five hours at 4 knots.	Adriatic Sea

[303] Hepper, *British Warship Losses in the Ironclad Era,* 1860-1919, 49.
[304] *Hepper, British Warship Losses in the Ironclad Era*, 1860-1919,82.
[305] Fock, *Fast Fighting Boats*, 25.

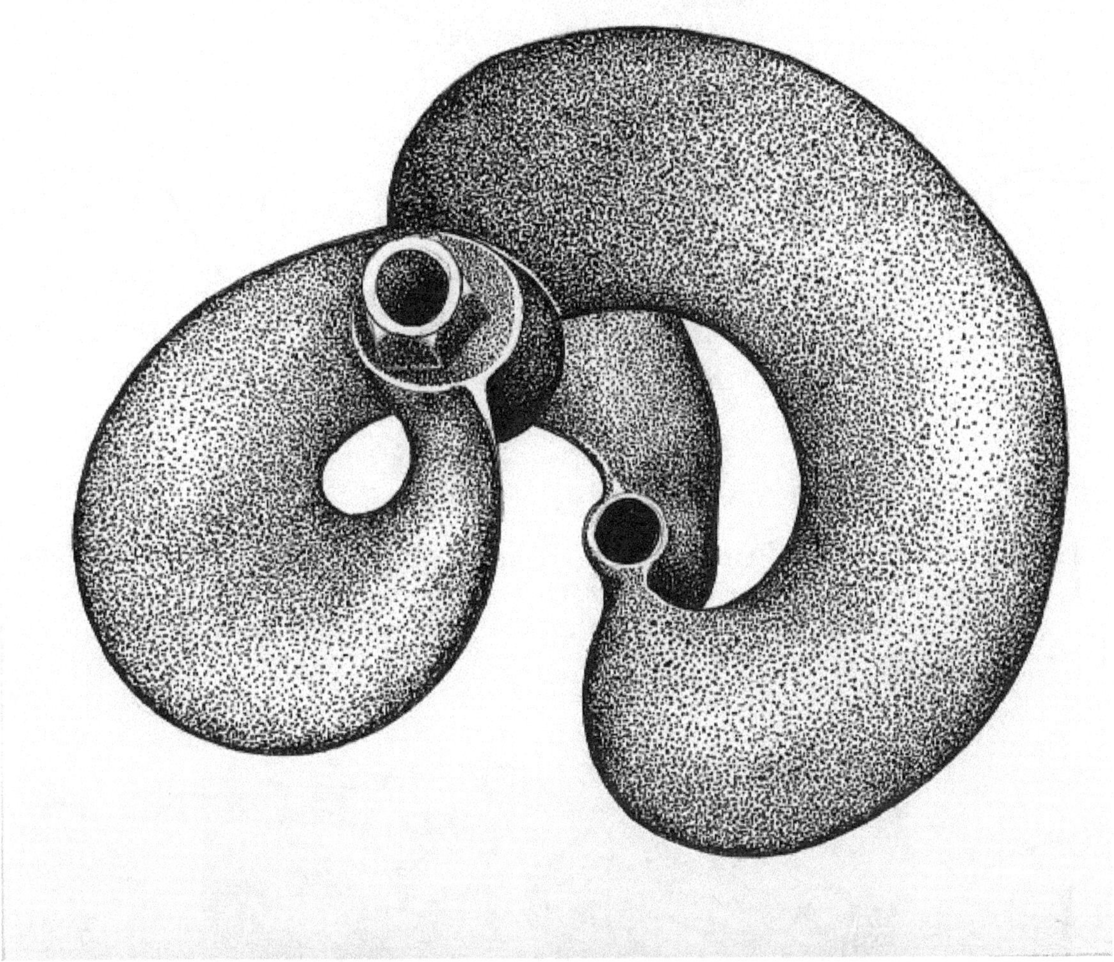

Figure 122. CMBs were experiencing problems with submarine nets and net obstructions in home waters as well as enemy territory. Thorneycroft addressed the problem with an innovative invention. The design prevented the propeller becoming entangle while not reducing speed and efficiency. (Illustration by author)

---

[306] Fock, *Fast Fighting Boats*, 34.

Figure 123. The Red Cross maintained a constant image for support of the soldiers and sailors of almost all nations regardless of which side they represented, Axis or Allied. (Library of Congress Prints and Photographic Division)

Figure 124.World War I became known for the huge emotional appeal to the general public. Posters like this one of wrecked sailors at sea stirred people of all nations to support their soldiers and sailors. (Library of Congress Prints and Photographic Division)

Date	Craft, Name, Event or Country	Event, Builder and Boat Specifications	City, River, Lake, Sea, Ocean
1916[307]	May 22 Italian Royal Navy *MAS 19*	The crew of *MAS 19* overpowered the crew of the Austro-Hungarian motorboat, *Leni*, assigned to the Trieste seaplane base. Austrian pilot Lieutenant Gottfried von Banfield [308]in his seaplane*, L-98*, recaptured the *Leni* the next morning by driving off the Italians with machine gun fire.	Trieste Adriatic Sea
1916[309]	June 7 Italian Royal Navy *MAS 5, MAS 7*	The first Italian motor torpedo boats to be fitted with side dropping gear for torpedoes. *MAS 5,* commanded by Tenente Melito and *MAS 7* commanded by Tenente Beradinelli. On the night of September 7, the boats attacked Austro-Hungarian ships at anchor in the roads of Durazzo and sank the steamer *Lokrum*.	Durazzo Harbor Albanian Coast Adriatic Sea
1916[310]	July 26 British Royal Navy *TB.9 1st Class TB*	The TB.9 had a collision with the British destroyer HMS *Matchless* in the dark. The *Matchless* rescued the crew of the TB.9 boat and the 9 boat sank (one man died). Shortly thereafter, the *Matchless* then had a collision with the destroyer HMS *Manley* near the Cork light ship. **Builder:** Woolsten 1907. **Displacement:** 247 tons; **Length:** 175'; **Beam:** 17'-6"; **Armament:** 2 x 12= pounder guns, three torpedoes.	St. George's Channel Irish Sea
1916[311]	June 25 Italian Royal Navy *MAS 5, MAS 7, MAS 9*	The two motor torpedo boats, with *MAS 9* commanded by Tenente Rizzo, again strike at Durazzo harbor and break through the obstructions, sinking the Austro-Hungarian steamer *Sarajevo*.	Durazzo Harbor Albanian Coast Adriatic Sea

307 Fock, *Fast Fighting Boats*, 47.
308 http://en.wikipedia.org/wiki/Gottfried_von_Banfield.
309 Halpern, Paul G. *A Naval History of World War I* (Naval Institute Press, 1994), 169.
310 Hepper, *British Warships Losses in the Ironclad Era,* 1860-1919, 66.
311 Halpern, *A Naval History of World War* I, 169.

Date	Craft, Name, Event or Country	Event, Builder and Boat Specifications	City, River, Lake, Sea, Ocean
1916[312]	September 6/7 Italian Royal Navy *MAS 5, MAS 7*	For the first time the MAS boats go on operations with torpedo boats (ships) *34 PN* and *38 PN* and sink the transport *Lokrum*, 924 tons, off Durazzo.	Adriatic Sea
1916[313]	September Italian Royal Navy MAS 20, MAS 21	In the confines of the islands and rocky inlets and bays, the Italians overcame the loudness of their motor torpedo boats. They installed two electric, battery driven, motors for "Silent Running." The twin 5-hp electric motors could move the boats along at 5 knots for 20 nm. This solved the noise problem for the attacks in the narrow Fasana Channel. **Speed:** 24 knots; **Displacement:** 12.30 tons; **Length:** 52'-6"; **Beam:** 8.6'; **Draft:** 3.9' ; **Armament:** 2 x 45 cm side dropping torpedoes, 2 x 6.5mm mg, four depth charges; **Engines:** two 225 hp, petrol, 2 x 5-hp electric	Fasana Channel Adriatic Sea
1916[314]	December 10 Austro-Hungarian Navy MAS 9, MAS 11	The Austro-Hungarian naval forces stationed large coastal battleships in Trieste harbor to bombard Italian batteries at Cortellazo at the mouth of the Piave river. The 2 motor torpedo boats, MAS 9 commanded by Tenente Rizzo, were towed by large torpedo boats into position at Trieste harbor and sank the Austro-Hungarian coastal battleship *Wein.* **Speed:** 25.2; **Armament:** 2-17.7" torpedo tubes, 1-47mm MG; **Length:** 52'-6"; **Beam:** 8'-8"; **Draft:** 4'-0"; **Engines**: 2 x 225 hp, petrol, 2 x 5 hp electric (20 nm , 5 hrs @ 4 knots); two screws; **Crew:** 8; **Displacement:** 12.3 tons.	Trieste Harbor Albanian Coats Adriatic Sea

312 Fock, *Fast Fighting Boats,* 47.
313 Fock, *Fast Fighting Boats,* 34.
314 Halpern, *A Naval History of World War I,* 170.

Date	Craft, Name, Event or Country	Event, Builder and Boat Specifications	City, River, Lake, Sea, Ocean
1917[315]	March 26/27 Italian Royal Navy MAS 1	The boat laid four mines north-west of the boom defense on Fasana channel off Palo Harbor. The *MAS 1* took credit for sinking the Austro-Hungarian submarine *U-5*.	Fasana Channel Adriatic Sea
1917[316]	America USN *Shearer 1-Man Torpedo Boat (Experimental)*	The unconventional design came from the Thames Towboat Company of New London, Connecticut on October 1, 1917. There is no proof that William B. Shearer was the designer of the one-man torpedo boat. He was a promoter, con-man, charlatan and great liar. In the 1930's he tried again to interest the navy in a larger version with three men and three torpedoes but did not find interest. He had many friends in high places of the U.S. Navy. **Speed:** 5 knots, semi-submerged; 9 knots, in calm waters; 7.5 knots in heavy weather; **Armament:** 1-18" torpedo; **Engine:** one-Scripps, 4-cycle 35 hp, petrol engine.	America
1917[317]	April 7/8 British Royal Navy *CMB-4, 5, 6, 9* Lt. Beckett	The four CMB's attacked and sank the German torpedo boat (ship) G-88. **Builder:** Thorneycroft, 40' Class **Speed:** 33.5 knots; **Length:** 40'-0"; **Beam:** 8.53'; **Armament:** 1 x 45.7 cm torpedo tubes, one twin.303 cal. MG; **Displacement:** 4.34 tons; **Engine**: 1 x 250/275 hp V-12 cyl. Thorneycroft, petrol; **Range:** 160 nm; **Crew:** 3.	Off Ostend North Sea

315 Fock, *Fast Fighting Boats*, 48.
316 Friedman, Norman, *U.S. Small Combatants* (Naval Institute Press, 1987) 16, 100-101.
317 *Fock, Fast Fighting Boats*, 30-31.

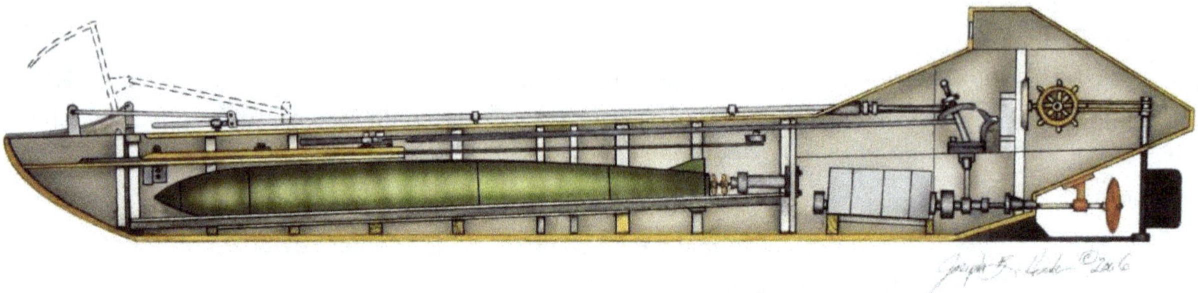

Figure 125. Shearer I-Man Torpedo Boat. Experimental. The man Shearer was a most untrustworthy and unsavory character in the quiet annals of U.S. navy history. Still, he always seemed to have friends in the higher echelons of the U.S. naval hierarchy. His design was built three times in spite of the conflict of purpose he brought to the table. His designs were hopeless to say the least. (Illustration by author)

Figure 126. Japanese torpedo boat *Kasumi* of the *Akatsuki* class. Date of photo not recorded. Laid down in 1900, completed in 1902 by. UK Yarrow. (United States Library of Congress's Prints and Photographs Division)

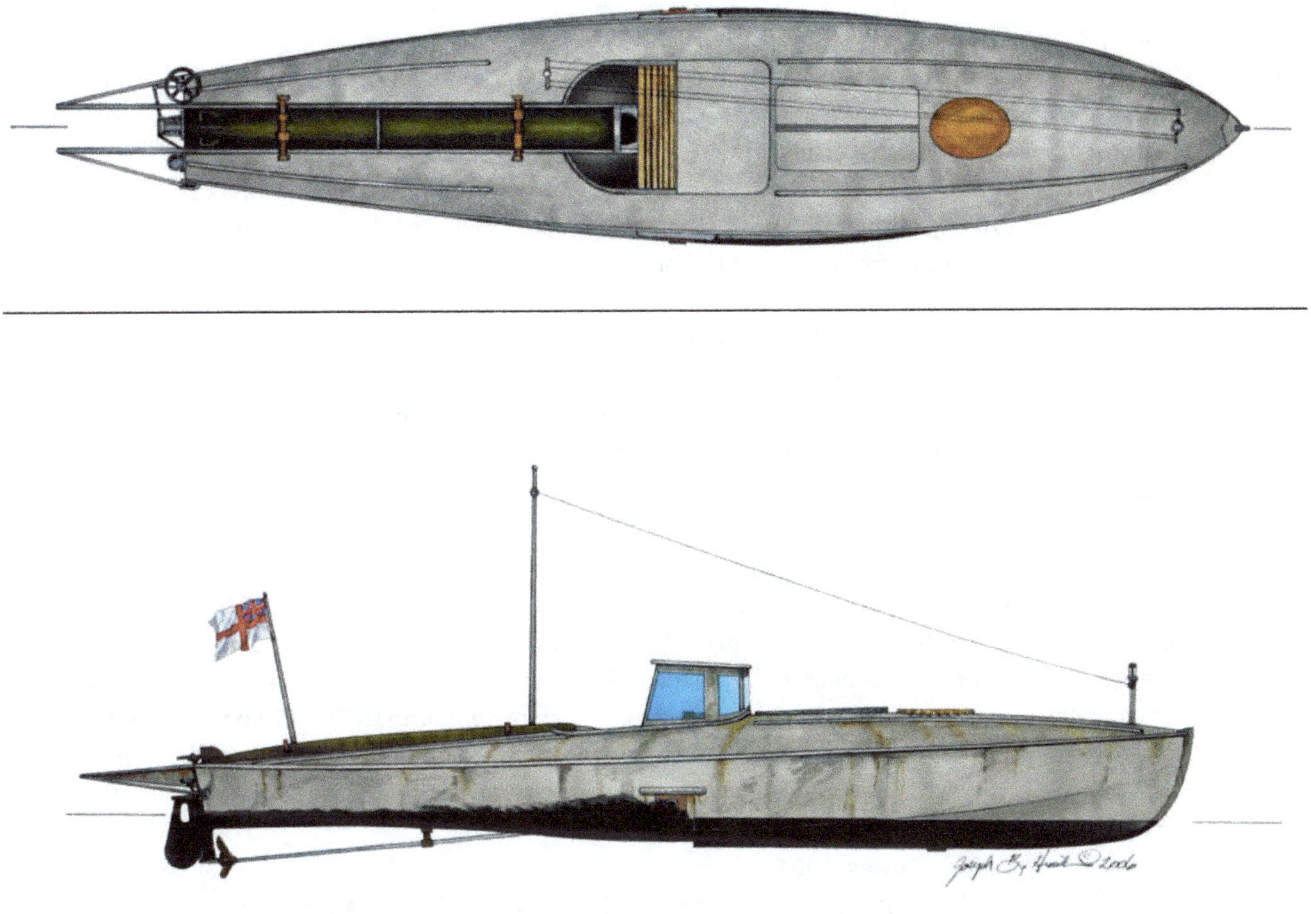

Figure 127. *CMB-4* (top and profile views) was one of the British boats that made a successful raid on the gigantic Russian fortress, Kronstadt, in August 1919 (see below). This small boat and a few others almost destroyed an entire Russian fleet while in its harbor. After this event, British naval doctrine included the torpedo boat in its overall naval strategy. (Illustration by author)

Date	Craft, Name, Event or Country	Event, Builder and Boat Specifications	City, River, Lake, Sea, Ocean
1917[318]	Germany Navy *LM Torpedo Boats*	Germany needed fast motor boats to tackle the British laid anti-U boat and anti torpedo boat net barriers at Zeebrugge and Ostend. They made 26 craft with a displacement from 6 to 7.5 tons. Airship engines were first used because of their availability and weight. These were Maybach, 6 Cyl, 210 hp, and petrol. Speeds ranged from 30.5 knots to 31.8 knots. The length was from 15 to 17 meters. **Builder:** Lürssen, Vegesack LM 1-2 **Speed:** 30.55 knots; **Displacement:** 7 tons; **Engines:** two Maybach 210 hp; **Length:** 15 m; **Beam:** 2.4 m; **Draft:** 1.08 m; **Armament:** 1 x 45 cm torpedo (bow), 1-3.7cm cannon; **Range;** 150nm @ 30 knots	Germany
1917[319]	June 10 Italian Royal Navy *MAS 15, MAS 21*	West of Premuda Island, Captain Rizzo in *MAS 15* with its torpedoes sank the most modern battleship of the Austro-Hungarian navy, the *Szent Istnan,* as it was making a break for the Otranto Straits barrage, 9 nautical miles north-west of Premuda Island. The torpedoes of *MAS 21* missed the battleship *Tegethoff* in the same battle.	Straits of Otranto Adriatic Sea
1917[320]	June 10 British Royal Navy *TB.17 1st Class TB*	In the late evening, *TB.17,* while patrolling the waters of the Thames estuary, is struck by the large merchant ship, *Kamouraska* The 17 boat was cut in half and sank. 3 men became missing and presumed drowned. The rest of the crew was rescued by armed trawlers in the area. **Builder**: Cowes, 1904 **Displacement:** 205 tons; **Length:** 165'; **Beam:** 17'-6"; **Armament:** 3 x3-pounder. guns, three-torpedoes	Shingles Bank Thames Estuary North Sea
1917[321]	June 19 British Royal Navy *CMB 1, CMB 9*	The boat was attempting to rescue downed RAF airmen off the coast of Dunkirk. There had been a dog fight between German and British airmen. German destroyers , the first on the scene, engaged and sank the *CMB 1. CMB 9* evaded the destroyers and escaped. **Builder:** Thorneycroft, Hampton 1916, 45' class	Off Dunkirk Dover Strait North Sea

318 Fock, *Fast Fighting Boats,* 53-58.
319 Hepper, *British Warship Losses, 1860-1919,* 92.
320 Hepper, *British Warship Losses, 1860-1919,* 92.
321 Hepper, *British Warship Losses, 1860-1919,* 93.

Figure 128. Wrecks of British cruisers after the Zeebrugge raid, April 1918. [322]

Figure 129. The Italian MTB MAS 15 has been ensconced in the national naval museum of Italy. This is one of the boats that had a huge impact on naval doctrine. This boat also began an era in which the world naval powers made the torpedo boat an important element of naval strategy. This one boat sank two battleships and other vessels that were thought to be "unsinkable." (Illustration by author)

[322] *Der Weltkrieg 1914-1918 in seiner rauhen Wirklichkeit. "Das Frontkämpferwerk,"* Munich, ca. 1925. http://en.wikipedia.org/wiki/File:Zeebrugge-Raid.jpg.

Figure 130. This photo speaks a thousand words. The sailor arising out of the confines shows the grim of his clothes and skin. Coal fired engines meant everything and everybody was constantly coated in coal dust. Still, the world's navies like their men to wear white as a uniform of the day. (Library of Congress Prints and Photographic Division)

Figure 131. Hans Frank (Admiral, retired, Deutsche Marine) translated the message of this poster. The first title is paying homage to German blockade runners. The second title says,"How our East-Africans received ammunition." The last line is "1 Mark, bounded 2 Mark" which is the title of the small magazine. (Library of Congress Prints and Photographic Division)

Figure 132. For a contrary view from the other side, see this ogre-like portrait of the German Kaiser. (Library of Congress Prints and Photographic Division)

Figure 133. In the scheme of overall history, the nation of Canada always seems to get left out or receive only a, "by the way…" notice. They have always done much more then that. They have sacrificed as much as some nations and more then others. The point is, they, like the Aussies and New Zealanders, have always been there for all of the great causes. (Library of Congress Prints and Photographic Division)

Date	Craft, Name, Event or Country	Event, Builder and Boat Specifications	City, River, Lake, Sea, Ocean
1917[323]	German Navy *Air-Propeller Hydroplane Torpedo Boat*	Engineer Brase built two-steel framed, single-step boats in May of 1917. The boat started construction at the Muggei Yards in Friedrichshafen and finished at the Imperial Yards at Kiel. One boat carried two Maybach HS Lu six-cylinder 240 hp engines with two 2.3 m propellers. **Speed:** 34 knots; **Range:** 400 nm; **Crew:** 4; **Armament:** 1-45cm torpedo (stern) 1-mg. The second boat carried three engines with three propellers. **Speed:** 40 knots; **Armament:** 1 x 45cm torpedo (stern), one MG; **Range:** 400 nm; **Crew:** 3.	Ostend North Sea
1917[324]	August 11 British Royal Navy *CMB 41*	Six British CMB's were assigned the task of crossing German minefields off the island of Vlieland. They were to attack German minesweepers, or other surface targets. The CMBs had just passed Terschelling Island, in Dutch waters, when they were attacked by first six German flying boats, from Borkum naval air station (composed of Friedrichshafen FF.49C and Brandenburg W.12's). Two more aircraft joined the fray followed by five from the Kampfstaffel Norderneyu. The CMBs shot down one Brandenburg W.29. Five of the CMBs were badly shot up and dead in the water. *CMB 41* had just enough power to make it to shore. 3 CMBs were sunk, two drifted into Dutch waters and then towed by Dutch torpedo boats into port. This was an early sign that aircraft would become one of the torpedo boat's most deadly foes. **Builder:** Thorneycroft, 45'	Ameland Lighthouse North Sea

[323] Fock, *Fast Fighting Boats*, 57-58.
[324] Halpern, *A Naval History of World War I*, 443.

Date	Craft, Name, Event or Country	Event, Builder and Boat Specifications	City, River, Lake, Sea, Ocean
1917[325]	August 17 German Navy *Hydroplane Attack Boat*	The first military use of this radical design came on this date. A small hydroplane flotilla of six boats attacked and sank the Russian minelayer *Penelope* off the coast of the Zerel Peninsula. The minelayer is later lost after hitting a mine, a faulty torpedo, or explosion from petrol carried onboard. *LM 1-2*: **Speed**: 30/55 knots; **Displacement:** 6 tons; **Armament:** 1 x3.7 cm gun; **Engines:** 3 x 210-hp Maybach; three screws; **Length:** 49.2'; **Beam:** 7'-6"; **Draft:** 3'-6". LM 3-4: **Speed:** 27/80 k, **Length:** 47.8', **Beam:** 7.9'1-3.7 cm gun. LM 5-6; **Speed:** 29/45k, **Beam:** 7.9', 1-mg, 1-45cm bow torpedo	Caspian Sea
1917[326]	September 27 British Royal Navy *CMB 8,* *CMB 13*	The 2 boats were assigned to rescue RAF airmen whose planes crashed into the North Sea. On the way to the site, the steering gear was carried away. The 13 boat took her in tow and at that time German shore batteries started shelling the 2 boats. **Builder:** Thorneycroft, 45' **Displacement:** 5 tons; **Length:** 45' **Beam:** 8'-6"; **Armament:** one torpedo	North Sea
1917[327]	November 2 British Royal Navy *CMB 11*	The boat may have caught fire and sank off the Dover coast. There is still speculation that the craft was involved in a collision. **Builder:** Thorneycroft, 45'. **Displacement:** 5 tons; **Length:** 45'; **Beam:** 8'-6"; **Armament:** one torpedo.	Strait of Dover

325 Fock, *Fast Fighting Boats*, 58.
326 Hepper, *British Warship Losses, 1860-1919,* 104.
327 Hepper, *British Warship Losses, 1860-1919,* 109.

Date	Craft, Name, Event or Country	Event, Builder and Boat Specifications	City, River, Lake, Sea, Ocean
1918[328]	April 8/12 Italian Royal Navy *Grillo*-Class *Barchini Saltori* (Jumping Boat)	The Italians built four *Grillo* Class "Jumping Boats" to penetrate Austrian-Hungarian boom and net defenses in their harbors. They made their combat début against Pola harbor on this date but had to abandon the raid. On August 12 they tried again and were forced to scuttle the *Grillo* boats, *Cavalleta* and *Pulce.* **Speed:** 2-5 knots; **Displacement:** 10.5 tons; **Length:** 12m; **Beam:** 2.6m; **Draft:** 0.48; **Armament:** 2x 45 cm side dropping torpedoes; **Engines:** 1 x - 10-hp electric propulsion motor, 1 15 hp electric motor for driving two claw-chains tracks each side of the craft, **Speed:** 5 to 5.5 knots per minute;, **Power:** 3-40 cell batteries.	Palo Harbor Adriatic Sea

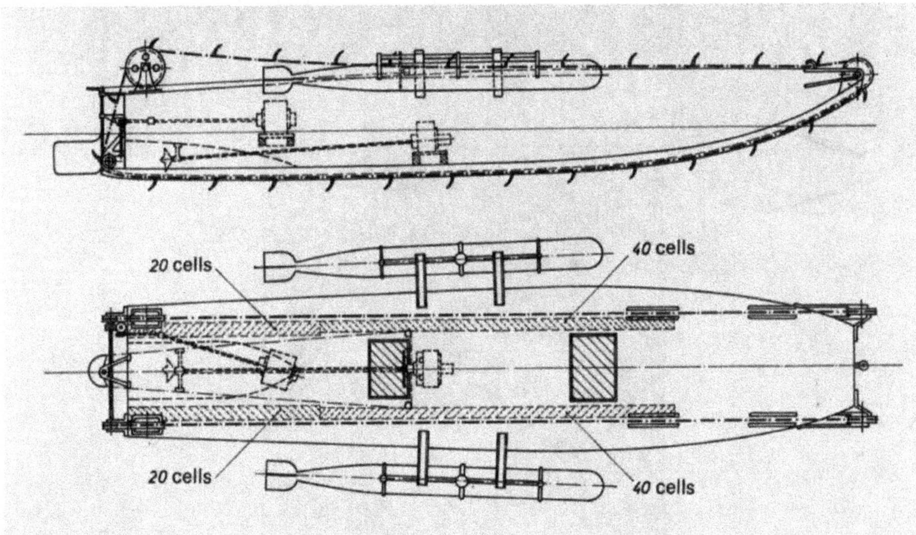

Figure 134.*Grillo*-class torpedo boat *Barchini Saltore* (Jumping Boat) This design became effective in solving the problems of torpedo craft entering harbors protected by nest and boom defenses. While invented by the Italians in WWI, it was the Austro-Hungarian navy who benefited most from the design. They captured two such boats and used then against the Italian navy. (National Archives)

[328] Halpern, *A Naval History of World War* I, 172; Fock, *Fast Fighting Boats*, 53-54.

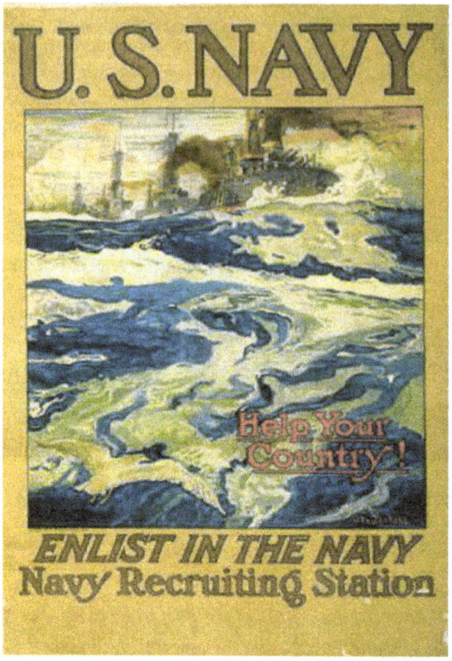

Figure 135. These U.S. Navy recruiting posters help convey the spirit of the era. (Library of Congress Prints and Photographic Division, Naval Historical Center)

Date	Craft, Name, Event or Country	Event, Builder and Boat Specifications	City, River, Lake, Sea, Ocean
1918[329]	April 12 British *Royal Navy* *CMB 18A*	This was one of 74 British warships that set out to attack German ports and surface vessels in Belgian ports. The planned smoke screen for the mission failed due to shifting winds. The fleet was ordered to turn back. In the darkness as the various craft attempted this maneuver, the 18 boat collided with another boat and sank. **Builder:** Thorneycroft, 60'. **Displacement:** 11 tons; **Length:** 60'; **Beam:** 11'; **Armament:** 1-torpedo.	North Sea
1918[330]	April 12/13 British Royal Navy *CMB 33A*	The CMBs attacked fortifications of the Napoleonic battleground of Eylau at Ostend. The 8.8 cm batteries of Eylau damaged *CMB 33A* as she ran aground on the shoals of Ostend Harbor. All of the crew drowned in the attempt. The German salvaged the craft and discovered secret papers and all equipment intact. Three bodies of the boats crew were later washed ashore. **Builder:** Thorneycroft, 60'. **Displacement:** 11 tons; **Length:** 60'; **Beam:** 11'; **Armament:** 1-torpedo.	Ostend North Sea
1918[331]	May 5 British Royal Navy *CMB 10*	In Dunkirk, the boat was undergoing routine maintenance when it caught fire and burned up. **Builder:** Thorneycroft, 45' Class	North Sea
1918[332]	May 12/13 MAS 99, MAS 100	The MAS boats attacked a convoy of Austro-Hungarian naval vessels of Durazzo and sank the transport *Bregenz.* The convoy was being protected by the destroyer *Dukla* and torpedo boats *TB-84* and *TB-98.*	Albania Adriatic Sea
1918[333]	May 16 Italian Navy *Grillo*	The attempt on May 6 ended quickly due to mechanical problems. The *Grillo* finally made it past obstructions this night but was discovered by the enemy forces and scuttled to prevent capture. The crew was taken prisoner.	Palo Harbor Adriatic Sea

[329] Hepper, *British Warship Losses, 1860-1919*, 127.

[330] Fock, *Fast Fighting Boats*, 31; Hepper, *British Warship Losses, 1869-1919*, 127.

[331] *Hepper, British Warship Losses, 1860-1919, 132.*

[332] *Fock, Fast Fighting Boats, 48.*

[333] *Halpern, A Naval History of World War I, 172.*

Date	Craft, Name, Event or Country	Event, Builder and Boat Specifications	City, River, Lake, Sea, Ocean
1918[334]	July 9 British Royal Navy *CMB 2*	The boat was totally destroyed by accidental gun fire at Portsmouth naval station. **Builder:** Thorneycroft, 45' Class.	England
1918[335]	July 18 British Royal Navy *CMB 50,* *CMB 47*	The two boats departed their parent ship, *Cleopatra*, and proceeded on their mission to interfere with German minelayers operating in the North Sea. In the dark, the boats became separated and rough weather set in. They set out to make South Dogger Bank lightship, taking on water and being pounded by ever increasing heavy seas. A Dutch trawler, *Jacoba,* stopped and agreed to take the crew on board but not to tow the *CMB 50* boat. Lieutenant Lewis scuttled his boat and it sank on the spot. **Builder:** Thorneycroft, 45' Class.	North Sea
1918[336]	August 11 British Royal Navy CMB Flotilla *CMB 40, 41, 42, 44, 47, 48*	At the mouth of the Ems River, six CMB's running in advance of a larger British naval force, in the Southern North Sea, to attack German minesweepers. They were attacked by eight German flying boats. The Germans lost two two aircraft in the boat-to-airplane duel but the British CMB's were badly shot up. *CMB 42* was ran to shore, set on fire and the crew picked up by a Dutch vessel.	North Sea
1918[337]	August 11 British Royal Navy *CMB 40*	This craft stayed with the 42 boat and met the same fate. Burning off shore and rescue by the Dutch ship. **Builder:** Thorneycroft, 45' Class	North Sea

[334] Hepper, *British Warship Losses in the Ironclad era 1860-1919*, 136.
[335] Hepper, *British Warship Losses in the Ironclad era 1860-1919*, 136.
[336] Fock, *Fast Fighting Boats*, 31; Hepper, *British Warship Losses in the Ironclad era 1860-1919*, 138.
[337] Hepper, *British Warship Losses in the Ironclad era 1860-1919*, 138.

Date	Craft, Name, Event or Country	Event, Builder and Boat Specifications	City, River, Lake, Sea, Ocean
1918[338]	August 11 British Royal Navy *CMB 41*	The boat was headed for the shore to be burned when she was attacked by German aircraft. Dutch authorities prevented the crew from burning their boat. The craft was returned to the British navy at the end of the war. **Builder:** Thorneycroft, 45' Class.	North Sea
1918[339]	August 11 British Royal Navy *CMB-47*	While under the command of Lieutenant Leslie Cockburn, the boat was shot up by German floatplanes off Terschelling Island. Two of his crew were wounded in the fight. The rest of the men beached and burned the boat. Builder: Thorneycroft, 45' Class.	Netherlands North Sea
1918[340]	August 11 British Royal Navy *CMB-44*	While being commanded by Lieutenant Alfred Swann, the boat was severely damaged in an attack by German aircraft off Terschelling Island. A Dutch destroyer saved the craft from being beached and captured by the Germans. The 44 boat stayed in Dutch hands for the duration of the war and was returned to the British navy at war's end. **Builder:** Thorneycroft, 45' Class.	Netherlands North Sea
1918[341]	August 11 British Royal Navy *CMB 48*	The boat was badly shot up off Terschelling Island and her engine disabled. She drifted towards shore and was taken in tow by a Dutch destroyer. She was later returned to the British Navy. **Builder:** Thorneycroft, 45' Class.	Netherlands North Sea

[338] Hepper, *British Warship Losses in the Ironclad era 18601919*, 138.
[339] Hepper, *British Warship Losses in the Ironclad era 1860-1919*, 139.
[340] Hepper, *British Warship Losses in the Ironclad era 1860-1919*, 139.
[341] Hepper, *British Warship Losses in the Ironclad era 1860-1919*, 139.

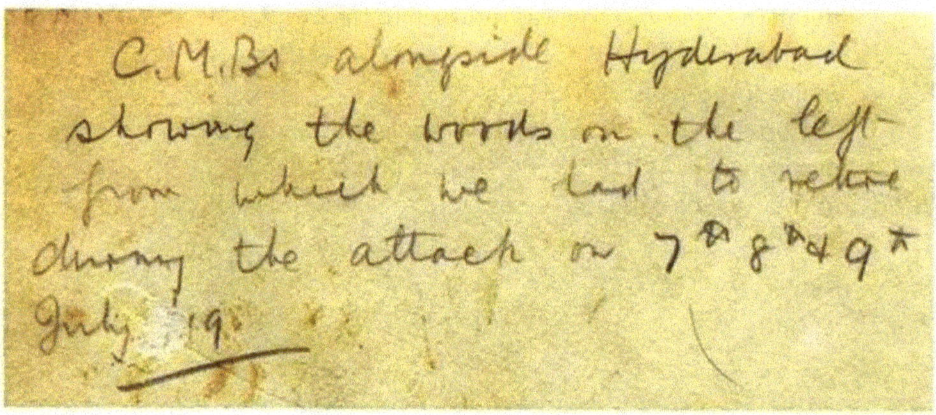

C.M.Bs alongside Hyderabad
showing the woods on the left
from which we had to retire
during the attack on 7th 8th & 9th
July 19

Figure 136. George Smith, Chief Yeoman of Signals, D.S.M, RN 1904-1928George Smith took many photos in his distinguished career and kept many hand written notes from his service to the Crown all over the world. He left an enormous collection of naval photos to his son, Gordon Smith who maintains one of the largest naval web sites in the world. (Collection of Gordon Smith, used with permission) The photo at top is described in the note at bottom as "CMBs alongside *Hyderabad* showing the woods on the left from which we tried to move during the attack on 7th, 8th, & 9th July 1919" with the North Russian Expeditionary Force on the River Dvina in 1919.

Date	Craft, Name, Event or Country	Event, Builder and Boat Specifications	City, River, Lake, Sea, Ocean
1918[342]	October 8 Italian Navy *Nibbio* Class Scout Torpedo Boats	A combined fleet of British light cruisers, Australian and British destroyers, Italian torpedo boats (ships), American submarine chasers, Italian *Nibbio* Class boats and MAS boats, as well as others, gathered together to destroy Durazzo Harbor and finish off the Austro-Hungarian forces. ***Nibbio:* Speed:** 18k; **Displacement:** 25.5 tons; **Length:** 80'-9"; **Beam:** 9'-10"; **Draft:** 3'-3"; **Engines:** one triple expansion steam, one screw; **Armament:** 2 x 14" bow torpedo tubes; **Crew:** 10.	Durazzo Harbor Adriatic Sea
1919[343]	June 17 British Royal Navy *CMB 4* 1st Attack Kronstadt Fortress	As a prelude to the planned full attack, Lieutenant Augustus Agar, Sub-Lieutenant John White Hampshire, RNR, and Chief Motor Mechanic Hugh Beeley, RNVR, made their way through the massive forts defenses. They attacked with a single torpedo and sank the Russian cruiser *Oleg*. **Builder:** Thorneycroft, 40' Class.	Kronstadt Fortress St. Petersburg Baltic Sea
1919[344]	July 31 British Royal Navy *CMB 67A*	The 67 boat was to be towed from England to the Baltic by the destroyer HMS *Venturous.* Heavy seas and bad weather made the trip miserable and the *67* boat parted her tow line and foundered for the last time north-east of Skagen lighthouse, Albaek Bay, Denmark. **Builder:** Thorneycroft, Oxford 1918, 60' Class.	Baltic Sea

[342] Halpern, *A Naval History of World War* I, 176.
[343] Agar, Augustus *The Baltic Episode*.
[344] Hepper, *British Warship Losses, 1860-1919*, 150.

Figure 137. (Collection of Gordon Smith, used with permission)

Figure 138. Another CMB deployed “against the Bolshevik Fleet.” (Collection of Gordon Smith, used with permission)

Date	Craft, Name, Event or Country	Event, Builder and Boat Specifications	City, River, Lake, Sea, Ocean
1919[345]	August 18 British Royal Navy *CMB 7AD & 4, CMB 24A, 31A 62BD, 72 ,79, 86, 88*	Due to the success of Lieutenant Agar's earlier attack on Kronstadt, Admiral Cowan called for 9 CMBs to attack the Fortress with a strong air arm to support the boats. The *67* boat was lost at sea on July 31st. and the *CMB 4* boat had been scuttled when damaged after the first raid. The CMBs ran at full power while running a gauntlet of fire from coastal batteries and ships as they forced their way into the confines of the harbor and wreaked havoc on the Russian fleet. At the end of the fight, the Russians had lost two battleships and one old cruiser that was acting as a submarine depot ship,. The docks at Kronstadt were largely destroyed. CMBs *7AD* and *4* were 40', *42A* and *31A* were 60'; *62BD, 72, 79, 86,* and *88* were 60'. CMBs sunk on this mission: *79A, 24A, 62BD, 67A.* Casualties: eight-dead, nine-POWs (three officers and six ratings) **Builder:** Thorneycroft, **Speed:** 39/40.4 knots; **Armament:** 2-53.3 cm torpedoes (rear), 4 x. 303 cal. MG; **Length:** 55'-0"; **Beam:** 11'-0"; **Draft:** 3.2'; **Engines:** 2 x375 hp Thorneycroft, petrol; 2-screws; **Displacement:** 12.5 tons; **Crew:** 11.	Kronstadt Fortress Baltic Sea
1919[346]	August 18 British Royal Navy CMB 24A	The 24 boat was on course to her target when the CMB 72 ran across her bows forcing the CMB 24 to take evasive action. She released her torpedoes and they both ran under their target ship, the Russian destroyer *Gavri'il,* sank the 24 boat with gunfire. **Builder:** Thorneycroft, 60'	Kronstadt Fortress Baltic Sea

[345] Agar, Augustus, Capt. VC, DSO *Baltic Episode A Classic of Secret Service in Russian Waters* (Naval Institute Press, 1963), 168-169.
[346] Hepper, *British Warship Losses, 1860-1919,* 151.

Date	Craft, Name, Event or Country	Event, Builder and Boat Specifications	City, River, Lake, Sea, Ocean
1919[347]	August 18 *CMB 62BD*	The *62* boat was maneuvering to make her attack when she collided with the *79* boat. They took the men on board and completed their attack on the *Petropavlovsk.* They made to get away and were hit by gunfire from shore installations and the destroyer *Gavri'il,* and the 62 boat caught fire and sank. **Builder:** Thorneycroft, 60'	Kronstadt Fortress Baltic Sea
1919[348]	August 18 *CMB 79A*	In the wild fighting, the *79* boat attacked her target ship, *Pamiat Azova (cruiser/*submarine depot ship) and sank her with torpedoes. As she beat a hasty retreat, she collided with the *62* boat and sank. **Builder:** Thorneycroft, 60'	Kronstadt Fortress Baltic Sea
1919[349]	August 26 *CMB 7*	On this night, as the boat ran at high speed for the coast to pick up an agent, Russian search lights from several forts on the shore forced the boat to take evasive action. Her tiller rope was shot away and the boat ran out of control onto a concrete breakwater near Fort No.5 off Kotlin Island. The crew pushed the disabled boat off, rigged a small sail and drifted with the current and wind. The next morning a Russian fishing boat towed the men and boat to a spot near their base at Terjioki. **Builder:** Thorneycroft, 45' CMB.	Baltic Sea
1919[350]	September 15 *CMB 28A*	From the Dvina River, the *28* boat was ordered to take a wounded officer to Archangel. Near the town of Suda, off the Bokachev River, the boat ran hard aground. To prevent use by the Soviets, she was destroyed by her crew. **Builder:** Thorneycroft, 60' **Displacement:** 11 tons; **Length:** 60'; **Beam:** 11'; **Armament:** 1-torpedo	.White Sea

347 Hepper, *British Warship Losses, 1860-1919,* 151.
348 Hepper, *British Warship Losses, 1860-1919,* 151.
349 Hepper, *British Warship Losses, 1860-1919,* 151.
350 Hepper, *British Warship Losses, 1860-1919,* 152.

Date	Craft, Name, Event or Country	Event, Builder and Boat Specifications	City, River, Lake, Sea, Ocean
1919[351]	September 27 *CMB 77A*	While stationed at Murmansk, the 77 boat was to be loaded on as deck cargo on the *SS Wisconsin Bridge*. The ship left port before the boat could be loaded. She was stripped of all war materials, then taken outside the harbor and sunk by gunfire from the *HMS Pegasus.* **Builder:** Thorneycroft, 60'.	Barents Sea
1919[352]	November 25 British Royal Navy *CMB 86 BD*	As she was being towed from Biiorko to Reval (Tallinn) she foundered and was lost. **Builder:** Thorneycroft, 60'.	Gulf of Bothnia Baltic Sea
1919[353]	December 28 British Royal Navy *CMB 72A*	At the end of hostilities, there was no way to ship the boat home and her deteriorated condition meant she could not be towed , she was scuttled off Reval (Tallinn) **Builder:** Thorneycroft, 60'	Gulf of Finland Baltic Sea
1919[354]	Finish Navy Finnish TB's	The Finns never received the motor torpedo boats they ordered. In their place, they used a mix of captured Russian boats. The small converted "torpedo boats" operated out of Biiorko Harbor, Finland. **Average speed**: 29 knots; **Armament:** 2 x 15" torpedo tubes, 2 x 3-pounder guns.	Gulf of Finland
1919[355]	Soviet Navy Fifteen captured British CMBs	The Soviets raise the 3 sunken CMBs *79, 24,* and *62* from Kronstadt Harbor and capture the surviving twelve boats as they were shipped by rail to Enzeli on the Caspian Sea. These boats become test beds for developing the Soviet MTB fleet.	Soviet Union

351 Hepper, *British Warship Losses, 1860-1919*, 153.
352 Hepper, *British Warship Losses, 1860-1919*, 154.
353 Hepper, *British Warship Losses, 1860-1919*, 154.
354 Fock, *Fast Fighting Boats*, 277-278.
355 *Soviet MTB's 1919-1969* Warship (Conway Maritime Press, New York), 123.

Date	Craft, Name, Event or Country	Event, Builder and Boat Specifications	City, River, Lake, Sea, Ocean
1920[356]	Italian Navy *MAS 397*	The next evolution of Italian motor torpedo boat design was the *MAS 397.* This line of boats was to be larger, heavier and better armed. Designed by the famous engineer Baggiest and built by S.V.A.N. Yards. Once the armament was reduced, the boat performed well. Constructions began as the armistice took effect and was finished in 1920. The *397* became the only model built. **Speed:** 28 knots; **Displacement:** 30 tons; **Armament:** 1 x 76 mm MG, 1-6.5mm MG, 2 x 450mm torpedo tubes, four depth charges; **Length:** 69.7'; **Beam:** 13.91'; **Draft:** 2.31'; Engines: 2 x 700/800 hp; 700 liters petrol; **Range:** 110nm @ 28 knots; **Crew:** 12.	Adriatic Sea
1923[357]	Soviet Union *G-5 "Tupolev" Torpedo Cutter*	Professor A.N. Tupolev, leading designer of all metal aircraft became the undisputed leader of the Soviet motor torpedo boat program. The Central Aero-Hydrodynamic Institute (CAGI) was awarded the contract to design the first Soviet built motor torpedo boats. The CAGI based much of their initial efforts on the captured British CMB's from the Kronstadt Raid. The design specifications were: **Radius:** 150 kilometers; **Speed:** 100 kh/h (55 knots); **Armament:** 2 18" torpedo tubes (rear), 1 x 7.62 mm MG; **Hull:** one version of aluminum and one of wood; **Crew:** 3.	Soviet Union
1927[358]	Soviet Union *SH-4 Torpedo Cutter*	The SH-4 was the last of a succession of designs that proved to be the basis for the Soviet motor torpedo boats of WWII. The hull was redesigned forward; there was a smaller conning tower, more careful design of construction joints, a new electrical wiring system and the addition of two torpedo troughs. **Speed:** 50 knots; **Displacement:** 13 tons; **Length:** 59.28'; **Beam:** 10.93'; **Draft:** 3.29'; Engines: 2 x 650-hp Wright T3 "Tornado," petrol; **Range:** 250 nm; **Armament:** 2 x 45.7 cm torpedo tubes, 2 x 7.6mm MG DA2; **Crew:** 6.	Soviet Union

[356] Fock, *Fast Fighting Boats*, 91-92.

[357] Budzbon, Przemyslaw *The Beginnings of Soviet Naval Power* (*Warship,* Naval Institute Press, 1971), 232.

[358] Budzbon, Przemyslaw *The Beginnings of Soviet Naval Power* (*Warship,* Naval Institute Press, 1971), 233-234.

Figure 139. The *G-5* was among the most unique torpedo boats of WWII. Built of Duralumin to the design of an eminent aircraft designer the boat had its share of built-in quirks. For instance, there was no gear box in the early models. The boats could not go slower then 18 knots. The engine was on or off. Docking was a constant trial and a bit dicey. (Illustrations by author)

Date	Craft, Name, Event or Country	Event, Builder and Boat Specifications	City, River, Lake, Sea, Ocean
1929[359]	Greece *T-1 &T-2*	Greece purchased two Thorneycroft CMB's to prepare them for fighting against their enemies in the Mediterranean Sea. Type: Thorneycroft 55' CMB.	Greece
1929[360]	Italian Navy *MAS 431*	The famous yards of S.V.A.N. turned out their first fully modernized motor torpedo boat that set a record for the fastest torpedo boat in the world. The boat's best feature was her petrol engines with a power to weight ratio of 1:37. **Speed:** 40 knots; **Range:** 124 nm @ 36 knots. **Length:** 16m; **Beam:** 4.68; **Draft:** 1.10m; **Displacement:** 39.8 tons; **Engines:** 3 x 500-hp Isotta-Fraschini, three screws; **Armament:** 2 x 17.7" torpedo tubes; 2 x 6.5mm MG.	Adriatic Sea
1929[361]	Germany Navy *Lürssen S 1*	The first *Schnellboote* built at the Lürssen Yards, Vegesack, Germany. The boats were built within the confines of the Treaty of Versailles. The new *Reichsmarine* relied upon the Allied authorization to resume building speedboats for foreign sales. **Speed:** 34.2 knots; **Range:** 350 nm; **Length:** 69.8'; **Beam:** 11.9'; **Draft:** 6.82; **Displacement:** 39.8/51.6 tons; **Armament:** 2 x 53.3cm bow tubes, two reloads, 1 x 7.92mm MG, 1 x 20 mm MG; 5 x 200 kg depth charges; **Engines:** three Daimler-Benz BFz,12-cylinder, 900-hp, one Maybach 100-hp, petrol;	Germany

[359] Fock, *Fast Fighting Boats*, 148.
[360] Fock, *Fast Fighting Boats*, 98.
[361] Fock, *Fast Fighting Boats*, 113.

Date	Craft, Name, Event or Country	Event, Builder and Boat Specifications	City, River, Lake, Sea, Ocean
1930[362]	Soviet Union *G-5, Series 7 Torpedo Cutter*	The G-5 model became the most produced MTB of the Soviet fleet. They had their quirks such as not being able to go below 18 knots (no maneuvering engine). Armor plating was scarce being mostly on the engine and fuel tanks. Photos and plans of these boats never show life rafts, rings or jackets. . 85 boats were built for the Black Sea Fleet. **Speed:** 45 knots; **Range:** 200 @ 45 k, 250 @ 34 k; **Displacement:** 14.84 tons; **Length:** 59.28'; **Beam:** 10.93'; **Draft:** 3.28'; **Armament:** 2 x 53.3cm torpedo tubes (rear), 1 x 7.6mm MG; **Engines:** 2 x 675-hp GAM 34, aviation petrol 1,050 liters; **Crew:** 7.	Baltic Sea Black Sea Caspian Sea
1932[363]	Germany Navy *Lürssen S 2-5*	A slight improvement over the *S-1* in working space on board and a higher speed. The other design improvements included superchargers for the engines and *"Lürssen-effect rudders,"* a pair of small rudders located next to the main rudder that lower the wake and enable the ship to go faster. For crew comfort, an oil cooking stove was added to supply hot water to the wheelhouse and living quarters. **Speed:** 33.8 knots; **Range:** 300 nm **; Displacement:** 46.5/58 tons; **Length:** 91.7'; **Beam:** 13.78'; **Draft:** 8.01'; **Armament:** 2 x 53.3cm bow torpedo tubes, two reloads, 1 x 12.7 mm MG, 1 x20mm MG; **Engines:** three Daimler-Benz Bfz ,12-cylinder, 1100-hp,oneMaybach S5, petrol.	Baltic Sea North Sea Norwegian Sea

[362] Fock, *Fast Fighting Boats*, 298-299.
[363] Fock, *Fast Fighting Boats*, 115.

Date	Craft, Name, Event or Country	Event, Builder and Boat Specifications	City, River, Lake, Sea, Ocean
1934[364]	Soviet Navy *G-5* Series 9	The code for the *SH-4* became *G-5* (*Glissiryushchyi*-No.5-Hydroplane No.5). Further improvements were made to the boats to allow for larger engines and general engineering improvements. Gross loaded weight down by half of one ton. **Speed:** 49 knots; **Range:** 200 nm @ 45.3k, 250nm @ 34.2 knots **Displacement:** 14.03 tons; **Length:** 62.66'; **Beam:** 10.93'; **Draft:** 2.13'; **Armament:** 2 x 53.3 cm torpedo tubes(rear), 2 x 7.6mm MG DA, Engines: 2 x 850hp GAM 34 BS, aviation petrol; **Crew:** 7.	Baltic Sea Black Sea Caspian Sea
Date	**Craft, Name, Event or Country**	**Event, Builder and Boat Specifications**	**City, River, Lake, Sea, Ocean**
1935[365]	French Navy *VTB-87*	The French were in a race with the German *Schnellboote* and Italian MAS MTB's. France had enemies on both of her shores. **Speed:** 52 knots (max), designed: 46 knots; **Range:** 250 nm; **Length:** 61.85'; **Beam:** 13.28'; **Draft:** 3.94'; **Displacement:** 22 tons; **Engines:** 2 x 2, 200-hp Lorraine, petrol, Aux: 2 x 80 hp, petrol; **Armament:** 2 x 45.7 cm side dropping torpedoes, 4 x 12.7 mm MG; **Crew:** 5.	France North Sea English Channel Bay of Biscay Mediterranean

[364] Fock, *Fast Fighting Boats*, 76.
[365] Fock, *Fast Fighting Boats*,72-74, 289.

Figure 140. French VTB-87 French boats were well built but the war prevented their continued construction. (French Fast Attack Boats, www.ww2Ships.com, James Davies)

Date	Craft, Name, Event or Country	Event, Builder and Boat Specifications	City, River, Lake, Sea, Ocean
1935[366]	German Navy *S 14-17*	The navy kept improving the model S boats from 1928 to the last days of World War II. Documents in the *Bundesarchiv* (military archives) testify to the sea keeping and reliability of all models of S-bootes to date. The specifications for this boat and the Italian MAS 424 boat below, when compared to the WWI-era boats described earlier, illustrate the dramatic improvements in naval engineering and torpedo boat capabilities during the interwar period. **Builder:** Lürssen Vegescack. **Speed:** cruise 37.7 k/max. 39.8k; **Range:** 750 nm; **Displacement:** 117 tons; **Length:** 113.6'; **Beam:** 16.5'; **Draft:** 4.6"; **Engines:** three MAN, Typr-11-Zu, 1 x 1 cylinder 4500/6150 hp Diesel; **Armament:** 2 x 53.3 cm bow torpedo tubes; 1 MG; 1 20 mm; **Crew:** 18-25.	Germany
1936[367]	Italian Navy *MAS 424*	The navy pursued their motor torpedo boat program from 1912 to 1945. The Italian torpedo boat building program was nearing satisfactory results with the MAS 400 line of boats by 1935. Due to alterations, the launch date moved to February 14, 1937. **Builder:** S.V.A.N. **Speed:** 53 knots max/37 cruise; **Range:** 350 nm @ 35 knots, 650nm@ 18 knots; **Displacement:** 22 tons ; **Length:** 60'-0", **Beam:** 11'-0", **Draft:** 2.82'; **Armament:** two stern-mounted 45.7 cm torpedoes, two quad 7.76 mm MG, four-depth charges; **Engines:** ; 3 x 1,600bhp, petrol, **Crew:** 9.	Adriatic Sea, Mediterranean Sea
1936[368]	Chinese Navy *CMB 3-4*	These boats were made for China but reverted to British ownership in 1939. They would become *MTB-26, MTB-27.* China also purchased Schellboote *S 2, S 5* and Italian *MAS 226* and *227.*	China

[366] Fock, *Fast Fighting Boats*, 118.
[367] Fock, *Fast Fighting Boats*, 98.
[368] Fock, *Fast Fighting Boats*, 67.

Date	Craft, Name, Event or Country	Event, Builder and Boat Specifications	City, River, Lake, Sea, Ocean
1936[369]	British Navy *MTB 1-12* *MTB 14-19*	German S-boats in the English Channel caused the Admiralty great concern. The boats were small, fast, and robustly built. These few boats bought time until the larger boats with heavier guns came on the scene. The Admiralty dispatched two flotillas of six boats each to the Mediterranean and Hong Kong. **Builder:** British Power Boat Company **Displacement:** 122 tons; **Length:** 60'-3"; **Beam:** 13'-3"; **Draft:** 2'-9"; **Engines:** 2- Napier Sea Lion 1,800 bhp, Petrol; **Speed:** 33 knots; **Armament:** 2 x 17.7" torpedoes (stern tubes) ; 2 mounts-for four quad mg 7.76 mm (mounted fore and aft), four twin .303-cal MG; **Crew:** 9.	England
1937[370]	British Navy *MTB 102* (Experimental)	First served in Malta as part of the Malta MTB Squadron then the evacuation of Dunkirk (8 trips), flagship for Admiral Wake-Walker, and carried both Churchill and Eisenhower to review the fleet for Operation Overlord in 1944. She has the record as the fastest MTB in the British navy at 48 knots. **Builder:** Vosper. **Speed:** 48 knots empty of weapons and 43 knots fully combat loaded; **Length:** 68'-0"; **Beam:** 14'-9"; **Draft:** 3'-9"; **Armament:** 2 x 21" torpedo tubes, four Vickers MG, two depth charges; **Engines:** 3 x 3,300-hp Isotta-Fraschini, petrol; **Crew:** 12 men.	English Channel
1937[371]	June 22 Malta MTB Flotilla	The British sent the MTBs *01, 02, 03, 04* and *06* to Malta as the 1st Maltese Flotilla. **Builder:** British Power Boat Co.	Mediterranean

369 Fock, *Fast Fighting Boats,* 66; North, *Royal Naval Coastal Forces,* 8.
370 MTB-102 http://www.powerset.com/explore/semhtm/MTB_102.
371 Richards, Brooks. Secret *Flotillas: The clandestine sea lines to France and French North Africa 1940-1944* (HMSO Publications Centre, 1996).

Date	Craft, Name, Event or Country	Event, Builder and Boat Specifications	City, River, Lake, Sea, Ocean
1937[372]	U.S .Army MacArthur's MTB Design	U.S. Naval Lieutenant, S. L. Huff was a member of General Douglas MacArthur's staff. MacArthur had Huff design a shallow draft fighting boat for the shallow waters of the Philippines. Once the plans fell into the hands of BuShips, BuOrd and Preliminary Design the boats died. Changes to specifications reduced the boat to a fishing boat and never built. **Speed:** 45 knots; **Range:** 500 nm @ 45 knots and 700 nm @ 10/15 knots; **Displacement:** Unknown; **Length:** 60'-70; **Beam:** 14'; **Draft:** 3'-6"; **Engines:** 3-4,500 hp, petrol; **Armament:** 2 mounts of 2 x .50 cal MG in armored turrets, twp x 20" torpedo tubes; 2-Depth Charges; **Crew:** 10.	United States
1937[373]	Soviet Union Type D-3	The torpedo cutters, D-3, were never built in large numbers. They were built of wood and never had the maintenance problems of the G-5's. The G-5's had to be pulled from salt water and allowed to soak in fresh water rivers or lakes for 7 to 10 days. The D-3 was a far more robust boat and nearly as fast. **Speed:** 42 knots; **Range:** 450 nm @ maximum speed; **Displacement:** 35 tons; **Armament:** 1 x 20mm SchWAK, one twin 12.7 mm DSch K, 1 x 12.7 mm aircraft MG, 2 x 53.3 torpedo tubes; **Length:** 68.90'; **Beam:** 12.12'; **Draft:** 3.94'; **Engines:** 2 x 850 GAM 34 BS, 1 x 675-hp GAM 34; **Crew:** 12/14.	Baltic Sea Black Sea

372 Friedman, Norman *U.S. Small Combatants* (Naval Institute Press), 110-111.
373 Fock, *Fast Fighting Boats*, 77-78.

Date	Craft, Name, Event or Country	Event, Builder and Boat Specifications	City, River, Lake, Sea, Ocean
1937[374]	Soviet Union *Andry Tupolev G-5 Series 10*	Andry Tupolev designed and produced the fastest motor torpedo boat of World War II using Aviation Duralumin. Stalin suspected Tupolev of being a spy and arrested the entire engineering and design arm of the Central Aerohydraulic Institute. After 18 months of torture and seclusion, in Siberia, a prison camp was made for the Tupolev workers. It was here the final design improvements of the G-5 torpedo cutter were ironed out. The team stayed imprisoned until 1943, after German forces begin to decimate the Soviet navy and air force. **Speed:** 49k max; **Range:** 200 nm at 36 knots cruise; **Displacement:** 16.26 tons; **Length:** 53.3'; **Beam:** 11.15'; **Draft:** 3.45'; **Engines:** 2 x 850 hp GAM BS; **Armament:** 2 x 53.3 cm torpedo tubes; 2 x 7.6 mm MG DA2.	Baltic Sea Black Sea
1937[375]	August 16 Chinese Navy *MTB-102* (*Shih*)	China bought several Thorneycroft 55' CMBs. The *Shih* first saw action against the IJN when she attacked the Armored Cruiser (OAC) *Izumo* at the Marco Polo bridge, over the Whangoo River, Shanghai. The attack failed but the Japanese saw the threat of the MTB in modern warfare.	Whangoo River Shanghai, China
1937[376]	August 24 Chinese Navy MTB *Kuan Wei, Kuan Fu*	These two boats were built by the British Power Company to be carried aboard the customs ship SS *Fu Hsing* to seek out and attack smugglers. **Builder:** British Power Boat Co., 45'.	China

[374] Fock, *Fast Fighting Boats*, 76-77, 298-299.
[375] Lengerer, Hans. *Motor Torpedo Boats of the Imperial Japanese Navy*, Part 1 (*Warship International* No.3, 2007), 263.
[376] Lengerer, Hans. *Motor Torpedo Boats of the Imperial Japanese Navy*, Part 1 (*Warship International* No.3, 2007), 264.

Date	Craft, Name, Event or Country	Event, Builder and Boat Specifications	City, River, Lake, Sea, Ocean
1938[377]	April Q-111 *(Luzon)*	At the insistence of General MacArthur the Thorneycroft company built the Q-111 for the Philippine government Offshore Patrol. **Speed:** 39 knots; **Length:** 65'; **Beam:** 15'; Engines: 3 x 600 hp Thorneycroft, petrol; three screws; **Armament**: 2 x 18" torpedo tubes, 3 x .50 cal. Vickers MG, one depth charge	Manila Philippines
1938[378]	U.S. Navy	Official designation by CNO, Admiral J.O. Richardson, classified torpedo boats (T) and (P) for patrol category. PT boats become an official arm of the U.S. Navy.	Washington, D.C.
1938[379]	French Navy VTB-10	In 1929, the French started in earnest to develop its own fast motor torpedo boat. By 1937, VTB-10 became the standard model for further design efforts. France placed orders with the British Power Boat Company for VTB 23-40 for delivery in late 1940.The delivery never took place due to Hitler's Blitzkrieg. **Builder:** British Power Boat Company. **Speed:** 42 knots; **Range** 500 nm @ 40 knots, 900 @ 15 knots; **Displacement:** 30 tons; **Length:** 70'; **Beam:** 19.7'; **Draft:** 5.5'; **Engines:** 3 x 1100-hp Rolls Royce Merlin; **Armament:** 2-53.3cm TT or 4-45.7cm torpedo tubes; 2 x 20mm automatic gun turrets (or) 4 x 12.7mm MG; **Crew**: 12.	France
1938[380]	Spanish Navy *LT-15 , LT-13, LT-12, LT-11, LT-14*	The Spanish Navy acquired 5 *S*-boat prototypes for their own use. The old *S* boats were given Spanish names and the Spanish Naval designation of "LT", *Lancha Torperdera* (torpedo boat). Spain would have a total of 17 S boats by the end of the bloody civil war that racked Spain in the late 30's. The boats were put to effective use and and gave the Germans a live-fire testing ground for the S boot. *S-1* became *Badajoz* (a city in SW Spain), *LT 15; S-4* became *Requete (Phalanx), LT 11; S-2 Falange (Revenge), LT 13 S-5* became *Toledo* (a major city), *LT 14; S-3* became *Oviedo* (a city in NW Spain), *LT 12*	Spain

377 Fock, *Fast Fighting Boats,* 70.
378 Friedman, *U.S. Small Combatants,* 116.
379 Fock, *Fast Fighting Boats,* 74-75.

Date	Craft, Name, Event or Country	Event, Builder and Boat Specifications	City, River, Lake, Sea, Ocean
1939[381]	Soviet Union *D-3* Torpedo Cutter	The D3 became the Soviet navy's choice for a larger motor torpedo boat without the maintenance difficulties of the Duralumin of the G-5. This design still employed the stepped hull. **Speed:** 42 knots; **Displacement:** 32/35 tons; **Armament:** 2 x 53.3cm side dropping torpedoes, 2 x 12.7 mm MG, four depth charges; **Engines:** 3 x 1,250-hp GAM 34 FN, petrol; three screws ; **Crew:** 12/14.	Russia
1939[382]	Finnish Navy	Two Thorneycroft boats built for the Finnish needed to be taken over by the British Navy and renamed MTB-67 and MTB 68. **Builder:** Thorneycroft, 55' CMB. **Speed:** 40.4 knots max; **Range:** 200 nm @ 39 knots ; **Displacement:** 17 tons **Length:** 55'; **Beam:** 11"; **Draft:** 3.3'; **Engines:** 2x 375-hp Thorneycroft; **Armament:** 2 x 53.3cm torpedo tubes; 2 x2 7.76 mm MG; **Crew:** 5.	Finland
1939[383]	August 7 Bulgarian Navy *F-3 World Speed Record Holder*	The Lürssen designed *F-3 (Schnellboote)* is the only military torpedo boat to attain a certified world speed record. On August 7, 1939, the boat driven by Gert Lürssen, grandson of the company founder, Fredich Lürssen. The F 3 reached a sustained speed of 68.14km/h or 36.79 mph over one mile. The feat was ratified by the Union International du Yachting Automobile- UIYA[384]. The F 3 was not disarmed until 1973. **Builder:** Lürssen **Speed:** 37 knots; **Displacement:** 68 tons; **Length:** 91.2'; **Engines:** 2 x 12-cylinder MB 500 Diesel @950bhp**; Armament:** 2 x 53.3 cm torpedo tubes, 1 x 20 mm rapid-fire cannon.	Germany

---

380 Yanez, Alejandro, *Schnellboote* in the Spanish Armada http://www.prinzeugen.com/Armada.htm. Accessed August 9, 2007, 1-3.
381 Fock, *Fast Fighting Boats*, 78.
382 Worth, Richard. *Fleets of World War II* (Da Capo Press, 2001), 19.
383 Dallies-Labourdette, Jean-Phillippe. *S-Boote German E-boats in action 1939-1945* (Histoire & Collections 2003), 15.
384 The UIYA was established in 1922. http://en.wikipedia.org/wiki/Union_Internationale_Motonautique.

Figure 141. The Spanish never declared war against any country but they were publicly pro-German. The Spanish bough the older S boats, numbers 1-5 in 1939. Later in WWII, they again purchased six more S boats plus the plans for the S boat construction. The Spanish Navy sent officers and sailors to train with German crews in the Baltic Sea. This meant that while Spain was not at war, their sailors were killing allied personnel in training exercises. They have never been held to account for their actions.

Figure 142. This very rare photo of a Finnish torpedo boat was taken from the website of Osmo Joronen. What has never been discussed is the number of times the Finnish government made advance payments to the American builders ELCO and Higgins and the British builders Thorneycroft and Vosper without ever taking delivery. Starting in1940, Finland ordered 23 boats from four countries, and only ever received five from Germany's ally, Italy. Finland's June 1941 alliance with Germany against the Soviet Union may have had something to do with this poor delivery record. (Used with permission of Osmo Joronen)

Figure 143. Lürssen-built Bulgarian torpedo boats underway.

THE SHIP KILLERS

THE DEFINITIVE ILLUSTRATED HISTORY OF THE TORPEDO BOAT

The Soviet TKA-125.

VOLUME IV: TIMELINE 1939 – 1940

BY JOE HINDS

NIMBLE BOOKS LLC

CONTENTS (INCLUDING FIGURES)

DEDICATION

John Lambert of LambertPlans.com is one of the very first historians I started working with in 1999. John has always been straightforward in his answers to my questions. He knew the importance that I attached to all of my illustrations being absolutely color correct. Like me, he has seen more than his fair share of MTBs presented in art form and as models with everything perfect but the boats colors.

John's generosity with his knowledge, photo files, and drawings has enhanced my efforts for every volume of *The Ship Killers*. I may get credit for producing my illustrations but it was largely John who helped me get the colors correct.

This volume is surely a small recognition for all the years of his wealth of wisdom. Thank you John and continued good fortune.

TIMELINE 1939 - 1940

Date	Craft, Name, Event or Country	Event, Builder and Boat Specifications	City, River, Lake, Sea, Ocean
1939[385]	September 1 1st SFltl.[386] Flotilla *S-23*	The *S-23* scored the first S-boat success of the war for Germany when, on patrol in the Gulf of Danzig, it sank the barrage pilot vessel, *Lloyd Bydgoski*, a former trawler. The trawler sank from 20 mm cannon fire after the boat's crew took to lifeboats. **Builder:** Lürssen, **Type:** *S-14.*	Gulf of Danzig Baltic Sea
1939[387]	September 5 2nd SFltl. *S-17*	In its first mission of the war the *17* boat, while on a scouting mission off Heligoland, Denmark, North Sea, became damaged due to rough weather. The boat was able to make it back to port for repairs. **Type:** *S-10.*	North Sea
1939[388]	October 9/10 US Navy *PT 9*	The Electric Boat firm (Elco) bought the license rights for a 70' PT boat from Scott-Paine of British Power Boat Co. The boat was delivered to America for a series of successful trials, off New London, Connecticut, set up by the U.S. Navy. Second trials were agreed to; at their completion, the Navy awarded Elco a contract for twenty-three boats.	New London, Connecticut Atlantic Ocean
1939[389]	November 16 British Navy *MTB 6*	Foundered off the Sardinia Coast and was lost. **Builder:** British Power Boat; **Displacement:** 22 tons; **Length:** 60'-3"; **Beam:** 13'-3"; **Draft:** 2'-9"; **Engines:** 3 x 1,800-bhp Napier Sea Lion, petrol: **Speed:** 33 knots max, 29 knots cruise; **Armament:** 2 x 18"torpedo tubes; 4 x 2 .303 MG: **Crew:** 9.	Mediterranean Sea
1939[390]	Philippine Navy	The five motor torpedo boats destined for the Philippine Navy were taken over by the British Admiralty and re-named MTB *327–331*. General MacArthur would sorely miss these boats in the Japanese invasion of 1941. **Builder:** Thorneycroft, 55' (CMB Type)	Philippines

---

385 Hümmelchen, G., "*German Schnellboote* (E-Boats)" in *Warship: Profile 31* (1979), 157.
386 *Schnelleboote* Flotilla.
387 Frank, *German S-Boats in Action in the Second World War*, 11.
388 Lambert, John, Al Ross, Jr. PhD. *Allied Coastal Forces of World War II, Vol.2* (Naval Institute Press, 1990), 104-105.
389 *Warships,* Supplement 65, Summer, 5.

Figure 144. *PT 10* in New York Harbor seen with flags flying on its first "Photo Op" day. (Library of Congress Prints and Photographic Division)

Figure 145. A poster encouraging rapid production of PT boats. (Library of Congress Prints and Photographic Division)

[390] *Jane's Fighting Ships*, 1921.

Figure 146. The Norwegian Navy did not realize that the MTB 6 was not sunk, but washed up on shore. (Gordon Smith, www.Naval-History.net)

Figure 147. The German navy started the war with a very reliable type of S-boat, the type *S-14.* Many of these boats survived the entire six years of combat.[391]

[391] Busch, F.O., *Insere Schnellboote im Kanal,* 22.

Date	Craft, Name, Event or Country	Event, Builder and Boat Specifications	City, River, Lake, Sea, Ocean
1940[392]	March 21 Fairmile *ML 100*	Noel Maklin created the Fairmile Marine Company from his automobile company, Fairmile Engineering of Fairmile, Cobham. The *ML 100* was 110' long and displaced 25 tons. This boat would start the line of famous "Dog Boats," the only Allied motor torpedo boat to stand up to the German S-boats.	England
1940[393]	April 7 Operation *Weserübung*	*S*-boats *19, 21, 22, 23,* 24 are part of the German naval invasion of Norway and assigned to Group 3. These same boats took part in the invasion of Poland. *S 21* and *S 19* collided in the fog and *S 21* was completely out of action for the remainder of the operation. **Type:** *S-14.*	Hardanger Fjord North Sea
1940[394]	April 19 2nd SFltl.	The flotilla boats sank the Norwegian torpedo boat, *Sael*, seized the minelayer *Uller,* and captured several other ships. **Type:** *S-14.*	Hardanger Fjord
1940[395]	May 10 2nd SFltl. *S-30, 31, 32, 33.*	Off the Dutch coast, the Flotilla came under heavy fire by a British cruiser *Birmingham* and four British destroyers. The *S-33* was rammed by a destroyer and sustained severe damage. The *S-31* torpedoed the destroyer HMS *Kelly.* She was under tow for ninety-one hours and safely taken back to Tyne, UK. **Type:** *S-14.*	Dutch coast
1940[396]	May 21 1st SFltl. *S-19, 21, 22, 23, 24.*	Boats of the flotilla sank the British mine destructor HMS *Corburn* off Le Havre. **Type:** *S-14.*	Off French coast
1940[397]	May 23 1st SFltl. *S-21, S-23.*	The two S-boats torpedoed the French DD *Jaguar* off Malo les Bains. **Type:** *S-14.*	Off French coast

392 Reynolds, *Dog Boats At War*, 3.
393 Rohwer, *Chronology of the War At Sea*, 18.
394 Hümmelchen, *German Schnellboote*, 158.
395 Hümmelchen, *German Schnellboote*, 158.
396 Hümmelchen, *German Schnellboote*, 158.
397 Rohwer, *Chronology of the War At Sea*, 24.

Figure 148. The Fairmile "D" type. The Fairmile line of MTBs and MGBs were the only vessels capable of fighting the German S-boats toe-to-toe. (Private collection of John Lambert).

Figure 149. All torpedo boat crews enjoyed the chance to show off for the camera. The S-boat design was sturdy and capable of standing up to heavy weather. (Private collection of Hans Frank)

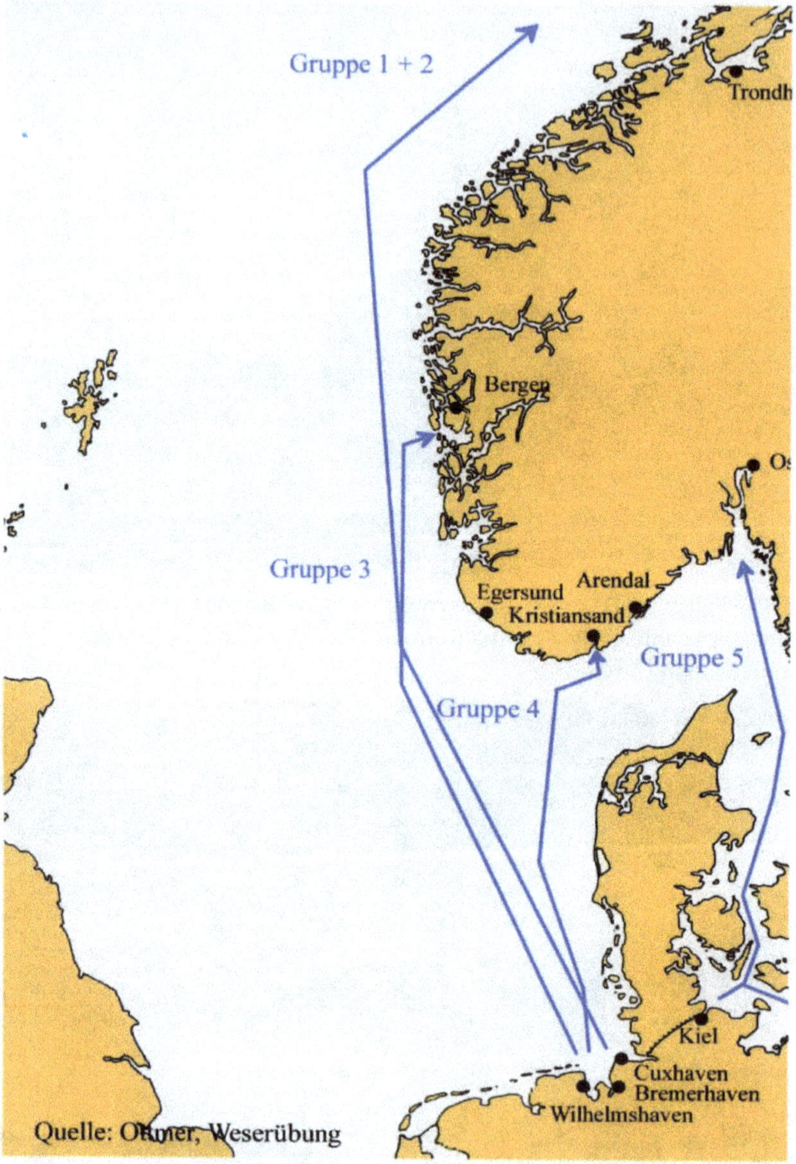

Figure 150. The German operation to capture Norway became a very costly adventure. The *Kriegsmarine* lost ten destroyers, the cruiser *Karlsuhe,* and heavy cruiser *Blucher.* The small boats were not immune; two S-boats collided and were out of action for a great deal of time.[398]

[398] Frank, *German S-Boats,* 13-15.

Figure 151. The S-boat crewmen feared the *Hunt* class destroyer (pictured: *HMS Kelly*), RAF fighters and fighter-bombers, and British radar. (Gordon Smith, www.Naval-History.Net).

Figure 152. *S-19* under way with the rest of 1st SFltl. [399]

[399] Frank, *German S-Boats,* 34.

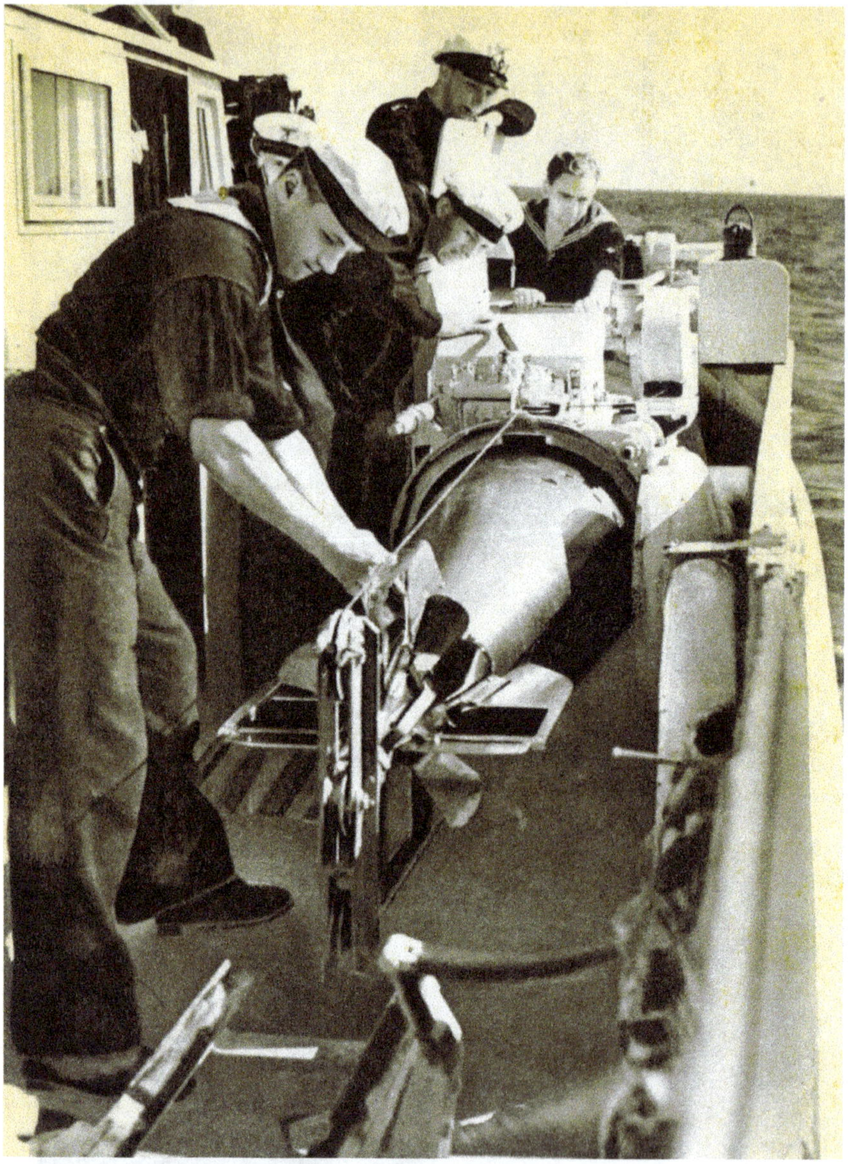

Figure 153. Torpedomen prepare their weapons before meeting the enemy. The French destroyer *Jaguar* and HMS *Destructor* are sunk this day by S-boats off LeHavre.[400]

---

[400] Busch, *Insere Schnellboote im Kanal,* 4.

Figure 154. S-boats leave port on their way to battle enemy ships in the English Channel.[401]

Figure 155. The RF *Sirocco* is sunk by torpedoes from the S-23 and S 26.[402]

[401] Busch, *Insere Schnellboote im Kanal,* 22.

Date	Craft, Name, Event or Country	Event, Builder and Boat Specifications	City, River, Lake, Sea, Ocean
1940[403]	May 25 *MTB 106* (ex-*DCTB 2*)	Assigned to 10th MTB Flotilla HMS *Beehive,* Felixstowe, used to evacuate British troops from Dunkirk. The last RN ship to leave Dunkirk, France. **Builder:** Thorneycroft 45' (Experimental).	English Channel
1940[404]	May 25 *MTB 107*	She is assigned to 10th MTB Flotilla, HMS *Beehive,* and Felixstowe. The 107 made eight trips to evacuate British troops from Dunkirk. **Builder:** Thorneycroft 45' (Experimental).	English Channel
1940[405]	May 26 1st SFltl.	They attack and damage the British destroyer HMS *Vega* and Dutch submarine *0-13;* neither boat sinks as the Germans claimed. **Type:** *S-14.*	North Sea
1940[406]	May 29 2nd SFltl. *S-30, S-34*	The *S-30* sinks the British destroyer, HMS *Wakeful,* off the coast of Nieuport (Nieuwpoort) Belgium. *S-34* sank the freighter SS *Abouki*r. **Type:** *S-14.*	Off Belgian coast
1940[407]	May 30 2nd SFltl. *S-24*	*S-24* was successful in hitting the French destroyer *Cyclone* with two torpedoes. The *Cyclone* makes it back to England, but she falls into German hands in the Port of Brest, France. **Type:** *S-14.*	Off Dunkirk Straits of Dover
1940[408]	May 31 2nd SFltl. *S-23, S-26*	*S-23* and *26* put torpedoes into the French destroyer *Sirocco* and it sinks immediately. **Type:** *S-14.*	English Channel
1940[409]	June 4 *MTB 107*	The *107* served as Air-Sea rescue boat during the Battle of Britain. Later, she detonated German laid acoustic mines, in theThames estuary. She was involved in the operation of the LRDG (Long Range Desert Group) and final fate unknown.	English Channel
1940[410]	June 11 *S 35, S-30*	Both boats attacked by RAF aircraft off Boulogne, France, and shot up badly. 7-dead, 22 wounded. **Type:** *S-30.*	English Channel

---

402 Uboat.net http://www.uboat.net/allies/warships/ship/6007.html.
403 Lenton, H.T. *British and Empire Warships of the Second World War* (Greenhill& NIP, 1998) 540.
404 Lenton, H.T. *British and Empire Warships of the Second World War,* 540.
405 Hümmelchen, *German Schnellboote,* 158.
406 Hümmelchen, *German Schnellboote,* 158.
407 Hümmelchen, *German Schnellboote,* 158.
408 Labourdette, *S-Boote,* 43.
409 Lenton, H.T. *British & Empire Warships of the Second World War* (Greenhill & NIP, 1998), 540.

Figure 156. The helmsman and radio operator stood long watches with little chance of relief.[411]

[410] Frank, *German S-Boats in Action in the Second World War,* 23.
[411] Frank, *German S-Boats,* 143.

Date	Craft, Name, Event or Country	Event, Builder and Boat Specifications	City, River, Lake, Sea, Ocean
1940[412]	June 19 *MTB 29*	(SIS Section D) The *MTB 29* assigned to reach the French shores, near Carantec, and retrieve the family of General de Gaulle. Unknown to the crew, the family had left earlier for Brest, and left on a merchant ship for England. **Builder:** Vosper, Built by Camper and Nicholson, Gosport. **Speed:** 42 knots; **Length:** 70', **Beam:** 14'-9"; **Draft:** 3'-3" forward, 5'-0" aft; **Displacement:** 37 tons; **Engines:** 3 x 1,200 hp Isotta-Fraschini petrol & 2 x Ford V-8 cruising engines with 24 Volt and 110 Volt, Ford 10 hp auxiliary engine. **Armament:** 2 x twin .50 cal MG, 2 x 21" torpedo tubes; **Crew:** 10; Range 400 nm @ 20 knots.	English Channel
1940[413]	June 19/20 1st SFltl.	The *S-19* and *S-26* sink the freighter *Roseburn* off Dungeness. **Type:** *S-14.*	English Channel
1940[414]	June 20 French MTB	The boats enlisted by the Secret Intelligence Service (SIS) to begin clandestine operations off the shores of France. Their first success was in landing agents and retrieving agents near Brest, France. **Builder:** British Power Boat Company, Hythe, Southampton Water. Designated as MA/SBS by Coastal Forces.	English Channel
1940[415]	June 21 2nd SFltl. *S-32*	The *32* boat was ordered to patrol in mid-Channel and listen for Allied convoys. While cruising at the lowest possible speed, the boat struck a mine. The mine blew off the front third, of the boat killing seven men immediately. After the rest of the crew was rescued, the *S-32* was scuttled by explosives on the spot. **Type:** *S-30.*	English Channel
1940[416]	June 21 *S-21, 32*	Mines off Boulogne sink both boats, 15 miles South of Dungeness. *S-21* is salvaged and repaired in 1940. **Type:** *S-14.*	English Channel

412 Richards, *Secret Flotilla: The clandestine sea lines to France and French North Africa 1940—1944* (HMSO Publication Centre, 1996), 304.
413 Labourdette, *S-Boote,* 43.
414 Richards, *Secret Flotillas*, 304.
415 Frank, *German S-Boats in Action in the Second World War*, 23.
416 Rohwer, *Chronology of the War At Sea*, 29.

Figure 157. *MTB 34*, a sister ship of *MTB 29*. (Private collection of John Lambert).

Figure 158. The *S-26* showing the effects of a prolonged fire fight at sea. Notice the dozens of small patches marking the many holes in the hull.[417]

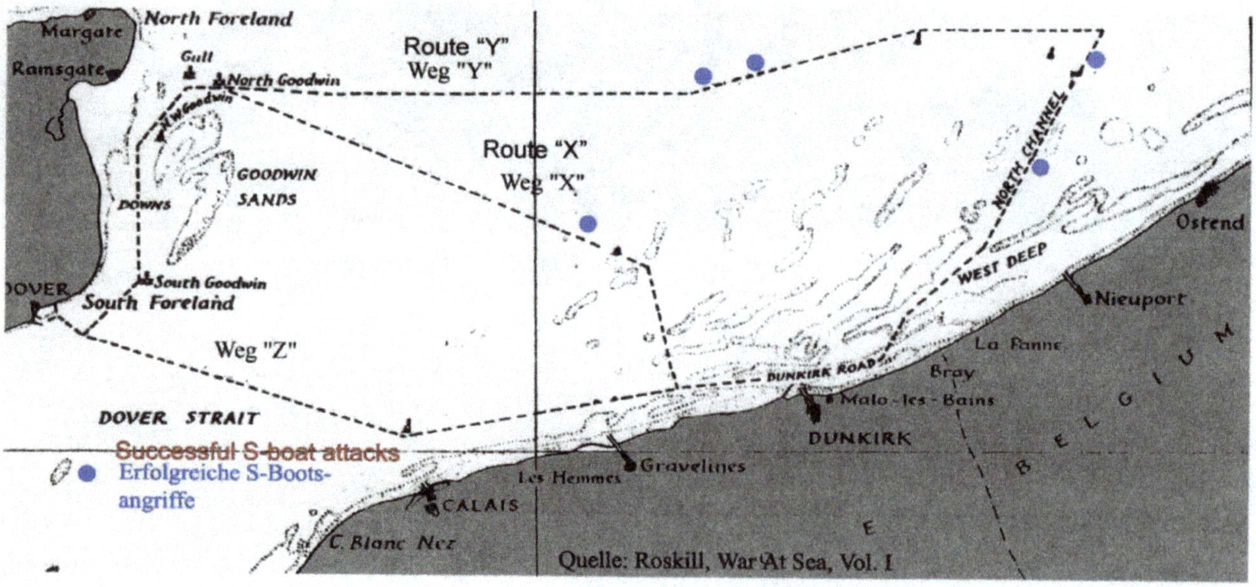

Figure 159. The *32* boat, while cruising at a low sped to listen for an oncoming convoy, strikes a mine and is scuttled on site.[418]

---

[417] Frank, *German S-Boats,* 98.

Date	Craft, Name, Event or Country	Event, Builder and Boat Specifications	City, River, Lake, Sea, Ocean
1940[419]	June 23/24 1st SFltl. *S-36,S-19*	Off the British coast near Dungeness, the *S-36* sinks the tanker *Albuera* and armed packet boat *Kingfisher*. The Germans refer to the English Channel as "The Shooting Gallery." **Type:** *S-14.*	English Channel
1940[420]	July 2 French MTB	(SIS Op AFB/SLO2) The crew failed to land the two agents carried aboard due to a faulty compass. **Builder:** British Power Boat Company.	Chebourg English Channel
1940[421]	July 4 1st SFltl. *S-14, 19, 20, 26*	The flotilla, operating out of Cherbourg, attacked Convoy 178 far, South of Portland, sinking the freighter *Elmcrest* and registering hits of two other ships. **Type:** *S-14.*	English Channel
1940[422]	July 10 2nd SFltl	Seven boats of the flotilla make for the Thames estuary to lay 12 EMC anchored mines and 12 explosive buoys.	Thames Estuary North Sea
1940[423]	July 12 1st SFltl. *S-23*	The *S-23* struck a mine while on mine-laying operations against convoy routes. The boat was scuttled and there was no loss of life. **Type:** *S-14.*	English Channel
1940[424]	July 17 *PT 9*	Built by Scott Paine British Power Boat Co. Given to Britain and renamed as RN *MTB 258,* 70'-Experimental	England

---

418 Frank, *German S-Boats,* 21.
419 Labourdette, *S-Boats,* 304.
420 Richards, *Secret Flotilla*, 304
421 Hümmelchen, *German Schnellboote*, 158.
422 Frank, Hans. *German S-Boats in Action in the Second World War* (Naval Institute Press, 2007), 26.
423 Frank. *German S-Boats,* 26.
424 Polmar, Norman, Samuel Loring Morison, *PT Boats At War World, World War II to Vietnam* (MBI Publishing 1999), 13.

Figure 160. *PT 9* (RN *MTB 258). (*Library of Congress Print and Photographic Division)

Figure 161. *Kriegsmarine* sailors had begun to call the English Channel "The Shooting Gallery" due to the many successes of the S-boat flotillas.[425]

---

[425] Busch, *Insere Schnellboote im Kanal,* 23.

Figure 162 (top and bottom). As American excitement about the looming war grew, so did the enormous propaganda efforts rise. The negative side of this was that propaganda replaced facts and the myths live on into this day. There was great resentment in the ranks of the PT crews against those who spun tales of the fantastic doings of the PT boats while neglecting the PT crews and their real stories. (Library of Congress Print and Photographic Division)

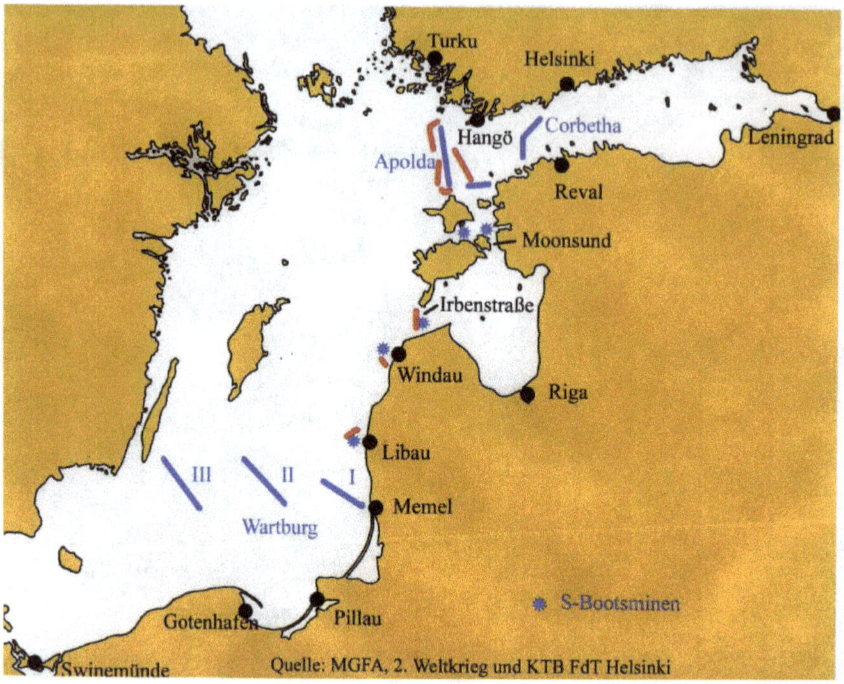

Figure 163. The *Kriegsmarine* moved quickly in the first days of Operation *Barbarossa and* blockaded the open ports of the Soviet Union and surrounding countries.[426]

Figure 164. The Soviet Torpedo Cutters (the type was known as MTBs to the rest of the world) had the unique distinction of being the only naval force to never issue life-saving gear to the MTB crews. They carried no life rings, rafts, mats, or cork vests. It was felt that when you went over the side, you were a goner. The only record of TKA sailors being rescued from the sea are by Germans and Finns.[427]

---

[426] Frank, *German S-Boats,* 44.

[427] Soviet Central Museum, http://www.cmaf.ru/eng/index_eng.htm

Date	Craft, Name, Event or Country	Event, Builder and Boat Specifications	City, River, Lake, Sea, Ocean
1940[428]	July 20 *PT 3*	Built by Fisher Boat Works, Detroit Michigan and sent to England and renamed *MTB 273*, 58' Experimental.	England
1940[429]	July 20 *PT 4*	Built by Fisher Boat Works, Detroit Michigan and sent to England and renamed *MTB 274,* 58' Fisher, Experimental.	England
1940[430]	July 25 1st SFltl.	Great controversy surrounds the sinking of the French passenger ship SS *Meknes* . The ship, carrying 1,300 French troops, was torpedoed off the Isle of Wright. One version of the story is that the soldiers were on their way to join up with Free-French troops in England. The other is that they were troops trying to join up with Vichy-Government forces in Southern France.	English Channel
1940[431]	July 26 1st SFltl. *S-19, 20, 27*	The three boats attack the remains of convoy CW8, and sink 3 convoy ships. The convoy fought back and did serious damage to the *S-21* and *S-27.* **Type:** *S-14.*	Southern coast English Channel
1940[432]	July 30 MTB	(SIS Ops-Free French-Operation A5/SLO1) The MTB attempted to land an emissary for General de Gaulle at St-Pabu, France. The first attempt failed and a second try, near Plougasnou, Morlaix succeeded.	St-Pabu, French coast
1940[433]	August 1 Soviet TKA Div.	Six TKA's attack German 1st M Flotilla in the Irben Straits and the M boats sank the TKA 122. **Type:** *G-5,* Motor Torpedo Cutter. **Displacement:** 14.84 tons; **Length:** 62.7'; **Beam:** 10.9'; **Draft:** 2.2'; **Engines:** 2 x 850 hp GAM34, petrol; **Speed:** 53 knots max, 48 knots cruise; **Range:** 250 nm; **Armament:** 2x53.3 cm torpedo tubes, 1 x 12.7 mm MG. **Crew:** 6.	Off Saaremaa Island, Latvia

428 Polmar, *PT Boats At War*, 13.
429 Polmar, *PT Boats At War*, 13.
430 Hümmelchen, *German Schnellboote*, 158.
431 Rohwer, *Chronology of the War At Sea*, 34.
432 Richards, *Secret Flotilla*, 304.
433 Fock, *Fast Fighting Boats*, 263.

Date	Craft, Name, Event or Country	Event, Builder and Boat Specifications	City, River, Lake, Sea, Ocean
1940[434]	August 7/8 1st SFltl. *S-20, 21, 25, 27.*	The flotilla sinks two ships, and damage two, out of convoy CW 9. The *S-21* and *27* suffered several wounded sailors from return gunfire. **Type:** *S-14*	English Channel
1940[435]	August 15 2nd SFltl.	The torpedo arsenal and adjoining workshops at Ostend were sabotaged and blew up in a massive explosion of 44 torpedoes. Four the flotillas boat were damaged *(S-24, 31, 35, 37)* and then sent to Germany for extensive repairs. This loss was felt strongly by the other flotillas.	Ostend North Sea
1940[436]	August 15 PT 6 (1)	She was made by Higgins Industries, New Orleans. The 58' Experimental became RN *MGB 68.*	America

Figure 165. When it came to bold propaganda, no country could top the US in boldness and exaggeration. (National Archives)

434 Hümmelchen, *German Schnellboote*, 158.
435 Frank, *German S-Boats*, 27.
436 Hyperwar: US Navy Torpedo Boats, 1.

Figure 166. The S-boats of the *Kriegsmarine* carried some of the most sophisticated equipment for targeting ship available at the time. The intricacies and accuracy of the equipment were astounding.[437]

[437] Frank, *German S-Boats*, 153.

Date	Craft, Name, Event or Country	Event, Builder and Boat Specifications	City, River, Lake, Sea, Ocean
1940[438]	September 3/4 MAS 536, 537	The two boats try to attack the two British destroyers HMS *Decoy* and HMS *Ilex*. The *Ilex* sinks MAS 537 in the Kaso Strait.	E. Crete Island, Sea of Crete
1940[439]	September 4 1st SFltl. *S-21, 18, 22, 54*	The flotilla attacks the Allied convoy coming from NE of Great Yarmouth. *S-21* sinks the freighters *Corbrook* and *New Lambton*, *S-18* sinks the freighters SS *Joseph Swan* and SS *Nieuwland*, *S-22* sinks the freighter SS *Fulham*, and *S-54* torpedoes a steamship. **Type:** *S-14* and *S-30.*	English Channel
1940[440]	September 7 2nd SFltl. *S-33, 36*	The two boats attack the British convoy FS 273. *S-33* sinks the Dutch freighter SS *Stad Almare* off Lowestoft. **Type:** *S-14.* **Displacement:** 100 tons; **Length:** 107.5'; **Beam:** 16.8'; **Draft:** 9.2'; **Engines:** 3 x 1,320 hp Daimler MB 502 diesel, **Speed:** 36 max, 30 cruise; **Range:** 800 nm, **Armament:** 2 x torpedo tubes , 2 x 20mm MG.	North Sea
1940[441]	September 24 *MTB 15*	Hit a mine in the Thames Estuary. **Builder:** British Power Boat Co. **Displacement:** 22 T: **Length:** 60'-3"; **Beam:** 13'-3"; **Draft:** 2'-9"; **Engines:** 3 x 1,800bhp Napier Sealion, petrol; **Speed:** 33 knots max, 29 knots cruise; **Armament:** 2 x18" torpedo tubes; 4 x 2 .303MG; **Crew:** 9.	North Sea
1940[442]	October 11 *MTB 22, 31*	The two boats attack and sink two German trawlers off Calais. **Builder:** Vosper. **Speed:** 40 knots max, 38 knots cruise; **Displacement:** 31 tons; **Length:** 70'; **Beam:** 16'-6"; **Draft:** 3'-3"; **Engines:** 3 x 1,200 hp Isotta-Fraschini, petrol, 2 x Ford V-8 cruising, petrol, 24-Volt and 110-Volt, 1 x 10 hp Ford auxiliary; **Range:** 400 nm at 20 knots; **Armament:** 2 x 21" torpedo tubes, 2 x .50 cal MG, 2 x 2 .303 MG; **Crew**: 12 .	Dover Strait

---

438 Rohwer, *Chronology of the War At Sea,* 37-38.
439 Rohwer, *Chronology of the War At Sea,* 39.
440 Rohwer, *Chronology of the War At Sea,* 39.
441 North, A.J.D. *Royal Naval Coastal Forces 1939-1945* (Almark Publishing Co.Ltd, 1972), 8.
442 Rohwer, *Chronology of the War At Sea,* 66.

Figure 167. *MTB 30.* (Private collection of John Lambert)

Date	Craft, Name, Event or Country	Event, Builder and Boat Specifications	City, River, Lake, Sea, Ocean
1940[443]	October 12 2nd SFltl. *S-37*	The 37 boat sank by a mine, with 13 dead and 11 survivors, off Orfordness. **Type:** *S-30*	North Sea
1940[444]	October 16 *MTB 106* (Experimental)	Used to detonate German acoustic mines in the Thames estuary. Destroyed after striking a mine off Sheerness, North Sea. **Builder:** British Power Boats Co. **Displacement:** 8 tons; **Length:** 44'-3"; **Beam:** 9'-6"; **Engines:** 1-Thorneycroft,Petrol; **Armament:** 1-18"torpedo tubes; 3-.303 mg; **Crew:** 4	North Sea
1940[445]	October 17/18 1st SFltl. *S-18, 24, 27*	The three boats attack a British convoy on the Southeast coats of England and the S-18 sinks the freighter *Hauxley* while *S 24, 27* sink two other freighters. **Type:** *S-14*, 107'	English Channel
1940[446]	October 21 *MTB 17*	Lost to a mine off Thames Estuary. **Builder:** British Power Boat Co., Hythe. **Speed:** 33 knots max; **Displacement:** 22 tons; **Length:** 60'-3"; **Beam:** 13'-3"; **Draft:** 2'-9"; **Engines:** 3 x 1,800 hp Napier, petrol; **Armament:** 2-18" torpedo tubes, 4 x twin .303-cal. MG; **Crew:** 9.	Belgium Coast North Sea
1940[447]	October 31 *MTB 16*	Hit a mine in the Thames estuary. **Builder:** British Power Boat Co. **Length:** 60'-3"	North Sea
1940[448]	October *TKA 13, 14, 15*	The boats attack and sink the Norwegian freighter *Renöy.* **Type:** *G-5.* **Length:** 62.7'	Baltic Sea

443 Frank, *German S-Boats,* 28.
444 *Warship* Supplemental 65, 8.
445 Rohwer, *Chronology of the War At Sea,* 45.
446 *Warship* Supplement 65, 9; North, *Royal Naval Coastal Forces,* 8.
447 *Warship* Supplement 65, 5.
448 Fock, *Fast Fighting Boats,* 266.

Figure 168. The Soviets used their TKA force like some people use razor blades. They would typically send out a dozen or more boats out (one occasion called for 30 TKs) , lose half and call the mission successful.[449]

[449] Soviet Central Museum http://www.cmaf.ru/eng/index_eng.htm.

Date	Craft, Name, Event or Country	Event, Builder and Boat Specifications	City, River, Lake, Sea, Ocean
1940[450]	November 19/20 1st SFltl. *S-38*	The *S-38* is rammed and sunk by the British destroyer, HMS *Campbell.* 18 German crewmembers are rescued and become POW's. **Type:** *S-38.*	North Sea
1940[451]	December 15/16 *S-58*	The *S 58* sinks the Danish freighter SS *N.C. Monberg* in a British convoy East of Great Yarmouth. **Type:** *S-30* **Displacement:** 102 tons; **Length:** 107.5'; **Beam:** 16.8'; **Draft:** 9.2'; **Engines:** 3 x 1,320 hp Daimler MB 502, diesel; **Speed:** 36 knots max, **Cruise:** 30 knots; **Armament:** 2 x 21" torpedo tubes, 2 x 20mm MG; 2x MG 34; **Crew:** 25.	North Sea
1940[452]	December 26 MTB 5	Lost to a mine off the Gunfleet. **Builder:** British Power Boat Co- 60'3"	North Sea
1940[453]	December 23/24 1st, 2nd, 3rd SFltl.	The three flotillas moved into attack convoy FN 366 while being escorted by the British destroyers HMS *Wolsey*, HMS *Verdun* and corvette HMS *Shearwater*. *S-101* and *59* sink the Dutch freighter *Stad Maastricht* and the *S-28* sinks the British armed trawler HMS *Pelton*. **Type:** *S-14*	English Channel

450 Frank, *German S-Boats,* 31.
451 Rohwer, *Chronology of the War At Sea,* 52.
452 *Warship*- Supplement 65, 3.
453 Rohwer, *Chronology of the War At Sea,* 53.

Figure 169. *MTB 5* had a short, exciting and violent life. Its history is typical of other MTBs in the world. They came on the scene of naval history, had a short violent combat life and sank beneath the waves leaving nothing behind but memories and a record of valor. (Gordon Smith, www.Naval-History.Net)

The Ship Killers: A Definitive History of the Torpedo Boat by Joe Hinds

Volume 1: Overview

Volume 2: Timeline 1585 – 1899

Volume 3: Timeline 1900 – 1939

Volume 3: Timeline 1900–August 1939

Volume 4: Timeline September 1939–1940

Volume 5: Timeline 1941

Volume 6: Timeline 1942

Volume 7: Timeline 1943

Volume 8: Timeline 1944

Volume 9: Timeline 1945

Volume 10: Appendices, Index, and Never-Before-Published Bonus Images

Complete set available as a unit in January 2010 (ISBN 978-1-934840-58-0)

If you would like to contribute a volume to The Ship Killers, *please contact the series editor, Joe Hinds, at Joe.Hinds@gmail.com, or the publisher, Fred Zimmerman, at wfz@NimbleBooks.com. Information about writing for Nimble Books is available under "Publish With Us" at www.NimbleBooks.com.*

THE SHIP KILLERS

THE DEFINITIVE ILLUSTRATED HISTORY OF THE TORPEDO BOAT VOLUME V: TIMELINE 1941

BY JOE HINDS

NIMBLE BOOKS LLC

CONTENTS (INCLUDING FIGURES)

DEDICATION

Gordon Smith is the kindest man I have ever known. For years he has been completely unflinching in offering aid for my research questions. He has done this in the midst of his creation of one of the largest and most prolific naval websites known to man.

He has stopped his own ongoing projects to look up some bit of obscure information I can't seem to find anywhere else. Gordon gave me access to his father's archive of naval photographs that go back to the beginning of the CMB building program and the earliest British MTB's. Gordon's influence and materials s that he released to me helped make my publishing efforts even more meaningful. The old photos of his father add a great human touch to Volume four in which they appeared.

I am forever grateful to have a magnificent friend such as Gordon.

TIMELINE 1941

Date	Craft, Name, Event or Country	Event, Builder and Boat Specifications	City, River, Lake, Sea, Ocean
1941[454]	January 1 *MTB 37*	Bombed in stocks before completion. **Builder:** Vosper, 72'-6"	England
1941[455]	January 20 *PT 33*	Grounded and scuttled in Subic Bay. **Builder:** Elco, *PT 20* class 77'	Subic Bay, Luzon, Philippines
1941[456]	January 31 *MTB 104, 107* (Thorneycroft Exp)	The small size of the two boats made them ideal for operation with the Landing Ships Infantry (LSI). They were carried aboard ships to meet the needs of covert operations. The *104* boat was assigned to Long Range Desert Group (LRDG) in North Africa. **Builder:** Thorneycroft (Experimental) **Speed:** 38 knots **Displacement:** 10 tons **Hull:** Stepped **Engines:** 1 x RY 12, petrol **Armament:** 2 x 18" torpedo tubes **Crew:**	Mediterranean
1941[457]	February 3 *PT 6 (2)*	Assigned for service in Canada. **Builder:** Higgins 81'0 Experimental, RN *MTB 270*	U.S.A.
1941[458]	February 5/6 2nd SFltl.	The *S-30, 54, 58,* and *59* make a sortie off the coast of England and the *S-30* sinks a British freighter.	North Sea

[454] *Warships Supplement 65,* 6.
[455] *Hyperwar: US Navy Torpedo Boats,* 8.
[456] Lenton, H.T. *British & Empire Warships of World War II* (Greenhill & NIP, 1998) 540.
[457]. *Hyperwar:US Navy Torpedo Boats,* 8.
[458] Rohwer, *Chronology of the War At Sea,*58.

Figure 170. *MTB 107.* (Private collection of John Lambert)

Figure 171. *MTB 29* at harbor waiting to go out on the next mission. (Private collection of John Lambert).

Figure 170. Twin Oerlikon cannon. These formidable weapons were popular with every navy. After the first year of battle, the British began to take serious measures to increase the fire-power of their MTBs and MGBs. It was apparent that they did not stand a chance in a fight with the S, M and R boats of the *Kriegsmarine.* [459]

[459] http://www.navy.gov.au/w/images/1347-9.jpg

Date	Craft, Name, Event or Country	Event, Builder and Boat Specifications	City, River, Lake, Sea, Ocean
1941[460]	February 14 *MTB 41*	Struck a mine in the North Sea and sank. **Builder:** White, Cowes: **Speed:** 39.89 knots maximum, 33.13 cruise; **Range:** 240 nm; **Displacement:** 38.5 tons **Length:** 73'; **Beam:** 18'; **Draft:** 2'-8.5" **Engines:** 3 x 1,120 hp Stirling Admirals, petrol; **Armament:** 1 x 6-pdr forward turret, 1 x 2 Oerlikon midships, 2 x 2 .303" MG, 2 x 18" torpedo tubes; **Crew:** `17	North Sea
1941[461]	February 20 *PT 7. PT 8*	Made by the Philadelphia Navy Yard as an experimental 81' boat for service in Canada and reassigned as RN *MTB 271, 272*	United States of America
1941[462]	February 25 1st SFltl. *S-30*	The *S-30* sank the British destroyer HMS *Exmoor.* **Type:** *S-14,* 107.5'	English Channel
1941[463]	February 25/27 *MAS 541, 546*	The British invasion of Casterlorizo Island, Operation Abstention, failed due to the superior force of Italian naval vessels including MAS 541 and MAS 546. **Type:** Baglietto, 61.7'	E. Rhodes Is. Dodecanese Islands Mediterranean
1941[464]	March 1 *PT 5*	Made by Higgins of New Orleans as experimental 81' boat and assigned number as RN *MTB 269.*	United States of America

---

460 North, *Royal Naval Coastal Forces*, 22.
461 *Hyperwar: US Navy Torpedo Boats,* 8.
462 Hümmelchen,*German Schnellboote,* 160.
463 Rohwer, *Chronology of the War At Sea,* 61.
464 HyperWar, *US Navy Torpedo Boats,* 8.

Figure 171. Rare photo of an S-boat cook. The German S-boats and English D boats were the only two MTBs of WWII that had cooking facilities and cooks.[465]

[465] Hans Frank, *German S-Boats*, 155.

Date	Craft, Name, Event or Country	Event, Builder and Boat Specifications	City, River, Lake, Sea, Ocean
1941[466]	March 7 *MTB 28*	Destroyed in fire, Portsmouth area. **Builder:** Thorneycroft **Speed:** 42 knots **Displacement:** 37 tons **Length:** 72' **Beam:** 16'-6" **Engines:** 3 x 3,600 bhp Isotta Fraschini, petrol **Armament:** 2 x 21" torpedo tubes, 4 x .50-cal MG	Portsmouth, England
1941[467]	March 7/8 1st & 3rd SFltl.	The two flotillas attack the north bound British convoys off Cromer, FN 26 & FN 29, with nine S-boats. The 1st flotilla sinks five ships and the 3rd sinks two ships. *S-28* rescues the master of the ship *SS Carduff*. *S-101* picked up men from the ship *SS Norman Queen.* 41 British sailors died. **Type:** *S-30* and *S-38*	English Channel
1941[468]	March 18 1st SFltl.	Six S-boats of the flotilla (S-*26, 29, 39, 55, 101, 102)* make a sortie into the Humber estuary and sink the French freighter Daphne II in convoy FN 34. **Type:** *S-14, 26* & *29; S-38, 39, 55; 101, 102*	English Channel

[466] *Warships Supplement 65*, 6.
[467] Frank, Hans. *German S-Boats in Action in the Second World War* (Naval Institute Press,2007), 32.
[468] Rohwer, *Chronology of the War At Sea*, 64.

Figure 172. *PT 32* on a day where her engine *didn't* catch fire.

Figure 173. The aftermath of the S-28 getting into a severe gun fight with British MTB and MGB's.[469]

---

[469] Frank, *German S-Boats,* 97.

Date	Craft, Name, Event or Country	Event, Builder and Boat Specifications	City, River, Lake, Sea, Ocean
1941[470]	March 25 Italian Barchini M.T.M.	The Italian commandos enjoyed a large success in Suda Bay, Crete when six Barchini (Punts) attacked and sank the cruiser HMS *York*, and the freighter SS *Piericles.*	Suda Bay, Crete, Mediterranean
1941[471]	April 6 MAS 213	The MAS 213 torpedoed the British light cruiser HMS *Capetown* off the harbor of Port Sudan. The cruiser is towed to port for repairs. The Italians scuttle *MAS 213, 204, 206, 210, 216.* **Type:** Baglietto, *MAS 201-217* **Speed:** 24 knots/ 4 knots; **Range:** 230 nm @ 16 knots, 20 nm @ 4 knots; **Displacement:** 13.4 tons **Length:** 52.5'; **Beam:** 8.7'; **Draft:** 3.6' **Engines:** -240 hp Fiat A12, petrol, 2 x 5 hp Rognini electric; **Armament:** 2 x 17.7" torpedo tubes, 1 x 6.5 MG, 10 x 37.5 pd depth charges; **Crew:** 8	Port Sudan Red Sea

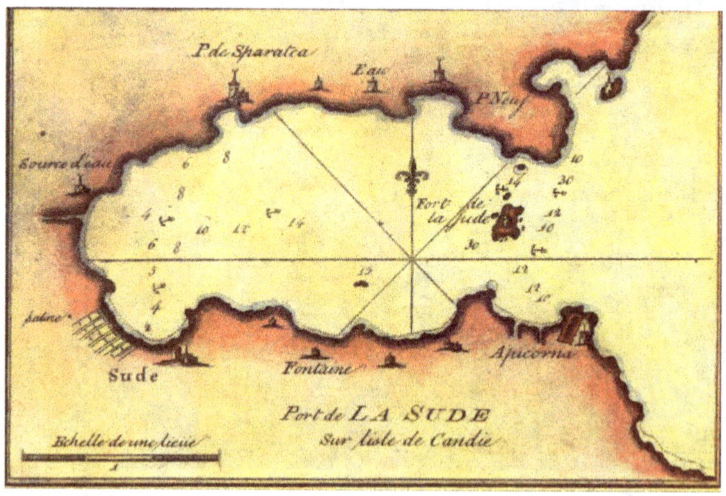

Figure 174. A historic map of Suda bay.

---

470 Suda Bay http://www.regiamarina.net/xa_mas/sudabay/sudabay_us.htm
471 Rohwer, *Chronology of the War At Sea,* 66.

Date	Craft, Name, Event or Country	Event, Builder and Boat Specifications	City, River, Lake, Sea, Ocean
1941[472]	April 17 2nd SFltl.	The flotilla attacked a British convoy off Great Yarmouth and two freighters and seriously damaged a third.	North Sea

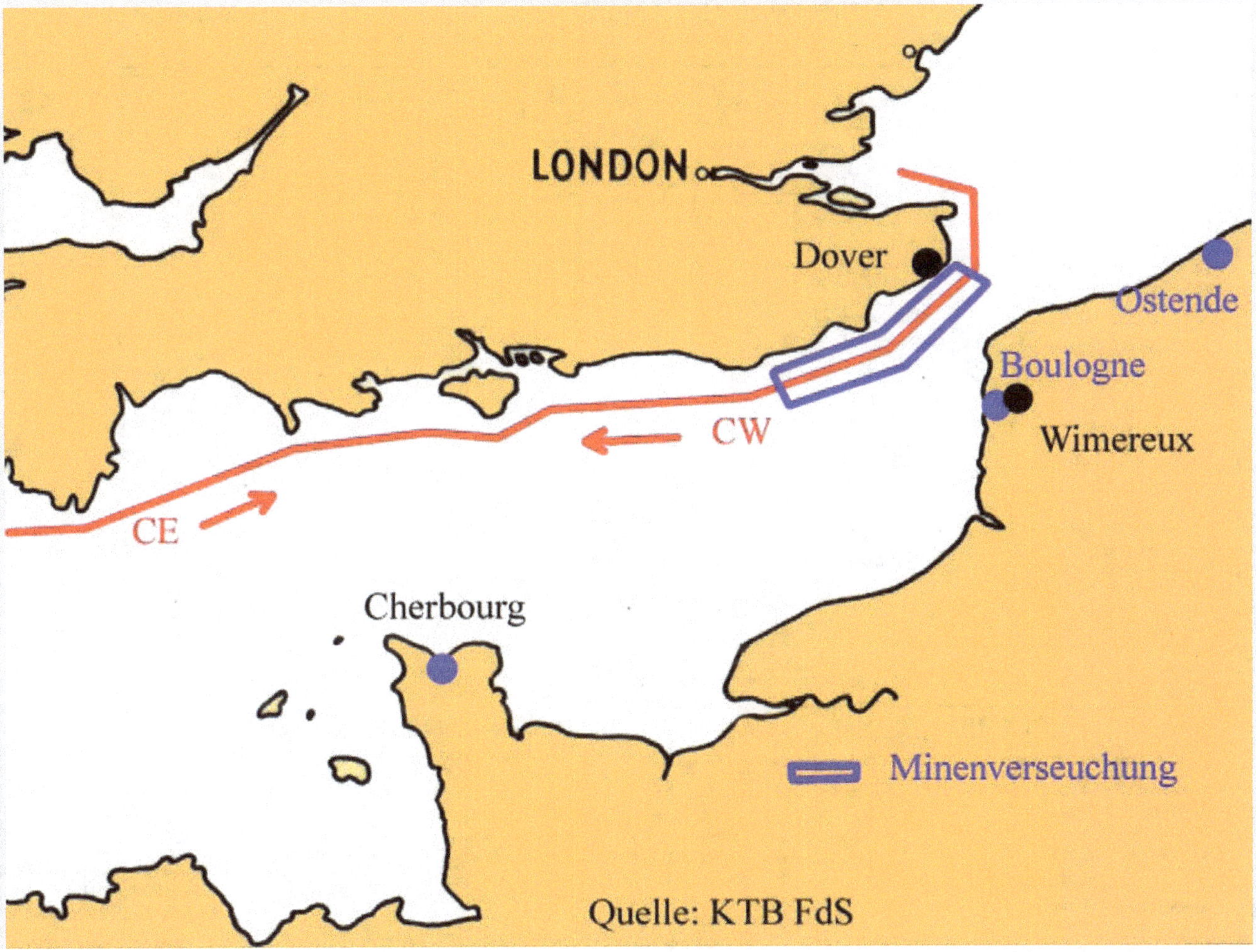

Figure 175. The successful S-Boat action off Great Yarmouth.[473]

---

[472] Rohwer, *Chronology of the War At Sea*, 68.
[473] Frank, *German S-Boats*, 53.

Date	Craft, Name, Event or Country	Event, Builder and Boat Specifications	City, River, Lake, Sea, Ocean
1941[474]	May 23 10th MTB Flotilla	Fighter bombers of the III./JG 77 attack Suda Bay and sank the MTBs *67, 213, 214, 216,* and *217.* Thorneycroft 55', CMB	Suda Bay Crete Island Mediterranean

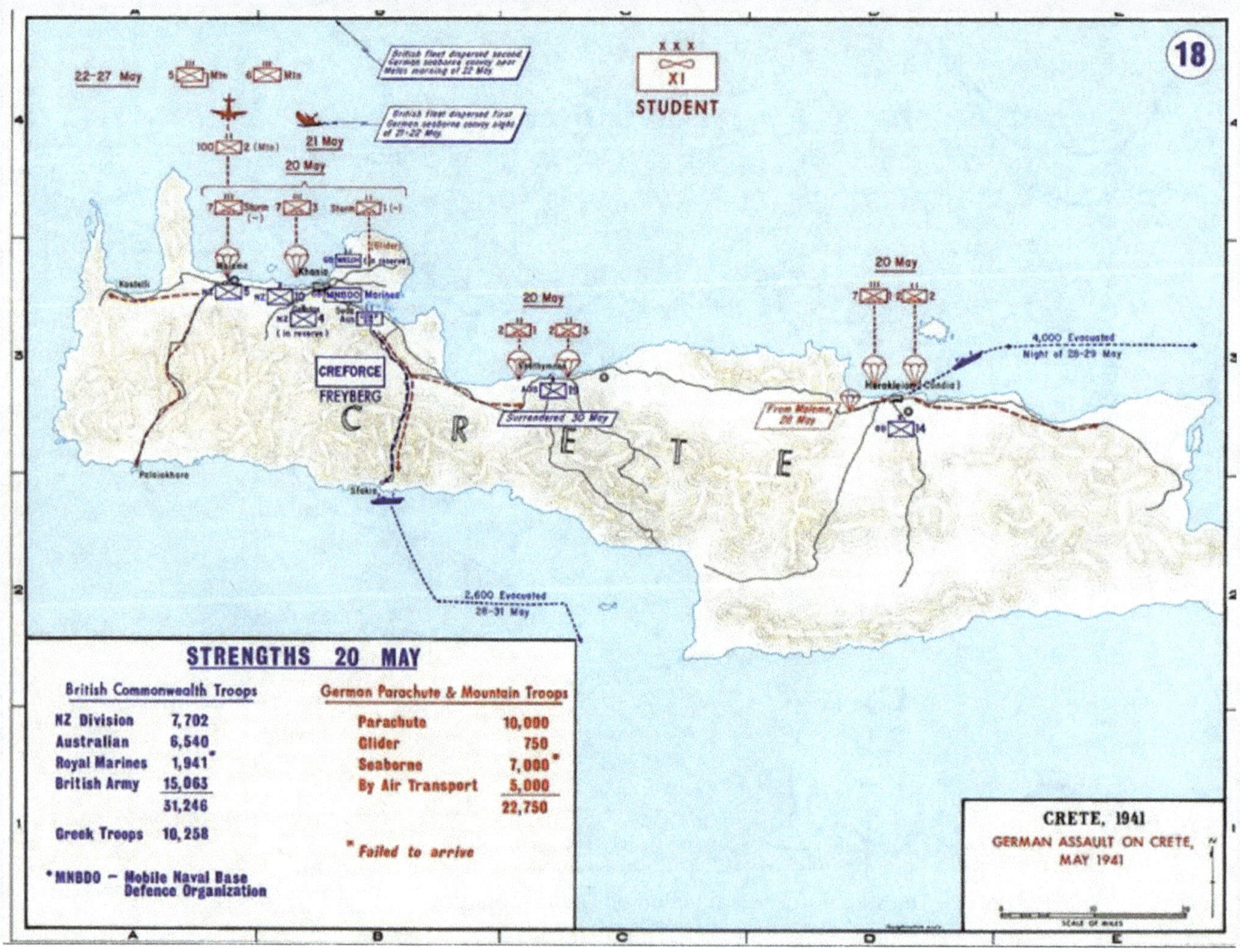

Figure 176. The campaign for Crete.

474 Rohwer, *Chronology of the War At Sea*, 75.

Date	Craft, Name, Event or Country	Event, Builder and Boat Specifications	City, River, Lake, Sea, Ocean
1941[475]	June 1 British Navy The 1st 'D' Class "Dog" Boat	The Fairmile 'D' MTB/MGB are the first coastal motor boat to be fully capable of going head-to-head with the Germans S-boats. The first builders were the Tough Brothers of Teddington, on the Thames. **Speed:** 34.5 knots; **Range:** 506 nm at maximum speed; **Displacement:** 105 tons **Length:** 115'; **Beam:** 20'; **Draft:** 4'9" **Engines:**4 x 12-cylinder 1,250 bhp supercharged 4M 2500 Packard 4M, petrol **Armament:** 2 x 1-6pdr 7cwt MK VII gun, 4 x .50 cal MG, 4 x .303 MGs, 2 x 20 mm Oerlikons on power mountings, 2 x 18" torpedo tubes Prototype, MTB/MGB *601-800, 5001-5029*	England
1941[476]	June 1 2nd SFltl. *S-59, S-60*	After mining the approaches to the Irben Straits, the *S-59* and S-*60* captured the Estonian steamer *Estonia.*	Irben Strait Gulf of Riga

[475] Lambert, John, Al Ross, Jr., PhD, *Allied Coastal Forces of World War II, Vol.1 ; Fairmile designs and US subchasers* (Naval Institute Press,1990), 90-91.
[476] Frank, Hans *German S-Boats* ,44

Figure 177. The introduction of the "D" MTB and MGB improved the odds faced by the Allied navies against the time tested S-boats.(Private collection of John Lambert).

Figure 178. *MAS-527.*[477]

[477] Ojor Onen, http://uralica.com/jornrn7.htm.

Date	Craft, Name, Event or Country	Event, Builder and Boat Specifications	City, River, Lake, Sea, Ocean
1941[478]	June 22-23 1st, 2nd, 3rd SFltl. *S-59, S-60*	A few minutes after hostilities opened between Germany and the Soviet Union the *S-59 and S-60* sank the ex- Latvian freighter Gasima off Gotland Island coast. **Builder:** Lürssen, **Type:** *S-30* **Speed:** 36 knots maximum, 30 knots cruise; **Displacement:** 102 tons; **Range:** 800 nm **Length:** 107.4'; **Beam:** 16'; **Draft:** 9.3' **Engines:** 3 x 2,000 hp Daimler MB 501, diesel; **Armament:** 2 x 53.3 cm torpedo tubes and 2-reloads, 2 x 20 mm; **Crew:** 22	Baltic Sea

[478] Hümmelchen, *German Schnellboote,* 161.

Figure 179. Gun maintenance was as important a task as engine maintenance. For all services, this was like life itself.[479]

Figure 180. This German book was the inspirational equivalent of the American story, *They Were Expendable*. This book was photographed in 1940 and printed early 1941. It remained a best seller for the duration of the war.

---

[479] Busch, Otto, *Unsere Schnellboote im Kanal*, 17.

Date	Craft, Name, Event or Country	Event, Builder and Boat Specifications	City, River, Lake, Sea, Ocean
1941[480]	June 22 1st SFltl. *S-103, S-39*	While underway, the *S-103* and *S-39* run hard aground in the fairway off Helsinki. The *S-103* was patched up enough to return to Germany for repairs. **Builder:** Lürssen; **Type:** *S-38* **Speed:** 38.5 knots max (39.5 for the *S-39),* 35 knots cruise; **Displacement:** 115 tons; **Range:** 700 nm **Length:** 114.6′; **Beam:** 16′; **Draft:** 9.5′ **Engines:** 3 x 2,000 hp Daimler 501 diesel; **Armament:** 2 x 53.3 cm torpedo tubes and 2 reloads, 2 x 20 mm; **Crew: 22**	Gulf of Finland
1941[481]	June 23 3rdSFltl. *S-35, S-60*	The *S-35* and *S-60* surprised the Soviet submarine S-3 and sank it after a lengthy sea battle. Both of the S-boat's last torpedoes were duds and they opened fire with all automatic weapons on board. The sub took off at high speed and the S-boats pursued the sub, which was wildly firing at the S-boats. *S-60* dropped two depth charges and crewmen tossed hand grenades onto the sub. This maybe the first time when a motor torpedo boat sank a submarine with hand grenades. The Germans rescue 20 Soviet sailors. *S-35* **Type:** S-10, *S-60* **Type:** S-30	Baltic Sea Gulf of Finland Mecklenburger Bucht

[480] Hans Frank, *German S-Boats*, 45.
[481] Hans Frank, *German S-Boats*, 44.

Date	Craft, Name, Event or Country	Event, Builder and Boat Specifications	City, River, Lake, Sea, Ocean
1941[482]	June 23 3rd SFltl. *S-44*	During a raid on the Hango coast, the S-*44* sank the Soviet guard boat *MO-238.* *S-44,* **Type:** S-38	Baltic Sea Gulf of Finland Mecklenburger Bucht
1941[483]	June 25/26 2nd &3rd SFltl. *S-106, S-43*	Off Moon Sound, the S-*106* and S- *43* ran into a Soviet minefield and both boats were lost. 19 crewmen died. The 3rd Sfltl. sank a Soviet destroyer. **Type:** *S-38*	Off Hiiumaa Coast Baltic Sea
1941[484]	June 25/26 3rd SFltl. *S-34,* *S-60, S-61*	On returning from a mine laying mission in the Irben Strait, *S-34* sank the Estonian steamer *Lizada* (flying under Soviet colors) and *S-54, 60,* and *61* launched torpedos into Windau (Ventspils) Harbor, Latvia. S-*60, 61, 54*: **Type:** S-30; *S-34*, **Type:** *S-14*	Baltic Sea
1941[485]	June 26/27 3rd SFltl. S-*59, 60, 35*	The *S-59* sank the Soviet destroyer *Storozevoi.* The *S-59* and *60* then sank the Soviet sub *S-10.* *S-35* and *60* sank the Soviet minesweeper, *T-208 (Skiv).* *S-59, 60:* **Type:** S-30; *S-35,* **Type:** S-14	Baltic Sea
1941[486]	June 26/27 2nd SFltl.	Boats of the flotilla captured a Soviet torpedo cutter, *TKA-47* off Backofen Island (Vzava), Latvia. **Type:** *G-5*	Baltic Sea Latvia

[482] Hümmelchen, *German Schnellboote,* 161.
[483] Frank, *German S-Boats*, 45.
[484] Hümmelchen, *German Schnellboote,* 161.
[485]Vercamer, *Naval Warfare in the Baltic Sea* http://www.feldgrau.com/baltsea.html (accessed January 3, 2007).
[486] Hümmelchen, *German Schnellboote,* 161.

Figure 181. The logo of the 2nd Schnellbootflottille.[487]

Figure 182. The flag of the Soviet Baltic Fleet.

---

[487] Schnellbooteflottille http://www.wlb-stuggart.de/km/sboot/sfl2.htm.

Date	Craft, Name, Event or Country	Event, Builder and Boat Specifications	City, River, Lake, Sea, Ocean
1941[488]	June 28 3rd SFltl.	Five S-boats took the Liepaja harbor, Latvia, and waited for the 291st Infantry Division, which arrived a day later. The loss of the port was a blow to the Soviets. They had hoped to save the two Latvian submarines and two Estonian submarines for their use in the Baltic Sea.	SW Latvia coast, Baltic Sea
1941[489]	July 21/24 1st US Plywood Derby	With war imminent, boat builders vied for US Navy contracts: the Big Show of design and engineering talent. Naval authorities and the Press examined the following boats. *PT 6,* 81' Higgins; *PT 8,* 81' Philadelphia Navy Yard; PT *20, 26, 30, 31, 33,* 77' Elco; *PT 69,*72' Higgins; *PT 70,* 76' Huckins; *MRB-2*, 70' Higgins. Elco's *PT 20* won the first race, at 38 knots, of 190 miles over smooth open water.	Off New London Connecticut

[488] Vercamer, *Naval Warfare in the Baltic Sea* http://www.feldgrau.com/baltsea.html (accessed January 3, 2007).
[489] Internet correspondence, Al Ross, Jr. PhD, August 5, 2007.

Date	Craft, Name, Event or Country	Event, Builder and Boat Specifications	City, River, Lake, Sea, Ocean
1941[490]	July 24 *MAS 532, 533*	An Allied large convoy, Operation Subsistence, left Gibraltar and suffered daily attacks by enemy aircraft. Off Pantelleria, *MAS 522* and *533* attacked and torpedo the transport *Sydney Star.* **Type:** Baglietto **Speed:** 44 knots maximum, silent running 8 knots **Displacement:** 28.3 tons **Range:** 360 nm at 44 knots, 1,100 at 6 knots **Length:** 18.7' **Beam:** 4.6' **Draft:** 1.47' **Engines:** 2 x 1,150 hp I.F. ASM 183, 2 x 70 hp Alfa Romeo 6-cylinder 2,500, petrol **Armament:** 1 x 13.2 mm MG, 2 x 450 mm torpedo tubes, side launched, 6 x 50 kg depth charges **Crew:** 10	Strait of Sicily Mediterranean

490 Rohwer, *Chronology of the War At Sea*, 88.

Date	Craft, Name, Event or Country	Event, Builder and Boat Specifications	City, River, Lake, Sea, Ocean
1941[491]	July 26 Operation Malta *MAS 451, 452*	Seventeen small craft led by *MAS 451* and *MAS 452* made a daring and fruitless raid on the Grand Harbour of Malta. All boats lost, 15 dead, including almost every officer, 20 men taken prisoner, and 19 wounded. **Builder:** Baglietto **Speed:** 42 knots maximum, silent running 8.9 knots **Displacement:** 25.5 tons **Range:** 330 nm, 836 nm at 8 knots **Length:** 59.06' **Beam:** 15.68' **Draft:** 5.1' **Engines:** 2 x -1,100 hp I.F. Asso 1000; 2 x 80 hp Alfa Romeo 6-cylinder 2,500, petrol **Armament:** 1 x 13.2mm MG, 2 x 450 mm torpedo tubes, side launched **Crew:** 11	Mediterranean Sea
1941[492]	July 27 3rd SFltl. *S-54, 55, 57, 58*	In the Gulf of Riga, Baltic Sea, S-54 sinks the Soviet destroyer *Smely.* The S-57 & 58 misses two German minesweepers at anchor off Arensburg. **Type:** *S-30*	Gulf of Finland Baltic Sea
1941[493]	July 28 3rd S-Boat Flotilla	The flotilla sinks the Soviet icebreaker *Lachplesis* by gunfire off Arensburg. **Type:** *S-30*	Gulf of Finland Baltic Sea

491 Allied Warships http://uboat.net/allies/warships/ship/5844.html and Gunnar Aselius, *The Rise and Fall of the Soviet Navy in the Baltic, 1921-1941*, 227.
492 Rohwer, *Chronology of the War At Sea,* 89.
493 Rohwer, *Chronology of the War At Sea,* 89.

Date	Craft, Name, Event or Country	Event, Builder and Boat Specifications	City, River, Lake, Sea, Ocean
1941[494]	August 1 Soviet TKA Div.	Six TKAs attack German 1st M Flotilla in the Irben Straits and the M boats sink the *TKA-122*. **Type:** *G-5,* Motor Torpedo Cutter. **Speed:** 53 knots maximum, 48 knots; **Displacement:** 14.84 tons; **Range:** 250 nm **Length:** 62.7'; **Beam:** 10.9'; **Draft:** 2.2' **Engines:** 2 x 850 hp GAM34, petrol; **Armament:** 2 x 53.3 cm torpedo tubes, 1 x 12.7 mm MG; **Crew:** 6	Off Saaremaa Island Cape Domesnaes, Gulf Of Riga Baltic Sea
1941[495]	August 2 *S-55, S-58*	In the Gulf of Riga*, S-55* and *S-58* make several attempts to strike the Soviet destroyer *Artem*. All attempts are unsuccessful. **Type:** *S-30*	Gulf of Riga Baltic Sea
1941[496]	August 11 4th SFltl. *S-49*	The flotilla attacks an Allied convoy off Dungeness and the *S-49* sinks the British freighter, SS *Sir Russell.* **Type:** *S-38*	North Sea
1941[497]	August 11/12 2nd US Plywood Derby	The second Derby ran over rough open water but still at 190 miles distance. Once again, an Elco boat won, *PT 21*, average speed of 25 knots. The boats entered were *MRB 2*, 70' Higgins; *PT 8*, 81' Philadelphia Navy Yard (Aluminum); *PT 21, 29,* 77' Elco; *PT 70,* 76' Huckins.	Off New London Connecticut Atlantic Ocean
1941[498]	August 14 *PT 6 (1)*	Higgins Industries produced this experimental 81' for consideration by the US Navy. It was re-named RN *MGB 68.*	United States

494 Rohwer, *Chronology of the War At Sea,* 89.
495 Rohwer, *Chronology of the War At Sea,* 90.
496 Rohwer, *Chronology of the War At Sea,* 92.
497 Internet correspondence, Al Ross, Jr. PhD, August 5, 2007.
498 Internet correspondence, Al Ross, Jr. PhD, August 5, 2007.

Date	Craft, Name, Event or Country	Event, Builder and Boat Specifications	City, River, Lake, Sea, Ocean
1941[499]	August 18/19 *S-58*	The *S-58* has a lucky day. She sinks the Soviet MS *T-51* (Pirmunas) laying at anchor off South entry to Moon Sound. **Type:** *S-30*	Gulf of Riga
1941[500]	August 20 4th SFltl. *S-48*	The *S-48* sinks the Polish freighter SS *Czenstochowa* and torpedoes a second ship of 2, 774 tons. **Type:** *S-38*	Off Cromer North Sea
1941[501]	September 7 4th SFltl. *S-48, 49, 50, 52 ,107*	The S-boats attack a small convoy off the coast of Norfolk and sink two small freighters. **Type:** *S-38*	English Channel North Sea
1941[502]	September 9 British MTBs	A British MTB force attacks a German convoy escorted by patrol trawlers, *V 202, V 208,* and sink the freighter (ex blockade runner) *Trifels* off Boulogne, France.	Strait of Dover
1941[503]	September 9 *MTB 56* RNorN	She joined a British MTB force to attack German shipping off the French Coast. She torpedoed and sank a 5,000 ton merchant ship. 75' Thorneycroft	English Channel
1941[504]	September 11/12 *TKA 11, 12*	The first attack by Soviet TKAs is against a German convoy off Petsamofjord. No hits are recorded. **Type:** G-5, 62.7'	Barents Sea
1941[505]	September 13 TKA (Unknown)	A torpedo from a TKA sinks the German patrol ship *V 308 Oscar Neynaber* from a TKA off Porkkala, near Gogland.	Gulf of Finland
1941[506]	September 14 TKA(Unknown)	The German motor minesweeper, *M-1707 (Lüneberg),* after being damaged by a mine and hit with a 130 mm shell from a Soviet shore battery and abandoned by its crew, was torpedoed and sunk by TKAs near Arensburg (Kuressaare) Saaremaa Island.	Gulf of Riga Baltic Sea

[499] Rohwer, *Chronology of the War At Sea*, 93.
[500] Rohwer, *Chronology of the War At Sea*, 93.
[501] Rohwer, *Chronology of the War At Sea*, 99.
[502] Rohwer, *Chronology of the War At Sea*, 99.
[503] Reynolds, *Dog Boats At War*, 29.
[504] Internet correspondence with Richard Worth, 2007.
[505] Internet correspondence with Richard Worth, 2007.
[506] Internet correspondence with Richard Worth, 2007.

Figure 183. A *V* class patrol boat.[507]

Figure 184. The British is the only country to use two national naval ensigns. The White Ensign is for combat vessels. The Red ensign is for all other vessels.

507 Michael Emmerich, German Naval History http://www.german-navy.de/kriegsmarine/ships/misc/vorpostenboot/index.html

Date	Craft, Name, Event or Country	Event, Builder and Boat Specifications	City, River, Lake, Sea, Ocean
1941[508]	September 14/15 *TKA 13, 14, 15*	The *TKA 13, 15*, attack and sink the Norwegian coaster *SS Mitinattsol* on the Petsamo-Kirkenes route. On the 15th, the *TKA 14* sinks the coaster Renöy. **Type:** *G-5*, 62.7'	Off Sar-Varanger, Finmark,Norway Barents Sea
1941[509]	September 15 *TKA 103*	She sank in a firefight with the German 31st M Flotilla off Revel (Tallinn). **Type:** *G-5*	Gulf of Finland
1941[510]	September 17 4th SFltl. S-*50, 51, 52*	The boats attack a British convoy off Cromer, sink the British freighter *SS Teddington*, and torpedo another ship causing some damage. **Type:** *S-38*	North Sea
1941[511]	September 22 Finnish Navy MTB	The motor torpedo boats Vinha and Syöksy made an attack on the harbor of Suursaari (Hogland in Russian) and sank the Soviet minesweeper *T-41/Sergej Kirow*.	Gulf of Finland
1941[512]	September 24 *TKA 24*	The *TKA 24* was first Soviet MTB to be captured by a German Trawler. **Type:** *G-5*	Gulf of Finland
1941[513]	October 1 *MTB 56* (RNorN)	The destroyer *Draug* tows the Norwegian torpedo boat *MTB 56* from Scapa Flow to Norway. 120 nm from Norway the MTB is let loose to attack a German convoy and she sinks the Norwegian tanker *Mv Borgny* south of Sorta. **Type:** Thorneycroft, 75'-6"	Off Sorta, Norway, North Sea

508 Rohwer, *Chronology of the War At Sea*, 93.
509 Fock, *Fast Fighting Boats*, 264.
510 Rohwer, *Chronology of the War At Sea*, 101.
511 Rohwer, *Chronology of the War At Sea*, 102.
512 Fock, *Fast Fighting Boats*, 266.
513 Rohwer, *Chronology of the War At Sea*, 105.

Figure 185. Loading torpedos on a Soviet *G-5* boat.

Date	Craft, Name, Event or Country	Event, Builder and Boat Specifications	City, River, Lake, Sea, Ocean
1941[514]	October 20 TKA *13, 61, 82*	*TKA 13* was lost after suffering damage off Lavassari, Russia, and *TKA 61* and *82* were lost to air attacks in the same area. **Type:** *G-5*	Gulf of Finland
1941[515]	October 6 *TKA 12, 14, 15*	The boats attack and sink the freighter *Bjørnungen* on the Petsamo-Kirkenes convoy route. **Type:** *G-5*	Off StureEkkeroy Finmark, Barents Sea

Figure 186. A Soviet *D-3* boat underway. [516]

[514] Fock, *Fast Fighting Boats*, 267.

[515] Internet correspondence with Richard Worth, July 12, 2007.

Date	Craft, Name, Event or Country	Event, Builder and Boat Specifications	City, River, Lake, Sea, Ocean
1941[517]	November 1 *MTB 108*	The boat burned and destroyed in the stocks while under construction. **Builder:** Vosper, 46'	England
1941[518]	November 3 Spanish Armada	Operational Instructions are issued for Spanish naval officers and NCOs to undergo training with the *Kriegsmarine* authority for Mine sweepers, U boats and *Schnellboots*. Their training takes place first in Könisberg, then to Helsinki, at the Army base of Kotka. The Spaniards work the mine barrage *Corbetha* between the Estonian, coast of Kallbada, and Porkkala, Finland. The Spanish presence is contrary to the rules of war at the time. They were in fact civilians allowed to kill Soviet aviators and other Allied crew members.	Gulf of Finland
1941[519]	November 20 *PT 1* & *PT 2*	Made by the Fogal Boat Yard for consideration by the US Navy, the two 81' Experimental craft were re-classified as Small Boats, SB.	Atlantic Ocean
1941[520]	November 19/20 *S-41, 53, 104, 105*	The flotilla attacked the British convoy off Great Yarmouth and sank 3 of the many ships, one being the British fleet oilier *HMS Methar*, by *S-104. S-41* received damage from gunfire and scuttled on site. **Type:** *S-38*	Great Yarmouth, English Channel

---

[516] Central Naval Museum, St. Petersburg, http://www.RussianMuseum.info/M153.
[517] *Warships Supplement 65*, 8.
[518] Yanez, Alejandro, *Schnellboote in the Spanish Navy* http://www.prinzeugen.com/Armada.htm.
[519] HyperWar: *US Navy Torpedo Boats*, 8.
[520] Hümmelchen,*German Schnellboote*, 162.

Date	Craft, Name, Event or Country	Event, Builder and Boat Specifications	City, River, Lake, Sea, Ocean
1941[521]	November 23/24 4th SFltl. *S-50, 51, 52, 109, 110*	The flotilla attacked a British convoy east of Orfordness with *S-109* sinking the British tanker *Virgilia* and Dutch tanker *Groenlo* with another freighter torpedoed and damaged. **Type:** *S-38*	Off Orfordness, Dover, North Sea
1941[522]	November 28/29 4th SFltl. *S 51, 52, 64*	In the same area of the North Sea, the 4th S-Flotilla boats sink three more British freighters. **Type:** *S-38*	Off Orfordness, Dover, North Sea
1941[523]	December 7 *PT 24*	Two gunners, GM1c Van Zyll De Jong and TM1c Huffman, of *PT 24*, believed to be the first to shoot down IJN dive-bomber near Berth S-13, Pear Harbor. **Builder:** Elco, *PT 20*-class 77'	Honolulu, Pacific Ocean
1941[524]	December 8 RON 3 PT *30, 34, 41*	The morning after the Japanese sneak attack on Pearl Harbor, the 84 men of RON 3 and their six PT boats became part of the small contingent fighting force to oppose the Japanese invasion fleet, Army, and air forces. **Builder:** Elco, Type PT 20-class, 77'	Kawit Island, Mindanano, Philippine Is
1941[525]	December 12 *PT 64*	Presumed sunk or scrapped by the USN 7th Fleet. **Type:** Elco PT 20-Class, 77'	Unknown

---

521 Hümmelchen, *German Schnellboote*, 163.
522 Hümmelchen, *German Schnellboote*, 163.
523 HyperWar: *US Navy Torpedo Boats*, 8.
524 Hinds, Squadron of Furies, *America In WW II*, Volume One, Number Three, 2005, 53-57.
525 Polmar, Norman and Samuel Loring Morison *PT Boats At War, World War II to Vietnam* (MBI Publishing, 1999), 116.

Figure 187. This photograph was taken at the U.S. Navy Motor Torpedo Boat Training Center, Melville, Rhode Island in late 1940. America, like Japan, waited too long to implement an aggressive MTB program. This lack of training would prove fatal in the early years of the war. (Library of Congress Print and Photographic Division)

Figure 188. The Rising Sun swiftly became feared throughout Asia and the Pacific.

Figure 189. Japanese naval poster with the flags of Germany and Italy flying alongside the Rising Sun. In late 1941, the Axis was a fearsome threat.

Figure 190. A U.S. Navy recruiting poster.

Figure 191. The PT boats' "Mosquito Fleet" logo was designed by the Walt Disney Studio. (U.S. Navy Historical Division)

Date	Craft, Name, Event or Country	Event, Builder and Boat Specifications	City, River, Lake, Sea, Ocean
1941[526]	December 15 *PT 33*	First PT boat lost in combat. Grounded 12-15-41 and destroyed 12-26-41 **Builder:** Elco, **Type:** PT 20-class, 77'	Cape Santtiago Luzon, PI
1941[527]	December 15 *MTB 68*	Off the coast of Libya, the *68* boat collided with *MTB 215* and was lost. **Builder:** Thorneycroft 55',CMB	Off Libya Mediterranean
1941[528]	December 16 *MTB 8*	One of the first British Power Boat Co. MTBs. Bombed by IJN Navy air force in Hong Kong Harbor **Builder:** British Power Boat Co, 60'	Hong Kong Harbor, South China Sea
1941[529]	December 19 *MTB 26*	Sunk by gunfire from IJN landing craft, Hong Kong Harbor. **Builder:** Thorneycroft, 55' CMB	Hong Kong Harbor

Figure 192. The one weapon the men of the small contingent of PTs most feared was the Japanese "Zeke" float planes.

---

[526] *Casualties, Navy and Coast Guard Ships*, WW II http://www.history.mil/faqs/faq82-1.htm, 8.
[527] North, *Royal Naval Coastal Forces*, 12.
[528] *Warship Supplemental 65*, 5.
[529] *Warship Supplemental 65*, 5.

Figure 193. The class of 55' CMTB, Coastal Motor Torpedo Boat, did well in both World War I and II.

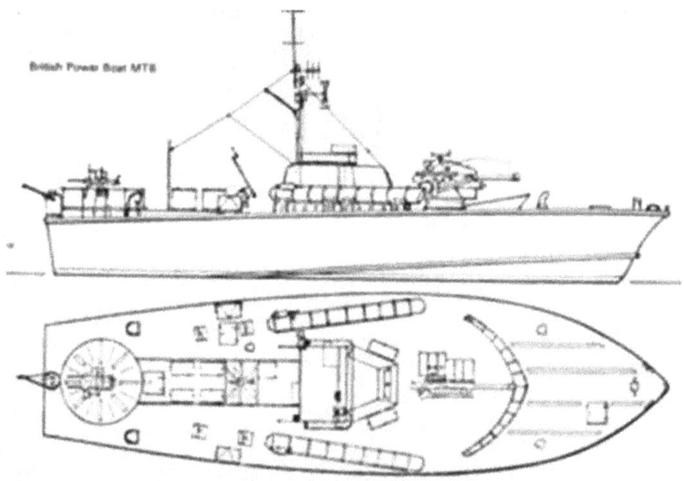

Figure 194. Profile and top view of British Power Boat MTB.

Date	Craft, Name, Event or Country	Event, Builder and Boat Specifications	City, River, Lake, Sea, Ocean
1941[530]	December 3d SFltl.	To disrupt or stop British convoys getting to the safety of the Grand Harbour, Malta, the flotilla concentrated in laying 73 mines off the entrance and approaches. This had an immediate effect in slowing the flow of freighters to Malta.	Mediterranean Sea
1941[531]	December 19 *MTB 12*	Sunk by gunfire from IJN landing craft, Hong Kong Harbor. **Builder:** British Power Boat Co., 60'	Hong Kong Harbor
1941[532]	December 26 *MTB 27*	Scuttled by crew to prevent capture, Mir's Bay (Hong Kong). **Builder:** Thorneycroft, 55' CMB	Hong Kong South China Sea
1941[533]	December 26 *MTB 9, 10, 11, 13*	Scuttled by crews to prevent capture by invading Japanese forces, Mir's Bay (Hong Kong) **Builder:** British Power Boat Co., 60'	Hong Kong Harbor

530 Hümmelchen, *German Schnellboote*, 163-166
531 *Warship Supplemental 65*, 5
532 *Warship Supplemental 65*, 5.
533 *Warship Supplemental 65*, 5.

Figure 195. A PT squadron.

THE SHIP KILLERS

THE DEFINITIVE ILLUSTRATED HISTORY OF THE TORPEDO BOAT

VOLUME VI: TIMELINE 1942

PT-59. *(Illustration by author)*

BY JOE HINDS

NIMBLE BOOKS LLC

CONTENTS (INCLUDING FIGURES)

DEDICATION

Roger is my friend from the Land of Oz (Australian) and has done an incredible amount of work on my volumes. He has proofed copy, run down conflicting information, and cheered me on when I was fed up with the whole thing.

His work and cheerful encouragement has kept me going in times of bad news and worse bad news. To him I cheerfully dedicate this volume to him and his lovely wife Deb.

PUBLISHER'S NOTE

In the process of editing a multi-volume work over many months, I have introduced certain inconsistencies, which I note here for the record, and to indicate that the author should not be blamed for these issues.

In the previous volumes, unfortunately, I hyphenated some, but not all boat numbers: *S-14* but not *PT 31*. I tried to be consistent by country: thus, *PT 31* but *S-14*.

I may have been wrong about this rule. According to U.S. Navy regulations cited by the U.S. Naval Historical Center,

> *4. Ships and service craft designated "active status, in service," except those described by paragraph 3 of this article, shall be referred to by name, when assigned, classification, and hull number (e.g., "HIGH POINT PCH-1" or "YOGN-8").— United States Navy Regulations, 1990, Article 0401*

Similarly, the Chicago Manual of Style, 15th edition, §8.124, presents the following guidance:

> *Names of specific ships and other vessels are both capitalized and italicized ....*
>
> *USS SC-530;*
>
> *Where necessary to avoid confusion between vessels of the same name—in a work on naval history, for example—the numbers should be included at first mention ...*
>
> *USS Enterprise (CVN-65) was already on its way to the Red Sea.*

Accordingly, in this volume, all boat names are italicized and hyphenated.

[1] http://www.history.navy.mil/faqs/faq63-1.htm.

In the preceding volume, I introduced a new "3 x 3 grid" that provides a much more readable display of boat specifications than the "run on paragraph" style used in volumes I through IV. I intend to maintain the grid style throughout the remaining volumes.

If you notice other inconsistencies or errors that should be addressed, by all means send an e-mail via wfz@nimblebooks.com.

TIMELINE 1942

Date	Craft, Name, Event or Country	Event, Builder and Boat Specifications	City, River, Lake, Sea, Ocean
1942[534]	January 17 *MTB-47*	Sunk during a firefight with surface vessels off Cape Gris-Nez. **Builder:** White, Cowes 72'	Straits of Dover
1942[535]	January 19 *PT-31*	Grounded and then scuttled to prevent capture. **Builder:** Elco Type *PT-20*-class 77'	Subic Bay Luzon, PI
1942	February 4 *MTB-308* (ex-*PT-50*)	Sunk by gunfire from Italian aircraft off Tobruk, North Africa. **Type:** Elco	Mediterranean

Figure 1. The English boats operating in the Channel were hard pressed at the beginning of the war to find tactics and develop an effective naval doctrine for its MTB force. Germany, Italy and the Soviet Union had pursued this mix since the early 1930's. Here, *MTB-47.* (John Lambert, Lambert-Plans.com)

534 "British RN, MGB, MTB, SGB, ML," http://www.naval-history.net/WW2BritishLosseseCoastal.htm.

535 "Casualties, Navy and Coast Guard Ships WWII," http://www.history.navy.mil/faqs/faq82-1.htm.

Figure 2. The US naval situation in early 1942 was dominated by the Japanese invasion of the Phillippines.

.

Figure 3. Posters that emphasized sacrifice did more to influence the American people then posters that demonized the enemy. (Library of Congress Print and Photographic Division)

Date	Craft, Name, Event or Country	Event, Builder and Boat Specifications	City, River, Lake, Sea, Ocean
1942[536]	February 14 MTB-776 "D"	Fairmile "D" boat lost to fire in Ostend Harbor during operations. **Builder:** Lady Bee, Southwick **Speed:** 32 knots max, 27 knots cruise **Displacement:** 105 tons **Range:** 506 nm **Length:** 115' **Beam:** 20' **Draft:** 4'9" **Engines:** 4 x 1,250 Packard 4m 2500 hp, petrol **Armament:** 6-pdr power turret forward and duplicate aft, 2 x twin .50-cal MG power turrets midships, 1 x twin Oerlikon 20 mm power turret midships, 2 x twin .303-cal. MG bridge, 2 x 18" torpedo tubes **Crew:** 30	Ostend Harbor North Sea
1942 [537]	February 18 *TKA-12, 13*	The TKA *12* and *13* attacked German vessels in Varangerfjord but score no hits. **Type:** *G-5*, 62.5'	Barents Sea, Arctic
1942[538]	February 19 2nd SFltl S-39, 53	The flotilla was returning to base after a night operation of both torpedoes and mining operations when the S 39 collided with S-53. The 39 boat was capable of making it home but the 53 boat was captured by British MTB's and scuttled by the crew. All crewmen were captured except the commander who was lost.	English Channel

536 *Warship Supplemental* 65, 22

537 Rohwer, *Chronology of the War At Sea*, 145.

538 Frank, *German S-Boats*, 51.

Figure 4. The "Dog" boats, as they were called, were the only MTB or MGB that could take on the *Kriegsmarine's* S Boats on an equal footing. Both classes of boats were of similar length, tonnage and crew members. Their fire-power was an equal match. (John Lambert, Lambert-Plans.com)

Figure 196. The Soviet *TKA-12, 13* were early model *G-5* types. (Central Naval Museum, St. Petersburg).

Figure 197. The S-boat crews lived better on their boats than crews in any other navy. They were dryer, and the sleeping compartment had steady heat from the recirculation of hot engine radiator water through out the living spaces on the boat. On the S boats, crews had the ability to get in out of the weather and thaw out. (Hans Frank)

Figure 198. One of His Majesty's Motor Torpedo Boats (see bottom right) in the Med.

Date	Craft, Name, Event or Country	Event, Builder and Boat Specifications	City, River, Lake, Sea, Ocean
1942[539]	March 11 2nd SFltl.	The flotilla sank the British freighter SS *Horseferry* east of Cromer.	North Sea
1942[540]	March 11 *PT-32, 34, 35,41*	Evacuation of General Douglas MacArthur. The four boats were loaded to the hilt with people, luggage and extra fuel for the trip from Corregidor to Cagayan on the island of Mindanao. The trip was exhausting: they traveled at night for 560 miles with Japanese forces on sea and land hunting them the whole time. The crew had to fight to keep the engines running as their fuel had been sabotaged with Paraffin wax. The trip was a success but cost two of the four boats n the little flotilla.	Luzon Sea, Sulu Sea
1942 [541]	March 13 *PT*-32	Destroyed off Tagauayan Island to prevent capture. **Type:** Elco *PT-20*-class 77'	Tagauayan Is. Philippines
1942 [542]	March 14 2nd SFltl.	The 2nd SFltl. attacked a convoy resulting in the sinking of the destroyer HMS *Vortigern.* Once the flotilla returned to their bunkers, they had to set sail again to rescue *S-111* , which had become separated from the flotilla in the fight. Three British MGBs gave battle and captured the S-*111*. It was being towed to England by the MGB force when the 2nd SFltl. drove off the MGB's and took the *S-111* under tow. RAF Spitfires arrived on site and in a 30-minute gunfight, inflicted many injuries on all boats. The *S-111* was cut loose and sunk by machine gun fire from the RAF planes. Of the entire crew of 111, 18 were missing and seven dead.	North Sea
1942 [543]	March 15 *S-104*	The *104* sank the destroyer *Vortigern* while it was escorting the convoy FS 749 off Cromer on the east coast of England.	English Channel

---

[539] North-East Diary 1939-1945 Roy Ropley & Brian Pears http://www.ne-diary.bpears.org.uk?inc/Inc/ISeq_26.html#D921

[540] Norman Polmar and Samuel Loring Morison, PT Boats At War, 26-27.

[541] *Casualties, Navy and Coast Guard Ships WWII.*

[542] Frank, *German S-Boats in Action*, 51.

[543] Europe- March 1942 http://www. Naval-history.net/WW2CampaignsCoastal-SpecialWarfare.htm .

Figure 199. MacArthur's route was totally infested with Japanese naval and army forces. The boats were overloaded with people luggage and spar drums of gasoline. It was an amazing journey in war torn Philippines.

Figure 200. Feelings against the Japanese military the Japanese people in general reached a fever pitch. The anger about the atrocities still persists to this day. Many cite the example of Germany making payments to captured Allied service men who worked in German manufacturing facilities. The Japanese still deny wrong doing and have never made a payment to any POW survivor who labored in a Japanese manufacturing plant. (Library of Congress Print and Photographic Division)

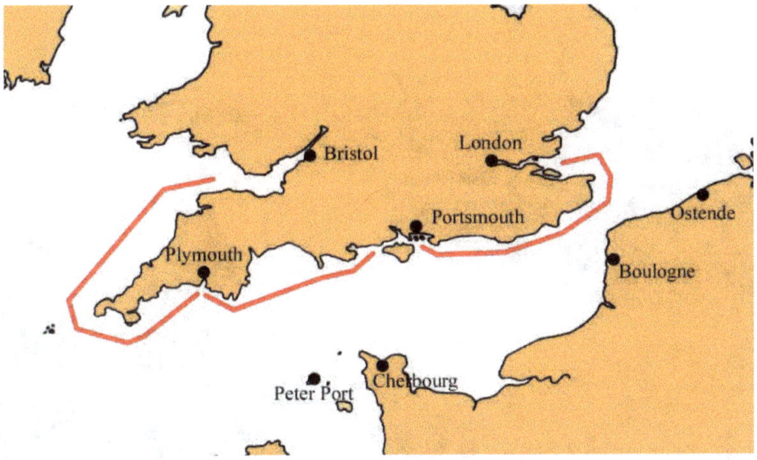

Figure 201. The route of the English convoy FS 749 as it came under attack by S-boats.[544]

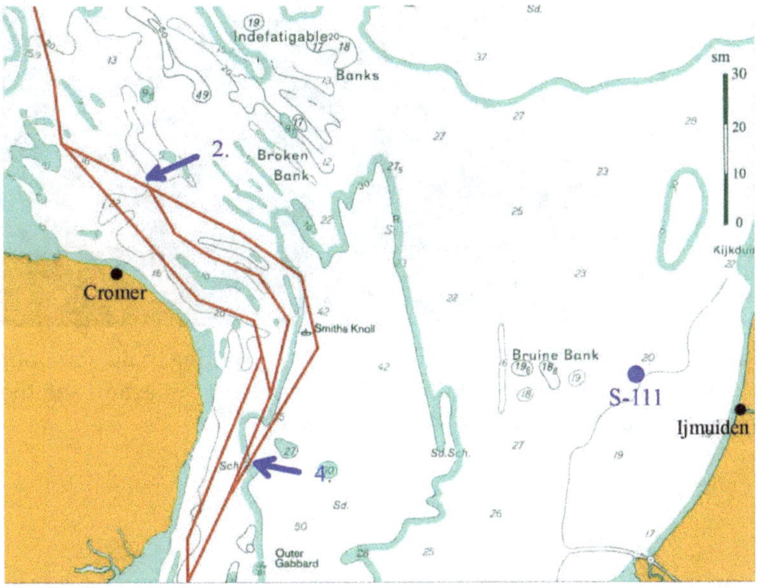

Figure 202. The S-111 was attacked by British MGBs (*MGB-87, 88* and *91*) and later British Spitfires. She was captured, the recaptured and then shot again by fighter aircraft. In this all day fire fight, she lost 14 men to the sea and the rest became POW's.[545]

[544] Hans Frank, *German S-Boats in the Second World War.*
[545] Hans Frank, *German S-Boats in the Second World War.*

Figure 203. The hardest part of any military mission is waiting to do your job. Heading out into the depths of Channel was a time for reflection and the wonder. You had to wonder on whose side is God.[546]

[546] Busch, *Unsere Schnellboote im Kanal*, 90.

Date	Craft, Name, Event or Country	Event, Builder and Boat Specifications	City, River, Lake, Sea, Ocean
1942 [547]	March 15/16 *PT-114, 129*	The two boats laid an ambush in a small inlet in Mia-Ama Bay, where they shot up and sank six Japanese barges loaded with troops and supplies. **Type:** Elco, 80'	Mia-Ama Bay Solomon Islands Solomon Sea
1942[548]	March 18 *PT-35, 41*	The two surviving PTs had to make another run carrying luggage, scared people and extra fuel to take the president of the Philippines and his family to Mindanao for their escape. This trip caused great damage to the *35* boat but she eventually recovered, only to be burned at the capture of Cebu City by the Japanese.	Luzon, Philippines
1942 [549]	March 28 *MTB-74*	For Operation Chariot, the boat was fitted with its torpedo tubes mounted on the forward deck. She became lost to gunfire from German shore batteries, at the mouth of the Lorie River, Southern France. **Builder:** Vosper: 72'-6"	Bay of Biscay Atlantic Ocean
1942 [550]	March 29 *MTB-215*	Presumed lost Mediterranean, but also listed as lost in June, 1944 in other sources. **Builder:** Thorneycroft, 55'	Mediterranean

[547] Keating, *The Mosquito Fleet*, 64, 65, 66.
[548] Polmar, *PT Boats At War*, 28-29.
[549] Lambert, *Allied Coastal Forces* Vol 1.
[550] *Warship Supplement* 65, 8.

Photo # NH 86429 Recruiting Poster, "Ouch!" -- Lieutenant J. D. Bulkeley, USN

Figure 204. PT commander John Bulkeley reached "rock star" status after word of his daring leadership style gave the American people the belief that Japan could be beaten. . He was the ideal man for the job in the Philippine waters. He knew the Japanese military from his time in China and he knew how they thought and planned.

Figure 205. *MTB-74* was one of a kind, built and fitted out for a single mission, Operation Chariot. She had the torpedo tubes, well forward on the bow to allow the torpedoes to be launched over the barrier nets at the harbor. She was sunk by German shore batteries. (John Lambert, Lambert-Plans.com)

Date	Craft, Name, Event or Country	Event, Builder and Boat Specifications	City, River, Lake, Sea, Ocean
1942 [551]	April 9 *PT-59*	The *59* boat was on routine training exercise from Melville, Rhode Island headed for practice torpedo runs near Narragansett Bay. An officer-in training accidentally turned the firing key for on one of the torpedo tubes and launched a torpedo. The torpedo passed Prudence island pier (which was loaded with depth charges), and struck the 4,000- ton navy supply ship, USS *Capella* (AK-13). The ship sank without loss of life., although her cargo was lost as well as a new float plane that was carried as deck cargo. **Type:** Elco *PT-20* class, 77'	Off Prudence Island Atlantic Ocean
1942[552]	April 9 *PT-34*	One of the six original PT-boats went down fighting. Petty Officer Al Ross shot down one IJN dive-bomber with his starboard side Lewis gun. Japanese float planes shot up the crew and then destroyed the 34 boat. **Type:** *PT-20*- class Elco 77'	Cauit Island, Mindanano, Philippine Islands
1942[553]	April 12 *PT 35*	The 35 boat had been brought to Cebu City with the hopes of getting her repaired and engines cleaned so she could keep fighting. The Japanese entered the city and the boat was burned to prevent capture.	Cebu CIty Luzon
1942[554]	April 15 *PT-41*	The last of the six original PT-boats finally met her end. After being trucked overland to Lake Lanao, Mindanano, she was destroyed to prevent capture **Type:** *PT-20*-class Elco 77'	Mindanano, Celebes Sea, Philippine Islands
1942[555]	April 25 TKA Div.	Two TKA's` sink the Soviet sub SC 401 off Komagnas. **Type:** *G-5*, 62.5'	Barents Sea Arctic

---

[551] "Briefing: PT 59's First Victory," StrategyPage.com http://strategypage.com/cic/docs/cic1396b.asp.
[552] Hinds, "Squadron of Furies, America IN World War II, 57.
[553] Polmar, PT Boats At War, 29.
[554] "Casualties, Navy and Coast Guard Ships WWII."
[555] Fock, *Fast Fighting Boats*, 266.

Figure 206. The *59* boat was John Kennedy's second command. At that time, the 59 was converted to a gunboat and all torpedo tubes were removed. Two twin barreled .50-cal MG were mounted forward on each side of the cockpit and four additional single .50 cals were added on each side of the cockpit from mishaps to stern.

//

Figure 207. *PT-34*. (Illustration by author)

Figure 208. The *34* boat was well shot up by Japanese float planes and ran aground on Cauit Island. Before being completely destroyed by further bombing, Petty Officer Al Ross shot down on of the attacking aircraft. The action would earn the *34* boat a place in the American propaganda effort as a "Poster" boat. Today, Petty Officer Ross's son is Dr. Al Ross, a leading historian of the PT boat.

Figure 209. The Japanese lost many inshore transports to the relentless PT boats.

Figure 210. The six PT boats that fought hard and long in hopeless conditions captured the heart of all America. They were relentlessly portrayed in movies, posters, and magazine articles like this one for *Life* Magazine. (Library of Congress Print and Photographic Division)

Figure 211. The Soviets used their MTB force (Torpedo Cutters) as razor blades. The boats never carried life saving equipment of any kind. Once you went over the side or your boat was sinking, you were a goner-the end. The only Soviet TKS crews rescued in the Baltic and Arctic Sea was made by Finnish or German crews. (Library of Congress Print and Photographic Division)

Date	Craft, Name, Event or Country	Event, Builder and Boat Specifications	City, River, Lake, Sea, Ocean
1942 [556]	May 7 3rd SFltl. *S-54, 31, 61*	Off Malta, under a moonlit night, the three boats sank the British *ML-130* with their 20mm machine guns. The S-boat crews recover only eleven men from the wreck. *S 54* & *61:* Type 30; S-*31:* Type S-*14*	Off Malta Mediterranean
1942[557]	May 7 *MAS-571*	The Soviets use their submarines to run 77 supply missions to Sevastopol. Near the end of this effort, the Italian MAS-571 torpedoes the Soviet sub *Shch 214.* **Type:** 500 Series **Builder:** Cantieri Riuniti dell' Adriatico (CRDA) Monfalcone	Black Sea
1942 [558]	May 10 3rd SFltl. *S 31*	After returning a night of heavy engagements at sea the *31* boat ran over a British mine and sank, taking 12 men with it. Earlier in the year, the *31* boat was the flotilla's "Lucky Boat." **Type:** S-14	Off Valletta, Malta, Mediterranean
1942[559]	May 12 4th SFltl.	British MTBs sink 2 of 4 torpedo boat escorts ,the *Seeadler* and the *Iltis.* The 4thSFltl. happens upon the survivors and rescues 83 crewmen from the two ships. They also pick up three surviving British crewmen from the *MTB-220,* which had been sunk in the action.	Straits of Dover
1942[560]	May 16 *MTB-338*	Lost to onboard fire and subsequent explosion. **Builder:** Canadian Power Boat Co. 70'	Trinidad, West Indies

[556] Royal Naval Vessels Lost At Sea, http://www.naval-history.net/WW2BritishLosses3Coastal.htm#ml
[557] Rohwer, Chronology of the War At Sea, 164.
[558] Frank, *German S-Boats in Action*, 18.
[559] Frank, *German S-Boats in Action*, 52.
[560] *Warship Supplement 65*, 11.

Figure 212. At times, the patrol boats from 4th SFltl.turned into rescue boats as they pulled their own and the enemy sailors from the water after a fire-fight.[561]

Figure 213. The panther logo of the 4th SlFtl.[562]

Figure 214. The coast of North Africa became a bloody battleground for the Allies and Axis coastal forces.[563]

---

561 Busch, *Unsere Schnellboote im Kanal*, 15.

562 4. Schnellbootflottille, http://www.wlb-stuttgart.de/seekrieg/km/sboot/sfl4.htm

Date	Craft, Name, Event or Country	Event, Builder and Boat Specifications	City, River, Lake, Sea, Ocean
1942[564]	May 17 3rd SFltl. *S 34*	While undertaking mine laying operations off Malta, shore batteries, directed by surface radar, opened fire and drove off all the German boats. The S-*34* received several hits, causing three dead, and the crew abandoned ship. The Luftwaffe flew in the next morning and sank the wreck. **Type:** *S-14*	Off Malta Mediterranean
1942[565]	May 18 *MAS-571*	In the continuing Black Sea operations, the *571* boat made a successful surprise attack on a convoy of German barges fully loaded with infantry troops while being escorted by six gunboats. Between machine gun fire and torpedoes, they sank four barges and damaged six other vessels. **Type:** Baglietto, C.R.D.A.,	Black Sea
1942[566]	June 4-7 Battle of Midway	Six months after the bombing of the American fleet in Pearl Harbor, the Battle of Midway was the critical defeat for the Japanese. This was also one of the few occasions where PT's are listed in the battle of order for a major battle. On the US side, there were 55 major warships and eight PT boats. The PTs were *20, 21, 22, 24, 25, 26, 27,* and *28*. The PTs were responsible for bringing down three Japanese Zero fighter aircraft and causing much damage to many more. The small boats rescued pilots and returned them to their own ship or other ships so that they could continue to fly in battle. These boats were of the original ELCO PT 20 Class, 77'.	Midway Atoll Pacific Ocean
1942[567]	June 13 TKA (Unknown)	Near Yalta, in the Black Sea, Soviet TKA's sank the Italian submarine CB-5. **Type:** *G-5*, 62.5'	Yalta, Black Sea

563 http://en.wikipedia.org/wiki/File:Britishmtb.jpg
564 Frank, *German S-Boats in Action*, 79.
565 "M.A.S. and Midget Submarines in the Black Sea 1942-1943", http://www.comandosupremo.com/Blacksea.html
566 Battle of Midway:4-7 June 1942 http://history.navy.mil/faqs/faq81-5.htm
567 Internet correspondence with Richard Worth, July 12, 2007.

Figure 215. Midway was one of the rare occasions where the PT boats were recognized the official Battle of Orders.[568] PT boats were part of the planned final defense for Midway Atoll.

Figure 216. *PT-25* served at Midway.

[568] http://www.history.navy.mil/photos/events/wwii-pac/midway/mid-1a.htm

Figure 217. Midway was the most decisive carrier battle of WW II.

Date	Craft, Name, Event or Country	Event, Builder and Boat Specifications	City, River, Lake, Sea, Ocean
1942 [569]	June 14 *MTB-259*	While on patrol, the *259* ran into bad weather somewhere in the Mediterranean and sank. **Builder:** Elco, 80' **Displacement:** 32 tons **Length: 8**1' **Beam:** 20' **Engines:** 3 x Hall-Scott 2,700-bhp, petrol **Armament:** 4 x 18" torpedo tubes, 2 x .50-cal. MG, 4 x depth charges	Mediterranean
1942 [570]	June 14 3rd SFltl. *S-54, 55, 56, 58, 59,60*	The flotilla attacked the Malta convoys Vigorous and Harpoon. The *S-55* sank the British destroyer HMS *Hasty* and the *56* boat put two torpedoes into the cruiser HMS *Newcastle*. **Type:** *S-30*	Mediterranean Sea
1942 [571]	June 15 *MTB-201*	Sank while under tow after seeing action in the Dover Straits. **Builder:** White, Cowes 72'-6"	Dover Straits North Sea
1942 [572]	June 19 1st SFltl.	The first German success in the Black Sea went to the *S-102* when it sank a Soviet freighter South of the Crimea.	Black Sea
1942[573]	June 21 3rd SFltl. *S 55 ,58*	As the British defenses of Tobruk collapsed, escaping vessels came under attack by the 3rd S-Flotilla. One motor sailing ship and six smaller craft sank or became captured. The *S-55* took several hits during the action. The S-flotilla's commanding officer and staff surgeon were killed. **Type:** *S-30*	Tobruk Mediterranean Sea

---

569 *Warship Supplement 65*, 10.
570 Coastal Forces and Special Units http://www.naval-history.net/WW2CampaignsCoastal-SpecialWarfare.htm.
571 North, *Royal Naval Coastal Craft*, 22.
572 Hümmelchen, *German Schnellboote*, 169.

Figure 218. MTBs at night.[574]

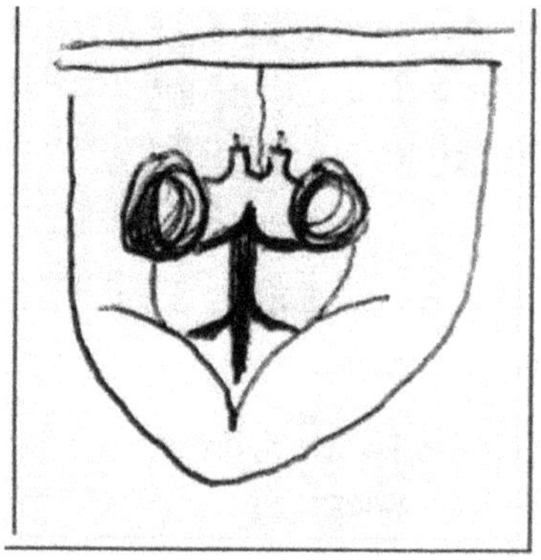

Figure 219. The sketch shows how the crew came about the design for their flotilla symbol. The German and Italians stuck to facsimile coats of arms while the American went for pin-up girls.[575]

[574] http://www.rodericktimms.royalnavy.co.uk/guest_book.html

[575] 1.Schnellboote-Flottile 1939-45 http://www.wlb-stuggart.de/seekrieg/km/sboot/sfl1.htm

Date	Craft, Name, Event or Country	Event, Builder and Boat Specifications	City, River, Lake, Sea, Ocean
1942 [576]	June 21 2nd SFltl.	The boats of the flotilla damaged the *TKA-47,* captured her, and turned her over to the Finish Navy. **Type:** *G-5,* 62.5'	Baltic Sea
1942[577]	June 24 1st SFltl.	Boats of the flotilla sank the Soviet at the Western entrance to the Gulf of Finland. TKA: Type *G-5,*62.5'	Gulf of Finland
1942[578]	July 3 1st SFltl. *S-28, 40*	The flotilla sank two Soviet gunboats fleeing from Sebastopol. *S-28* and *S-40* suffered heavy damage and many casualties. *S 40:* **Typ**e: S-*38; S-28* **Type**: S-*14*	Black Sea
1942[579]	July 4 *MTB-601* "D"	Sunk in combat with German surface ships off Dover. **Builder:** Fairmile "D" boat, 115'.	Dover Straits North Sea
1942[580]	July 9 2nd SFltl. *S 48, 50, 63, 67, 70, 104,109*	The flotilla attacked an Allied convoy near Lyme Bay and sank one Dutch freighter and three Norwegian freighters plus the armed trawler HMS *Manor (FY-333).* The *S-67* sank the British tanker *Pomella.*	Southern England English Channel
1942 [581]	July 7/9 TKA Div.	A force of 30 TKAs, with Soviet Guard boats, landed troops on Someri Island, the military base at Itäppaa. Finnish coastal batteries sank 8 TKAs and Finnish gunboats, Hämeenmaa, and Uusimaa sank seven more TKAs. The Soviets total loss was: *TKA* (Motor Torpedo Cutter) *13 ,31, 42, 83, 101, 113, 123 and MO* (Gunboats) *201, 211, 212, 225, 301, 306, 308, 314.*Of the landing force 149 men surrendered and 126 died. Others held out for weeks. **Type:** *G-5,* 62.5'	Someri Island Gulf of Finland

576 Fock, *Fast Fighting Boats,* 263.
577 Fock, *Fast Fighting Boats,* 263.
578 Hümmelchen, *German Schnellboote,* 166.
579 North, *Royal Naval Coastal Craft,* 50.
580 Royal Naval Vessels Lost At Sea 1939-1945 http://www.naval-history.net/WW2BritishLossesbyDate2.htm
581 Fock, *Fast Fighting Boats,* 263.

Date	Craft, Name, Event or Country	Event, Builder and Boat Specifications	City, River, Lake, Sea, Ocean
1942[582]	June 22	The Italian 12. Squadriglia *MAS* operated four boats, MAS-*526, 527, 528,* and *529* as part of a joint operations force on the inland waters of Lake Ladogain the summer and autumn of 1942. In winter 1942, the MAS boats retired, and the Soviets resumed the truck convoys over Ladoga's thick ice made famous in *The 900 Days,* Harrison Salisbury's classic account of the siege (and starvation) of Leningrad.	
1942[583]	July 13 1st TKA Brigade, 2nd Div.	The TKA brigade, in conjunction with the Soviet air arm, sank a German assault boat, slightly damaged 25 vessels by gunfire and shrapnel in a German convoy off Dünamünde (Bolderia). **Type:** *G-5,* 62.5'	Gulf of Finland
1942 [584]	July 17 *MTB-316* (ex-*PT-58)*	The boat was sunk with all hands by the Italian cruiser *Scipione Africano* off Reggio in the Strait of Messina. **Builder:** Elco Bayonne, N.J. **Speed:** 42 knots **Displacement:** 34 tons **Length:** 77' **Beam:** 19' **Engines:** 3 x 1,200 hp Packard, petrol; **Armament:** 2 x 21" torpedo tubes, 1 x 20 mm Cannon, 2-twin .50 cal. MG	Strait of Messina.
1942[585]	July 20 *MTB-339, 342*	Mines laid by the U-66 on 20 July 1942, damaged a U.S. Coast Guard Cutter on July 29, 1942. The MTB 338, 342 were damaged on 2 August 1942, from these mines in Port Castries. **Builder:** Canadian Power Boats Co., 70'	West Indies Lesser Antilles St Lucia Is.

582 Antonen, Harri. Joint Naval Forces On Lake Ladoga In The Summer And Autumn 1942, http://www.geocities.com/finnmilpge/fmp_los_k42.html

583 Fock, *Fast Fighting Boats*, 263.

584 *Warship Supplement 65*, 11.

585 HMS MTB-342 (Motor Torpedo Boat) -Allied Ships hit by U-Boats, http://uboat.net/allies/merchants/ship.html?shipUID=1989

Figure 220. A *MAS* boat on Lake Ladoga in 1942.{Joe: source?}

Figure 221. MTB-451 in convoy.

Figure 222. The "Dog Boats" quickly made a good name for themselves in standing to a good fight with the R and S boats in the Channel. (John Lambert, Lambert-Plans.com)

Date	Craft, Name, Event or Country	Event, Builder and Boat Specifications	City, River, Lake, Sea, Ocean
1942[586]	July 21 3rd SFltl.	The flotilla sank the *TKA 71,* East of Arensburg Island (Oesel Island, Estonian). **Type:** Mixed models	Gulf of Riga Baltic Sea
1942 [587]	July 24 *MTB-601*	The boat blew up in Dover after action in the Dover Strait. **Builder:** Tough Brothers, Teddington Wharf Fairmile "D" **Speed:** 30 knots **Displacement:** 105 tons **Range:** 506 nm at maximum speed **Engines:** 4 x 1,250 hp supercharged V-12 Packards, petrol **Armament:** 2 x 21" torpedo tubes, 1 x 6-pdr. 7 cwt Mk VII (forecastle), 4 x .50-cal. Vickers MG on 2-Mk V mounts, 4-.303-cal. Vickers gas MG in 2-twin mounts, 2 x 20mm Oerlikons on twin Mk V mounts	Dover Strait North Sea
1942[588]	July 27 *MTB-344* "The Little Pisser"	Inducted into the Small Raiding Force (SRF) for clandestine operations on the north and south coast of France. Boat became part of 1st MTB-Flotilla, Newhaven. For obscure reason the 344 was first painted "Ocean Green" and later "Mountbatten Pink." **Builder:** Thorneycroft, Hampton **Speed:** 40 knots **Displacement:** 17 tons **Length:** 60' **Engines:** 2 x 1,300 hp Thorneycroft, petrol **Armament:** 2 x 18" torpedo tubes, 4 x .303-cal. Vickers gas MG	S.E. coast France Mediterranean

---

[586] Fock, *Fast Fighting Boats*, 263.
[587] Reynolds, L.C. *Dog Boats At War; A History of the Operations of the Royal Navy D Class Fairmile Motor Torpedo Boats and Motor Gunboats 1939-1945* (Sutton Publishing, 1999), 273 and Lambert, John *Anatomy of the Ship. The Fairmile 'D' Motor Torpedo Boat*, (Conway Maritime Press, 1985), 12.
[588] Richards, *Secret Flotilla.*

Figure 223. The *344* boat was an experimental Thorneycroft built for use in running agents in and out of France. She carried no torpedoes or depth charges.[589]

[589] Combined Operations http://www.combinedops.com/Op_Aquatint.htm

Date	Craft, Name, Event or Country	Event, Builder and Boat Specifications	City, River, Lake, Sea, Ocean
1942[590]	July 26 3rd SFltl. *S-54*	The S-*54* sank the Soviet destroyer *Smelyi* and other boats of the flotilla sank the PK 238 north of the Gulf of Riga. **Type:** S-*30*	Gulf of Riga Baltic Sea
1942[591]	July 29 MTB Sqd. 2	The squadron was still in the Panama Canal Zone when it was alerted to move to the South Pacific. The 14 PTs were split to form a new "Expendable Squadron" flying the original Squadron 3 colors. Beginning on this date, the first boats, *PT-38, 46, 48* and *60* were loaded onto on Navy oilers, then towed to a release point past Espiritu Santo, from whence they sailed onward to Tulagi, adjacent to Guadalcanal, arriving October 12.. The second group of PTs: *37, 39, 45* and *61* went by freighter to Noumea and then to Tulagi, arriving October 25.	Tulagi Is. Solomon Is.
1942[592]	July 30 MTB Division 1	Four PT boats originally assigned to the South Pacific had their orders changed. Four boats, *PT-22, 24, 27,* and *28* were sent to Alaska by cargo ship. Here the Navy quickly learned that the new PT boats were no match for the rough weather of the Bering Sea. All four were destroyed by weather by January 14, 1943. Most other PTs assigned to Alaska met the same fate.	Aleutian Islands Bering Sea
1942 [593]	August 2/3 *MAS-573, 568, 569*	The three Italian boats attack the Soviet heavy cruiser *Molotov* (*Kiro* class) and the destroyer *Kharkov* (*Leningrad* class). The *568* fired two torpedoes at the cruiser and one hit. The destroyer gave chase, but all the MAS-boats dropped their depth charges in their wake, causing considerable damage to the destroyer and more to the cruiser.	Black Sea

[590] Reynolds, *Dog Boats at War*, 263 and Lambert, *Anatomy of the Ship*, 12.
[591] Polmar, Norman and Samuel Loring Morison, *PT Boats At War*, 53.
[592] Polmar, Norman and Samuel Loring Morison, *PT Boats At War*, 50.
[593] *M.A.S. and Midget Submarines in the Black Sea 1942-1943* http://www.comandosupremo.com/Blacksea.html.

Figure 224. The area of operations around Riga.

Figure 225. The "Expendables" squadron split into two units and elected to carry over the original RON 3 pennant.

Figure 226. The first PT's assigned to Alaska soon learned that their boats were not meant for Alaskan weather. (Library of Congress Print and Photographic Division)

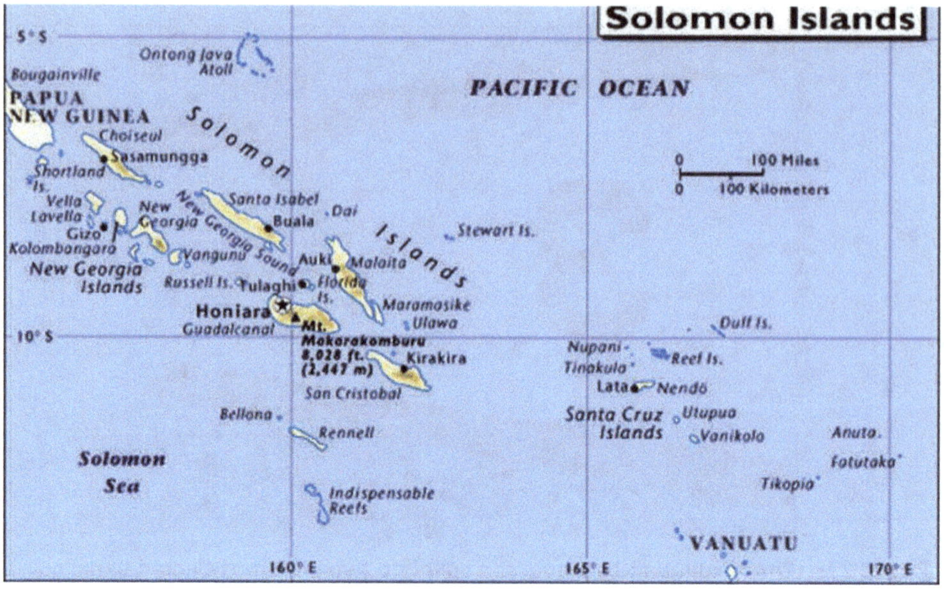

Figure 227 (top and bottom). The map does not begin to show the total complexity of the island chains. There were in fact tens of thousands of small islands that never could be shown on any but the largest scale maps. This is why the MTB squadrons of the U.S.Navy were truly in their element of ideal operations. BIG ships could never go where the PTs went (top: U.S. Naval Historical Department; bottom: PT King)

Figure 228. *MAS-549.* Italian MAS boats and other small craft stuck to limited naval doctrines that proved to be effective. The boats operated only to their advantage and avoided unnecessary gun battles. The method of hit and run worked well in preserving the boats and their crews.[594]

Figure 229. Areas in the central Mediterranean where MAS boats operated.[595]

---

594 http://www.regiamarina.net/arsenals/ships_it/mas_vas_ms/mas_vas_ms_us.htm

595 Map: Gordon Smith www.histori-navy.com Smith

Figure 230. The exposed crews of the Italian MAS boats suffered mightily from aircraft attacks.[596]

Figure 231. A Soviet TKA. (Central Naval Museum, St. Petersburg)

[596] http://www.regiamarina.net/arsenals/ships_it/mas_vas_ms/mas_vas_ms_us.htm

Date	Craft, Name, Event or Country	Event, Builder and Boat Specifications	City, River, Lake, Sea, Ocean
1942[597]	August 3 5th SFltl. *S-112*	This is the first S-boat to be fitted with Rotating Aerial Radar, FUM072. The radar sets, acquired from the Luftwaffe, had uninspiring results. The antennas were far too large and cumbersome to be of any real value for the meager results expected in detection of an enemy destroyer at 10 km. **Type:** S-*38*	Germany
1942[598]	August 8 *MTB-218*	Ran over a mine in a firefight with German surface vessels and sank. **Builder:** Vosper 72'-6"	Dover Straits North Sea
1942[599]	August 10/11 *MGB-335*	The boat was captured before secret documents could be deep sixed and the Germans learned that the English MGB flotillas in the North Sea were composed of twelve MGB flotillas of six boats each. Most important was the disclosure that radar units were being put into service on the MGBs and MTBs. The Germans countered with heavier guns, the 4 cm Flak 28 (Bofors).	North Sea
1942[600]	August 13 *MAS-554, 553, 557*	The British convoy Pedestal, composed of 29 warships and 14 merchant ships, was attacked by an Axis force of 33 warships, a huge air armada, and two Italian Ms Motor Torpedo Boats on the approach to Malta. In three days of attacks, the Allies lost 2 aircraft carriers and their air arms carried, 2 cruisers, a destroyer, 8 merchant ships and damages to other convoy ships. The 2 Italian MTB's, Ms-16 and Ms-22 sank the cruiser HMS Manchester. The Axis naval forces lost 2 ships and 3 suffered some damage.	Mediterranean

597 Whitley, *German Coastal Forces*, 39.
598 *Warship Supplement 65*, 9.
599 Whitley, *German Coastal Forces*, 36.
600 The Supply of Malta, http://www.naval-history.net/xAH-MaltaSupply01.htmOperation Pedestal, http://en.wikipedia.org/wiki/Operation_Pedestal#August_13

Date	Craft, Name, Event or Country	Event, Builder and Boat Specifications	City, River, Lake, Sea, Ocean
1942[601]	August 15 TKA (Unknown)	TKA's are credited with the sinking of the German ferry barge, *MFP F-138A,* near Kazantip Cape. **Type:** *G-5* ,62.5'	Black Sea
1942[602]	August 18 *MTB-43*	Lost in combat with German coastal forces off the North East coast of France. **Builder:** White, Cowes, 72'	North Sea
1942 [603]	August 18 *MTB-237*	Sunk during combat with enemy surface forces off Cape Barfleur. **Builder:** Camper-Nicholson, Vosper 72'-6"	English Channel
1942[604]	August 26 TKA (Unknown)	A German ASW whaler, *UJ 1216* (Star XXI) sank by a TKA torpedo hit off Vikalla Flat near Bol'shoj Tuters Island. **Type:** *G-5,* 62.7'	Barents Sea
1942 [605]	August 28/29 Training SFltl. *S-20, 22, 103*	The training flotilla started poison gas experiments with dim reviews. It was a waste of the S-boat's training time. They tried again in October 6/7 off Bornholm Island, Denmark and the trials were again a waste of time. The crews became too sick to remain on duty after just a few hours exposure to the gas. *S-20, 22, 103* Aug. 28/29; S-24, 50, 103 Oct. 6/7	Off Bornholm Island, Southern Baltic Sea

601 Internet correspondence with Richard Worth, July 12, 2007.
602 *Warships Supplement 65*, 6.
603 *Warships Supplement 65*, 9.
604 German Auxiliary Fast Attack ASW,
http://oceania.pbworks.com/German%20Auxiliary%20ASW%20Fast%20Craft
605 Rohwer, *Chronology of the War At Sea*, 191.

Date	Craft, Name, Event or Country	Event, Builder and Boat Specifications	City, River, Lake, Sea, Ocean
1942[606]	August 29 MAS-228	On the coast of Egypt, in the Daba area, British destroyers HMS *Aldenham, Hurworth, Croome* and HMS *Eridge* bombarded shore targets. *MAS-228* attacked and severely damages the *Eridge,* which is towed to Alexandria and declared a constructive total loss.	Off Daba, Egypt Mediterranean
1942[607]	September 5 1st SFltl S 27, 72	During night operations against the Soviets, the *S-27* was struck and sunk by a torpedo from *S 72*. 12 crewmen were lost with the boat and the rest seriously injured. This included the flotilla surgeon.	Ivan Baba Black Sea
1942[608]	September 9 *MAS-567, 569, 572, 573*	The Italian Naval base at Yalta was attacked by a large force of Soviet fighter-bombers. The attack laid waste to the harbor and many small craft were sunk, including *MAS-571, 573* and a barge. The *MAS-567, 569 and 572* suffered much damage.	Black Sea
1942 [609]	September 12 TKA Div.	The first attack of the TKA's operating from Pumanski Bat base on Fisher Peninsula, made unsuccessful attack on Axis convoy off Petsamo Fjord.	Barents Sea

---

606 HMS Eridge, http://www.naval-history.net/xGM-Chrono-10DE-Eridge.htm.

607 Frank, *German S-Boats*, 94.

608 *M.A.S. and Midget Submarines in the Black Sea 1942-1943* http://www.comandosupremo.com/Blacksea.html.

609 Internet correspondence with Richard Worth , July 12, 2007.

Date	Craft, Name, Event or Country	Event, Builder and Boat Specifications	City, River, Lake, Sea, Ocean
1942 [610]	September 13/14 *MTB-344* "Little Pisser" (Experimental)	As part of the Small Scale Raiding Force (SSRF) the *344*-took11 commandos to the beach near St. Laurent (Omaha Beach, Normandy) with the assignment to bring back a German prisoner, reconnoiter the area for invasion planning purposes and return to England. The men ran into a German ambush: the commando casualties were three-dead, two-seriously wounded, four captured, two unhurt. Of the captured men, two were executed later executed. The 344 boat limped back to England with a damaged transmission. **Builder:** Thorneycroft, Hampton (Experimental) **Speed:** 40 knots **Displacement:** 17 tons **Length:** 60' **Armament:** 4 x Vickers MG (no torpedoes or depth charges carried on mission) **Engines:** 2 x 1,300 hp Thorneycroft, two auxiliary "silent engines" **Crew:** 6, plus 11 commandoes	Normandy Beach English Channel
1942	September 13/14 15th MTB-Flotilla *MTB-308, 310, 312, 314* (ex-*PT-50, 52, 54, 56*)[611]	The boats of the flotilla had an operational agreement with the American PTs for a joint attack on Tobruk and Benghazi. *MTB-308, 310 and 312* were lost in combat and to attacks by Axis aircraft. The grounded *314* became the German *RA 10* after its capture. The boat was sunk by the RAF on September 2, 1943. **Type:** Elco	Tobruk to Beghazi coast North Africa
1942 [612]	October 2 5th SFltl. *S-65, 77, 82, 112*	The flotilla boats sank the armed trawler *HMS Lord Stonehaven* off Eddystone Light. **Type:** *S-38*	SW Rame Head, Eddystone Rocks

---

[610] Stewart, Nigel "Operation Aquatint 12/13 1942...An SSRF Operation" http://www.combinedops.com/Op_Aquatint.htm.
[611] Lambert, John. *Allied Coastal Forces,* Vol.1, 119.
[612] Allied Warships of WWII-ASW Trawler HMS Lord Stinehaven http://uboat.net/allies/warships/ships/6414.html

Figure 232. PT boat crews were almost always on some type of duty. Aside from the engines the next most important job would be working on the guns. (Wikipedia)

Date	Craft, Name, Event or Country	Event, Builder and Boat Specifications	City, River, Lake, Sea, Ocean
1942[613]	October 3 6th SFltl. *S-69, 71, 75*	The three boats sink two merchant ships off Cromer. **Type:** *S-38*	Off Cromer, North Sea
1942[614]	October 6 TKA (Unknown)	TKAs sink the German ferry lighter *MFP F-131A* near Anapa on the Northern Black Sea Coast near the Sea of Azov. Type *G-5,* 62.5'	Black Sea
1942[615]	October 14 MTB-236	The German raider *Komet* headed out for its second cruise and took the risky English Channel route. British destroyers and MTBs that were lying off Cherbourg gave chase and the *MTB-236* sank the raider. **Builder:** Camper Nicholson,Vosper 72'-6"	English Channel
1942[616]	October 14 6th SFltl. *S-69, 71, 75*	Once again, this trio of S-boats sank two merchant ships off Cromer. **Type:** *S-38*	Off Cromer North Sea
1942[617]	October 14 RON 3 *PT-46, 48, 60, 38*	The boats of RON 3 have a good shooting night when they sink a IJN light-cruiser that is part of the "Tokyo Express" running supplies to Guadalcanal. **Type:** Elco *PT-20* class, 77'	Off Guadalcanal Solomon Islands Coral Sea

[613] Hümmelchen, *German Schnellboote*, 169.
[614] Internet correspondence with Richard Worth, July 12, 2007.
[615] Coastal Forces and Special Units http://www.naval-history.net/WW2CampaignsCoastal-SpecialWarfare.htm, 5
[616] Hümmelchen, *German Schnellboote*, 169.
[617] Keating, *The Mosquito Fleet*, 18-20.

Figure 233. *PT-46* (National Archives).

Date	Craft, Name, Event or Country	Event, Builder and Boat Specifications	City, River, Lake, Sea, Ocean
1942[618]	October 26 *MTB-620, 623* (RNoR)	In the harbor of Askvoll, Norwegian coast, south of Flore, the *620* sank the German supply ships *Harvestehude* and the 623 sank the *Hertha.* Both ships were at anchor. Fairmile "D", 115' **Builder:** *620* Dickie, Bangor & Tarbert, *623* Wallasea Bay Yacht Station **Speed:** 30 knots **Range:** 600 nm at 18 knots **Crew:** 26 **Engines:** 4 x 1,250 hp Packard, petrol **Armament:** twin 20 mm Oerlikon AA, twin Colt 12.5 mm MG, 6-7.6 MG, Vickers "Pom-Pom", 2 x 6-pdr cannon	North Sea
1942[619]	October 28 TKA	The ferry lighter *MFP F-470C,* while trying to avoid attacking TKAs, beached herself near Anapa and was later salvaged. **Type:** *G-5,* 62.5'	Black Sea
1942[620]	October 31 *MTB-87*	She was lost when she struck a mine in the North Sea **Builder:** Vosper, 72'-6"	North Sea
1942[621]	November 19 5th SFltl. *S-68, 77, 82, 112, 115, 116*	The flotilla attacked an Allied convoy off Eddystone Light, sinking three freighters, and the armed Trawler, HMS *Ullswater.* **Type:** *S-38*	SW Rame Head, Eddystone Rocks, English Channel

618 The Norwegian MTB Flotilla in Shetland http://shetlopedia.com/The_Norwegian_in_Shetland, 2.
619 Internet correspondence with Richard Worth, July 12, 2007.
620 *Warship Supplemental* 65, 7.
621 Hümmelchen, *German Schnellboote*, 169.

Figure 234. The main service weapon of the PT boats was the .50 cal machine gun. It was a heavy caliber and required constant maintaince. The barrels could burn out quickly in a major fire-fight and needed to be changed frequently. Each boat carried several spare barrels for the 50's. (National Archives)

Figure 235. A boat gunner in *Life* Magazine, 1941. (LIbrary of Congress Print and Photographic Division).

Figure 236. The *MAS* boat crews had the advantage of years of training and a navy with a strong naval doctrine in the use of MTBs. The Italians had been using and perfecting the MTB since 1915.[622]

[622] Commando Supremo http://www.comandosupremo.com/

Date	Craft, Name, Event or Country	Event, Builder and Boat Specifications	City, River, Lake, Sea, Ocean
1942[623]	November 23 *MAS-543, 547*	The Italian transport Numidia, loaded with 1,100 Allied POWs, leaves Tripoli for Palermo, Sicily, escorted by the torpedo boat *Ardito*, two MAS-and two MS boats. During transit, British fighter planes attack the transport, leaving 2 dead and 11 wounded.	Mediterranean
1942[624]	December 3 *S-115*	The *115* boat found the destroyer HMS *Penylan* amidst the convoy PW 257 off Start Point and sank her with torpedoes.	English Channel
1942[625]	December 9 *PT-59*	While laying in wait to machine-gun Japanese barges in the shallows near Cape Esperance, Lieutenant Jack Searles and his crew spotted a Japanese submarine approaching shore. It was being unloaded by barges and trapped in the bay. The 59 boat fired two torpedoes which both struck amidships, sinking the submarine in seconds. The sub turned out to be the 2,180-ton blockade runner, *I-3*. **Type:** Elco *PT-20*-class, 77'	Off Guadalcanal Solomon Sea
1942[626]	December 11 *PT-44, 110*	The two PT boats laid in waiting for enemy ships to make another trip down The Slot. The two boats attacked 3 IJN destroyers and sank the destroyer *Teruitsuki*. The *44* boat was chewed up in a gunfight with the other two destroyers and burned, then sank. Two men survived and nine were lost. **Type:** 44-Elco *PT-20* class, 77'; 110: *PT-103* class, 80'	The Slot Solomon Sea

623 Rohwer, *Chronology of the War At Sea*, 214.
624 "Coastal Forces and Special Units" http://www.naval-history.net/WW2CampaignsCoastal-SpecialWarfare.htm.
625 *United States Submarine Losses World War II* (Reissued with an Appendix of Axis Submarine Losses, fully indexed. Naval History Division Office of the Chief of Naval Operations, Washington, 1963), 175, 217.
626 Polmar and Morison, *PT Boats At War*, 56.

Figure 237. The PT boats that went out to take on the capital ships of the "Tokyo Express" were the only vessels available for fighting in tight spaces. They had tremendous advantage in their speed but more importantly their shallow draft.

Figure 238. Crew of *PT-59.*

Figure 239. The men of the first MTB squadrons in the Pacific learned by doing and making mistakes, while accomplishing much. The U.S. Navy never developed a fighting doctrine for its PTs until 1942. The U.S. Navy had envisioned the PT boat as a harbor protection vessel, mail boat, scout boat and not much more. (Library of Congress Print and Photographic Division)

Date	Craft, Name, Event or Country	Event, Builder and Boat Specifications	City, River, Lake, Sea, Ocean
1942 [627]	December 10 *PT-36, 37, 40, 43, 44, 48, 50, 109*	In a desperate move by the US command, they sent 8 PT-boats *(PT-36, 37, 43, 40, 44, 48, 50* and *109)* to turn back the night's "Tokyo Express" of 8 destroyers. The PTs' aggressive maneuvers, wild machine gun firing, and many torpedo launches unnerved the Japanese flotilla commander and the destroyers retreated back up the Slot.	"The Slot" Guadalcanal Coral Sea
1942[628]	December 12 *PT-44*	This night the Tokyo Express made another run to drop supplies and men for their Guadalcanal garrison. The *44* boat managed to put two torpedos into IJN destroyer *Teruzuki* just before her shells from the destroyed the 44 boat. Two crew members survived and nine died. **Type:** Elco *PT-20* class, 77'	Off Savo Island Guadalcanal Is., Coral Sea
1942[629]	December 12 4th SFltl. *S-48, 63, 80, 110, 117*	The flotilla had one of its best nights; they sank five Allied freighters off Lowestoft.	Off Lowestoft North Sea
1942[630]	December 18 *MTB-30*	*MTB-30* ran over a mine in the North Sea and sank. **Builder:** Vosper 71'-3/4"	North Sea
1942[631]	December 24 PT-122	On Christmas Eve, off Buna Island New Guinea, an IJN submarine, *I-4 (Ueno)* was spotted. The *122* boat fired two torpedoes at 1,000 yards, and minutes later, two more at 500 yards. Shortly afterwards, the submarine blew up in one massive explosion. A second submarine fired at *122* and missed. The *122* boat was given credit for one kill. **Type:** Elco PT-103 class, 80'	Off Buna Battlefield, Guadalcanal Coral Sea
1942[632]	December *TKA-86*	The *TKA-86,* class *G-5,* received a rocket launcher for the new ROFS-82' missiles. The naval launch system had 12-missile rails based on the Army version. The *TKA-86* was the first missile boat of the Soviet navy posted near Novorossiysk.	Black Sea

627 "This Day in Naval History – Dec 10" http://www/news.navy.mil/search/display_history.asp?story_id=281.
628 Keating, *The Mosquito Fleet*, 40 and "This Day in Naval History-Dec. 08" http://www.news.navy.mil/search/display_history.asp?story_id=281.
629 Rohwer, *Chronology of the War At Sea*, 218.
630 "British RN, MGB, MTB, SGB, ML," http;//www.naval-history.netWW2BritishLosses3Coastal.htm.
631 Keating, *The Mosquito Fleet*, 56. and United States Submarine Losses, Naval History Division, 1963, 175
632 Przemyslaw, Budzbon. *The Beginnings of Soviet Naval Power.* (Naval Institute Press, 1977), 240.

The Ship Killers

The Definitive Illustrated History of the Torpedo Boat

Volume VII: Timeline 1943

The crew of a Soviet G-5 boat.

By Joe Hinds

Nimble Books LLC

CONTENTS (INCLUDING FIGURES)

DEDICATION

To Michael Emmerich.

http://www.german-navy.de/smb/rcmodels/index.html

I wanted to put Michael's website right at the top of the page because it is such a-well made site. It appeals to the historian, to the scale model builder and to people who are never satisfied with a little bit of information. Michael has helped me on everything German for my book series. I find information in books, famous and obscure, and I can always count on him to be the third source of verification.

He has traded a few hundred emails over the years and I am richer for it and the end results of his accuracy show up in my books.

TIMELINE 1943

Date	Craft, Name, Event or Country	Event, Builder and Boat Specifications	City, River, Lake, Sea, Ocean
1943[633]	January 1 *S-58*	She sank the HMS *Horatio*, minesweeper, off the North African coast. **Type:** *S-30*	Mediterranean
1943[634]	January 1 *MTB 105*	The *105* was taken on board as an LSI or Landing Ship Infantry. Her small size of 45' made the boat a perfect, immediate small water-taxi, submarine hunter, and landing craft for secret operations. Taken in tow by an escort vessel and later scuttled. **Builder:** Thorneycroft 45'	North Sea
1943[635]	January 2 30th MTB Flotilla *MTB-618, 623, 625, 626, 627, 631*	In Operation Cartoon the flotilla made a three pronged attack, with 7 MTBs, against German forces defending a mining facility on the island of Stord. The *618* and *623* boats made a diversionary attack against a German lookout station 25 miles north of Stord. The *620, 625*, and *631* made their way in the fjord and set fire to a herring oil factory, laid mines in the main channel, and shot up and grounded the merchant vessel SS *Isle L.M. Russ.* The primary assault was made by 2 MTBs (*626* & *627*) with 50 Norwegian and 50 British commandos. These troops destroyed machinery, facilities, stores and gun emplacements. In their return to base the *625* and *631* shot down a German Ju-88 bomber. **Builder:** Fairmile "D", 115' *618* Harris, *623* Wallasea, *620* Curtis, *624 & 625* Robertson, *626* Tough Brothers , *631* King.	Norwegian Sea
1943[636]	January 6 PT-*22, 24*	Damaged in collision during rough weather in Dora Harbor, Unimak Island. **Type:** Elco *PT-20* class 77'	Aleutians Bering Sea

633 Hümmelchen, *German Schnellboote*, 167.
634 *Warship Supplement* 65, 8 and Richards, *Secret Flotillas*, 363-369.
635 Reynolds, L.C. *Dog Boats At War: A History of Operations of the Royal Navy D Class Fairmile Motor Torpedo Boat and Motor Gunboats 1939-1945 (Sutton Publishing, 1999)*, 37.
636 Rohwer, *Chronology of the War At Sea*, 224.

Figure 240. New crewmen were trained in all weapons carried by the S-boats.[637]

Figure 241. The SS *Fidelity* was a "Q" ship or spy ship used for several clandestine operations. She carried her own seaplanes and MTB force along with British and French commandos.[638]

---

[637] Busch, *Unsere Schnellboote im Kanal*, 53.
[638] http://www.uboat.net/allies/merchants/ship.html?shipID=2560.

Figure 242. Mines and mine technology were prominent items in the German war chest.[639]

[639] Frank, *German S-Boats in Action,* 154.

Date	Craft, Name, Event or Country	Event, Builder and Boat Specifications	City, River, Lake, Sea, Ocean
1943[640]	January 9 S-*104*	An RAF mine laying operation started with great intensity against the coast of France, Holland, Bay of Biscay. The *S-104* hit an aerial laid mine in the English Channel and sinks. **Type:** *S-38*	English Channel
1943[641]	January 10 *PT*-27, *29*	Suffered severe storm damage in Dora Harbor, Unimak Island. **Type:** Elco *PT-20* class 77'	Aleutians Bering Sea
1943[642]	January 10/11 *PT-112, 43*	Japanese destroyers were running supplies and reinforcements to Guadalcanal through a route called "The Slot." *PT-45, 39, 48, 115, 112, 43, 40, 59, 46,* and *36* were set up in groups of three along the route. Six boats fired 21 torpedoes and did some damage to the enemy destroyer *Hatsukaze*. *PT-112* and *PT-43* were sunk by Japanese naval gunfire. PT *112*, **Type**: Elco *PT-103* class, 80'. *PT-43* **Type**: Elco *PT-20* class, 77'	'The Slot' Solomon Islands Coral Sea
1943[643]	January 11 *PT-112*	The *112* was sunk by gunfire from an IJN destroyer off Guadalcanal. **Type:** Elco *PT-103* Class, 80'	Solomon Sea
1943[644]	January 12 *PT*-28	Ran aground in Dora Harbor, Adak Island, and abandoned on January 14th. **Type:** Elco *PT-20* class 77'	Aleutians Bering Sea
1943[645]	January 14/15 PT Squadrons	Nine Japanese destroyers made another supply run down the Slot and were met by US Marine Corps SBD Dive-Bombers and 13 PT boats. Seven PTs fired 17 torpedoes and registered no hits. The boats in the fight were: *PT-36, 37, 38, 39, 40, 45, 46, 47, 48, 59, 109, 115, and 123*. Many of the PT's were shot up, and of the 17 torpedoes fired no hits were substantiated. **Type:** PT 20 Class 77' and PT 103 Class 80'	Guadalcanal Solomon Islands Coral Sea

640 Rohwer, *Chronology of the War At Sea*, 224.
641 HyperWar: *US Navy Torpedo Boats (PT-Boats), 1940-1945* http://www.ibiblio.org./hyperwar/USN/ships/ships-py.html , 2.
642 Rohwer, *Chronology of the War At Sea*, 224.
643 Lambert, *Allied Coastal Forces*, Vol. 2, 140.
644 HyperWar: *US Navy Torpedo Boats* (PT-Boats), 2.
645 Rohwer, *Chronology of the War At Sea*, 224.

Figure 243. Higgins boats and a tender.

Figure 244. The shot-up remains of PT-43 show the grim outcome when little boats take on ships with big guns.[646]

---

[646] http://www.pt171.org/PT171/index.php?date=1944-12-03.

Figure 245. As news of the inhumane treatment of American and Allied POW's spread war posters took on a tone of personal vengeance towards the Japanese. (National Archives)

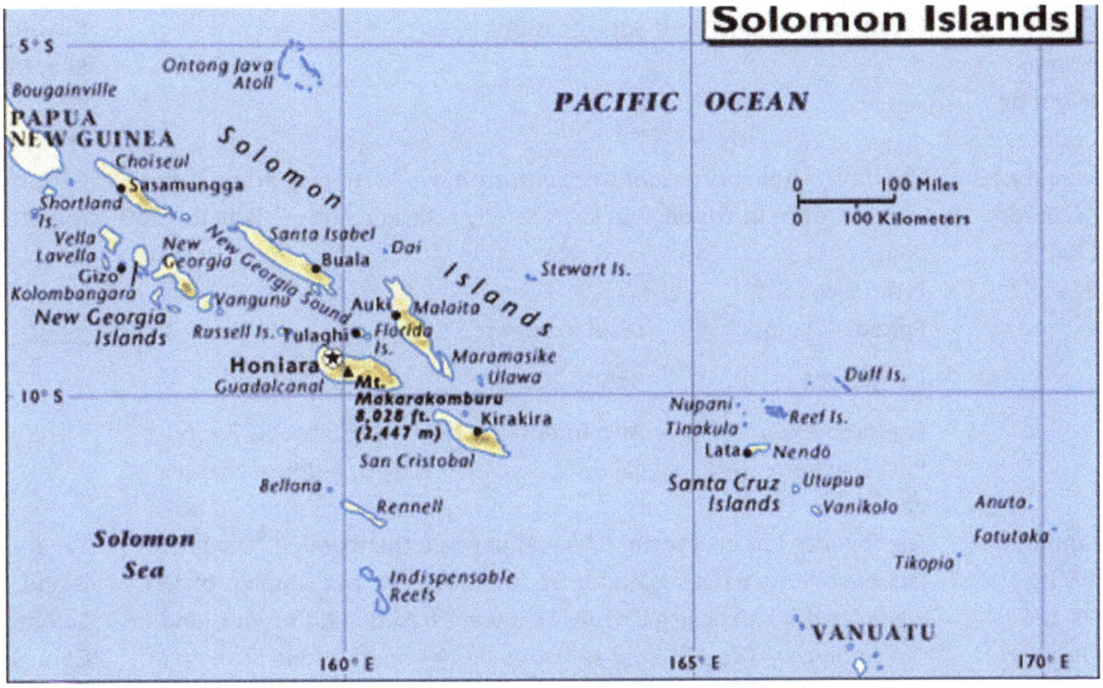

Figure 246. The infamous "Slot" ran from Guadalcanal and between Santa Isabel and the New Georgia Group and Choiseul..[647]

Figure 247. PT-120, "the Roaring Twenty." (National Archives)

---

[647] http://www.ibiblio.org/hyperwar/USMC/USMC-C-NSol/index.html.

Date	Craft, Name, Event or Country	Event, Builder and Boat Specifications	City, River, Lake, Sea, Ocean
1943[648]	January 19 *MTB-260, 264, 313*	The Italian submarine *Santorre Santarosa*, while running ammunition to Axis troops in Tripoli, was successfully attacked and sunk by the *260* boat. **Type:** Elco , 81' **Speed:** 42 knots **Displacement:** 32 tons **Length:** 81' **Beam:** 20' **Engines:** 3 x 2,700 hp Packard, petrol (AV gas) **Armament:** 2 x 18" torpedo tubes, 2 x .50 cal. MG, 6 x .303-cal. MG, 4-depth charges	Off Libyan coast Mediterranean
1943[649]	January 17/18 *PT-120* "Roaring Twenty"	The favorite targets for the *PT boats* became the wooden "Daihatsu" Japanese Army barges used for troop movement and supplies of all types for the Japanese garrisons that were based in different island in the Solomons. The *120* boat sank two barges and left one burning off Sananandа Island. **Type:** Elco *PT-103* class,80'	Sananandа Island Solomon Islands Coral Sea
1943[650]	February 1-2 *PT-123*	The *123* had the distinction of being wiped out by an early-model IJN glide bomb near Savo Island, off Guadalcanal. Four crewmen died. **Type:** Elco *PT-103* class, 80'	Solomon Sea
1943[651]	February 1 *PT-128*	Japanese aircraft off Guadalcanal destroyed the *PT-128*. **Type:** Elco *PT-103* class 80'	Guadalcanal Coral Sea
1943[652]	February 1 *PT-111*	The *111* boat was destroyed by the Japanese destroyer *Kawakaze* off Guadalcanal. **Type:** Elco *PT-103* class 80'	Guadalcanal Coral Sea
1943[653]	February 1 *PT-37*	A Japanese destroyer sank the *37* boat off Cape Esperance. **Type:** Elco *PT-20* class 77'	Guadalcanal Coral Sea

[648] *United States Submarine Losses World War II*, 179, 210.
[649] Keating, *The Mosquito Fleet*, 45, 58.
[650] Keating, *The Mosquito Fleet*, 47.
[651] Lambert, *Allied Coastal Forces Vol.* 2, 140.
[652] Lambert, *Allied Coastal Forces Vol.* 2, 140.
[653] Lambert, *Allied Coastal Forces Vol.* 2, 121.

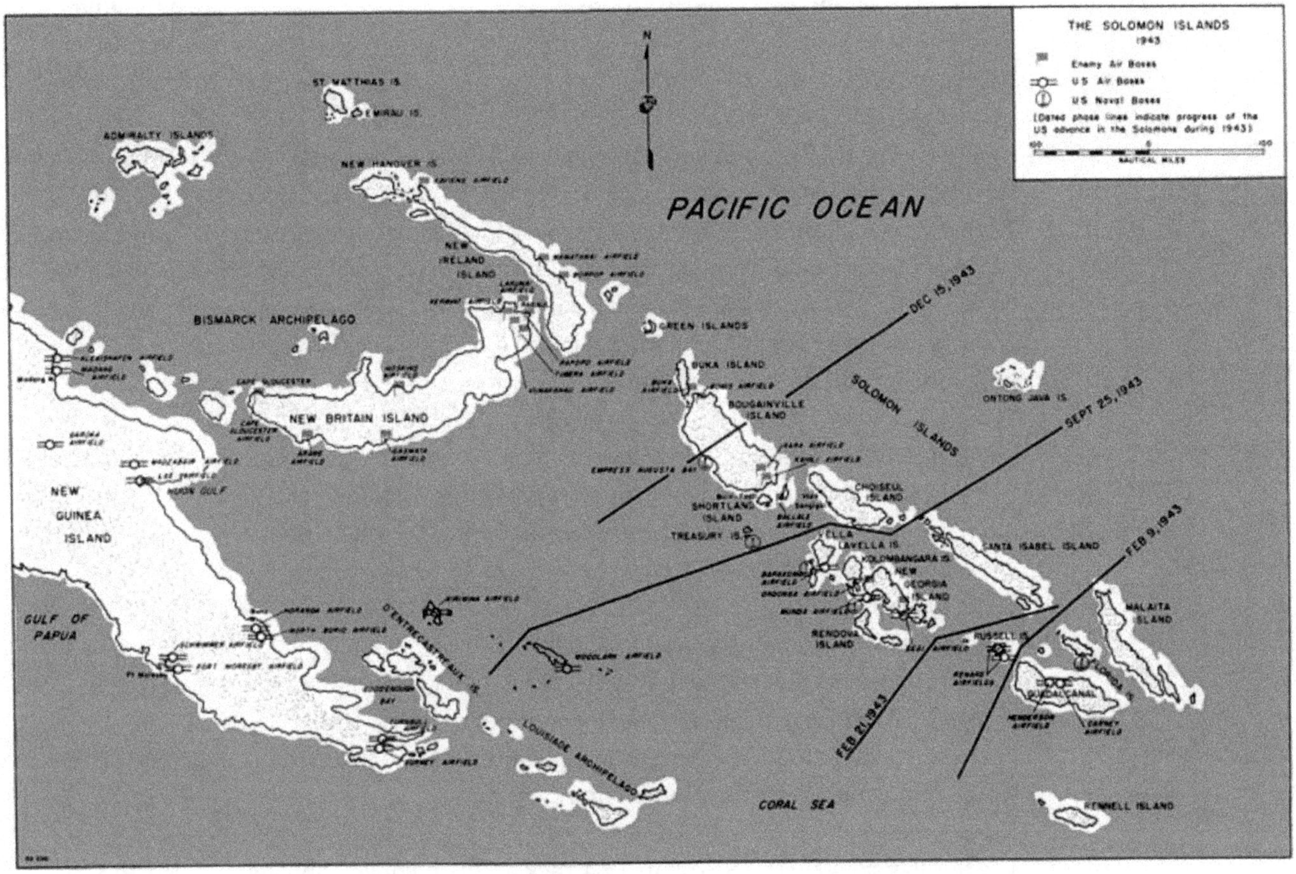

Figure 248. The Japanese were widely dispersed over the Southwest Pacific as 1943 began. Their situation quickly began to unravel. (National Archives)

Date	Craft, Name, Event or Country	Event, Builder and Boat Specifications	City, River, Lake, Sea, Ocean
1943 [654]	February 1/2 "The Slot" 11 PT boats	American minelayers sprinkled 300 mines north of Guadalcanal off Savo Island. Eleven PT boats (*36, 37, 39, 47, 48, 59, 109, 111, 115, 123, 124*) attacked 20 IJN DD's. The IJN destroyer Mekigumo was severely damaged by a mine or torpedo from a PT boat. The ship was scuttled on the spot. Two more IJN DD's and a submarine were also damaged. The PTs lost three boats, (*37, 115, 123*) with nine men killed, seven men missing and six seriously injured. **Type:** *PT 20* Class 77' and *PT 103* Class, 80'	Off Savo Island The Slot Coral Sea
1943[655]	February 15/16 *MTB-61, 77, 82*	The British MTBs attacked a four-ship convoy escorted by the Italian torpedo ships *Monsone* and *Sirio* and the corvettes *Gabbiano* and *Antilope*, south of Marettimo Island west of Sicily. The British boats were unsuccessful this night. **Type:** Vosper, 72'-6"	Strait of Sicily Mediterranean
1943[656]	February 17/18 2nd, 4th, 6th SFltl. S-71	The flotillas were working on mining operations southeast of Lowestoft and northeast of Great Yarmouth. The flotillas are attacked by the British corvette HMS *Kittiwake* and the destroyers HMS *Montrose* and HMS *Garth.* The *Garth* shelled the *71* boat until it stopped moving, and then rammed and sank her. **Type***: S-38*	North Sea
19[657]	February 22/23 2nd, 8th SFltl.	The flotillas sailed with 15 S-boats to attack a British convoy and were driven off by the destroyers HMS *Garth* and HMS *Southdown.* The *S-94* and S-128 collided with each other and were lost.	North Sea
1943[658]	February 24 *MTB 262* (ex-*PT-13)*	Bombed by enemy aircraft off Cape Serratt, the *262* was sunk. **Builder:** Elco *PT*-10 Class, 70'	Northern Tunisia Mediterranean
1943[659]	February 26/28 5th *S*-Boat Flotilla	The four boats of the flotilla, *S-65, 68, 81,* and *85,* attacked a British convoy in Lyme Bay, on the South-Eastern coast of England, and sank the motor vessel *Modavia, LCT-381*, and the armed trawlers *Harstead* and *Hailsham.*	English Channel

---

[654] Bulkley, *At Close Quarters*, 103.
[655] Rohwer, *Chronology of the War At Sea*, 231.
[656] Rohwer, *Chronology of the War At Sea*, 233
[657] Whitley, *German Coastal Forces*, 69
[658] Lambert, *Allied Coastal Forces*, 110
[659] Rohwer, *Chronology of the War At Sea*, 234.

Figure 249. *MTB 234.* (Illustration by author)

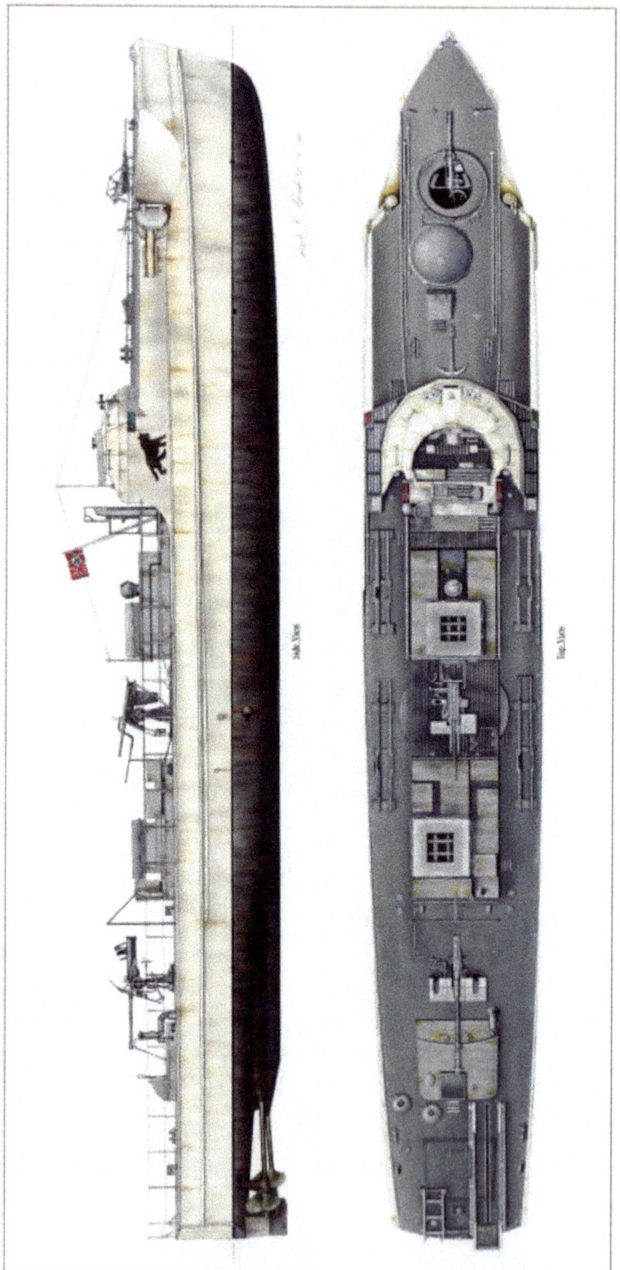

Figure 250. The S-boat type *S-100* was the culmination for years of engineering and naval architecture. The *S-100* boats were not to be matched any other MTB afloat with the exception of the British "Dog Boats." Even then, the S-boats superior speed and cruising range remained uncontested. (Illustration by author)

Figure 251. Palermo was a much-needed repair facility for the S-boats. Any damage to this port put the entire Mediterranean S-boat operation into jeopardy.[660]

[660] Frank, *German S-Boats At War,* 158.

Figure 252. The many S boat flotillas carried a tradition of having very distinctive badges that they wore on their uniforms and their boats. The devices of the 2nd (top) and 6th SFltl (bottom) were among the most colorful.[661]

[661] Frank, *German S-Boats at War,* 168.

Date	**Craft, Name, Event or Country**	**Event, Builder and Boat Specifications**	**City, River, Lake, Sea, Ocean**
1943[662]	February 28 *S-35*	*S-35* was sunk by a mine North-West of Bizerta, all hands lost. **Type**: *S-10*	Off Tunisia Coast Mediterranean
1943[663]	March 1 *S-56*	The boat was lost in an Allied air raid of fighter-bombers by a bomb in Palermo Harbor. **Type:** S-30	Sicily Tyrrhenian Sea
1943[664]	March 2 *PT-143, 150*	Just off New Guinea in the Bismarck Sea, the Japanese transport *Ogawa Maru* suffered an attack by modified B-25's from Australia and was severely damaged. The *143* and *150* boats attacked her with one torpedo each and sank her. **Type:** Elco *PT-103* class 80'	Bismarck Sea South Pacific
1943[665]	March 4 2nd & 6th SFltl. *S-70, 74, 75*	The *S-70*, during a night transit, ran into a British mine and sank with all hands. Two boats of the 6th SFltl, *S-74* and *S-75,* were attacked by two waves of Spitfire fighters. The *74* boat sustained serious damage and the *75* boat burned in a fire and had to be blown up. All three boats sustained many dead and wounded sailors. *S-74, 75:* **Type:** *S-38*	English Channel
1943[666]	March 3/4 *PT-66, 67, 68, 121, 128, 143, 150*	In the aftermath of the Battle of the Bismarck Sea, the battle waters were littered with the flotsam and jetsam of the ships who died. Mixed in with this wreckage were thousands of Japanese soldiers and sailors trying to swim to nearby islands. Orders from the Navy high command were for the PTs to prevent these potential reinforcements from reaching shore. Australian aircraft and PT boats were ordered to wipe out the survivors. **Type:** Elco *PT*-103, 80'	Bismarck Sea
1943[667]	March 7/8 *S-114, 119*	*S-114* and *119* collided. The *114* boat rescued the crew of the *119* and then scuttled her. *S 114, 119:*Type *S*-38	North Sea

[662] Rohwer, *Chronology of the War At Sea*.235.
[663] Frank, *German S-Boats in Action in the Second World War* (Naval Institute Press, 2007), 83.
[664] Keating, *The Mosquito Fleet*, 62.
[665] Frank, *German S-Boats In Action*, 56.
[666] Frank, *German S-Boats In Action*, .57.
[667] Rohwer, *Chronology of the War At Sea*, 236.

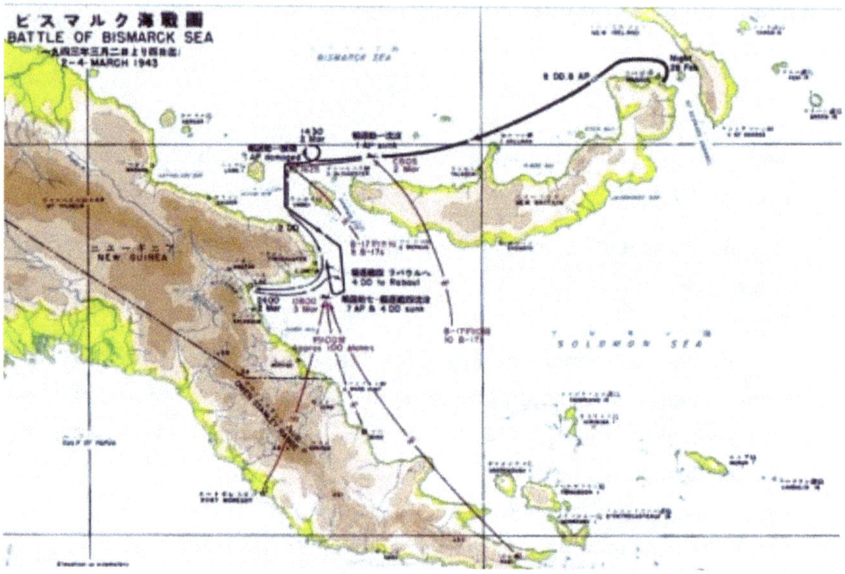

Figure 253. PT boats saw heavy action in the Battle of the Bismarck Sea.[668]

Figure 254. The use of semaphore signals never lost their usefulness during the war in spite of the increased quality of marine radios.[669]

[668] http://en.wikipedia.org/wiki/File:Battle_of_the_Bismark_Sea.jpg.
[669] Frank, *German S-Boats at War*, 145.

Figure 255. Beaching an Elco boat. The PT boat retained its glamour throughout World War II. (National Archives)

Date	Craft, Name, Event or Country	Event, Builder and Boat Specifications	City, River, Lake, Sea, Ocean
1943[670]	March 10 *MTB 622 "D"*	A German destroyer off Terschelling Island sinks her. **Builder:** Brooke Marine, Fairmile "D" 115' **Speed:** 30 knots **Displacement:** 108 tons **Length:** 115' **Beam:** 31'3" **Engines:** 4-Packard 4, 800 bhp **Armament:** 2 x 6-pdr (single mounts) 3-20mm (one twin, one single mount) 4 x .50 cal. MG, 4 x.303 MG, 2 x 21" torpedo tubes	Netherlands North Sea
1943[671]	March 12 S-55	The 55 boat sank the British DD, HMS *Lightning,* in the Sicilian Straits between Sicily and Tunisia. **Type:** ***S***-30	Mediterranean
1943[672]	March 12/13 *MTB 617, 624, 628*	Near Terschelling Island, the three boats attacked a three-ship German convoy with several escort vessels. They sank two ships with torpedoes, the *Liege* and *Hermond*. **Builder:** *617* Osborne, Littlehampton; *624* Dorset Yacht Co., Hamworthy; *628*:-Upham, Brixham	North Sea
1943[673]	March 13/14 *MTB-631* "D"	The Royal Norwegian "D" boat was stranded on the coast of Norway and captured by the Germans. The boat was inducted into the Kriegsmarine and becomes, S-631. **Builder:** King, Burnham on Crouch Fairmile "D" 115'	Norway North Sea

---

670 *Warship* Supplement 65, 18.

671 Hümmelchen,*German Schnellboote*, 167.

672 Reynolds, L.C. *Dog Boats At War*, 25.

673 *Warship Supplement* 65, 18; Reynolds, *L.C. Dog Boats At War*, 39 and Lambert, *The Fairmile 'D' Motor Torpedo Boat*, 11.

1943[674]	March 13/14 *MTB-619, 631*	The *619* and *631* waited for convoy targets off the port of Floro. That night the boats moved into the harbor and *619* sank the German freighter *Optima* with two torpedoes. The *631* became grounded and the crew was rescued by the 619. **Builder:** Dorset Yacht Co., Hamworthy Fairmile "D" 115'	Norway North Sea
1943[675]	March 15/16 *PT-114, 129*	The two boats successfully ambushed and destroyed six Japanese barges loaded with men and supplies in Mia-Ama Bay, Gulf of Huon. **Type:** Elco *PT-103* class 80'	Papua New Guinea Solomon Sea
1943[676]	March 17 *PT-119, 63*	These boats were destroyed by fire in Port Tufi, New Guinea because of a refueling accident. *PT-119,* **Type:** Elco *PT-103* class 80'; *PT-63,* **Type:** Elco *PT-20* class.	Coral Sea
1943[677]	March 22/23 *MTB-265, 311*	The two boats torpedoed and sank a German "F Lighter off Plane Island, Tunisia. *MTB-265,***Type:** Elco 81', *MTB-311,* **Type:** Elco 77'	Tunisia Mediterranean
1943[678]	March 22/23 *MTB-629, 628*	The two Dog boats moved in to reinforce a contingent of short MTBs that were attacking a German convoy off Terschelling. All were driven off by the escort's gunfire. The two Dog boats were slightly damaged but it later proved that they severely damaged an armed trawler. **Builder:** *629* Dickie; *628* Upham, **Type:** Fairmile "D", 115'	North Sea

Figure 256. Security concerns were very real.

674 Reynolds, L.C. *Dog Boats At War,* 39.
675 Keating, *The Mosquito Fleet,* 65-66.
676 Keating, *The Mosquito Fleet,* 65- 66 and Lambert, *Allied Coastal Forces,* 140.
677 Lambert, *Allied Coastal Forces,* Vol. 2, 110.
678 Reynolds, *Dog Boats At War,* 25.

Figure 257. PT boat on the run (National Archives).

Figure 258. The P-40 aircraft was made famous in China and continued to be a fighter pilot's favorite airplane.[679] Unfortunately, it proved effective against its own side's PT boats on March 27, 1943 (see entry for that date).

[679] http://en.wikipedia.org/wiki/Australia

Figure 259. There have been small boat sailors in every navy in every era. (National Archives)

Date	Craft, Name, Event or Country	Event, Builder and Boat Specifications	City, River, Lake, Sea, Ocean
1943[680]	March 28/29 2nd SFltl. *S-29*	Seven flotilla boats started out to attack an Allied convoy but were driven off by *MGB-321, MGB-333,* and the British destroyers HMS *Blencathra* and *Windsor.* The S-29 was damaged in a collision and abandoned. **Type:** S-14	North Sea
1943[681]	April 1 British MTBs	British MTBs off Terschelling, Holland, sank a German Patrol Trawler.	North Sea
1943[682]	April 1 *MTB-266, 311*	The boats torpedoed and sank two freighters off Cani Rocks, Tunisia. *MTB-266,* **Type:** Elco 81'; *MTB-311,* **Type:** Elco 77'	Tunisia Mediterranean
1943[683]	April 2 *MTB 267* *(ex-PT-18)*	The craft became damaged due to heavy weather and was sunk by depth charges and gunfire from MTB-13. *267* Elco, 81'; *13* British Power Boat Co., 60'	Mediterranean
1943[684]	April 2 MTB 63, 64	Both lost in collision off Benghazi, Libya. **Builder:** Vosper, Portsmouth & Portchester, 72'-6"	Libyan coast Mediterranean

Figure 260. The Mediterranean Seas were a new experience for the MTB crews.[685]

680 Rohwer, *Chronology of the War At Sea*, 241.
681 Rohwer, *Chronology of the War At Sea*, 242.
682 Lambert, *Allied Coastal Forces*,Vol. 2, 110.
683 Lambert, *Allied Coastal Forces*, 114.
684 *Warships Supplement 65*, 7.

Figure 261. An S-boat from 2nd Schelleboote Flotilla.[686]

Date	Craft, Name, Event or Country	Event, Builder and Boat Specifications	City, River, Lake, Sea, Ocean

[685] http://wpcontent.answers.com/wikipedia/commons/thumb/a/a1/Britishmtb.jpg/300px-Britishmtb.jpg.
[686] Frank, *German S-Boats at War,* 156.

1943[687]	April 10 *MAS-501, 503*	A flight of American B-24s attacked the Italian Naval base at La Maddalena, Sicily and during the raid the *MAS 501* and *MAS 503* were sunk along with the heavy cruiser *Trieste*. The cruiser *Gorzia* was damaged. **Type:** Baglietto **Speed:** 48.5 knots maximum · **Displacement:** 22 tons · **Range:** 400 nm at 42 knots, 1,000 nm at 6 knots **Length:** 55.8′ · **Beam:** 14.4′ · **Draft:** 3.9′ **Engines:** : 2 x 2,000 hp Isotta-Fraschini, petrol ;auxiliary engines: 2 x 80 hp, petrol CRRO D8 **Armament:** 2 x 45.7 cm side dropping torpedo tubes; 1 x 13.2 mm MG, 6 x50 kg depth charges **Crew:** 10	Mediterranean
1943[688]	April 13/14 5[Th] SFltl.	Off the Lizard, the flotilla's boats *S-65, 81, 82, 90, 112, 116,* and *121* attacked the British convoy PW 323, which comprises six freighters, two destroyers, and three armed trawlers. *S-90* torpedoed the *Eskdale* and the ship is later finished off by the *S-112*. *S-65* and S-121 torpedoed the freighter *Stanlake* and *S*-90 and *S*-82 finished it off.	North Sea English Channel
1943[689]	April 14 *MTB-640 MGB-644*	Off the North African coast near Bizerta, the two British boats attacked 3 S-boats, causing severe damage to one and some damage to the other two. **Builder:** *640* Wallasea Bay Yacht Co; *644* Tough Brothers Ltd. Fairmile "D," 115′	Mediterranean
1943[690]	April 14/15 2nd, 6th SFltl.	The British destroyer HMS *Westminster* and corvette, HMS *Widgeon* attacked the two flotillas. The S-83 received minor hits in the firefight. Torpedoes sank the British armed trawler HMS *Adonis.* **Type:** *S 38*	North Sea English Channel
1943[691]	April 17 1st SFltl. , IV MAS Flotilla, 3rd MM Flotilla	Operation *Neptun* was the operational plan for the three flotillas to break the Soviet Myschako bridgehead. The nightly operations resulted in the sinking of many Soviet barges, small vessels, and pontoons. The *S-47, 51, 102, 72,* and *28* were credited with many kills. These engagements included many Soviet patrol cutters and torpedo cutters.	Black Sea

---

[687] Rohwer, *Chronology of the War At Sea*, 244.
[688] Hümmelchen,*German Schnellboote*, 170; Rohwer, *Chronology of the War At Sea*, 245.
[689] Reynolds, *Dog Boats At War*, 51 and Lambert, John *The Fairmile 'D' Motor Torpedo Boat* (Conway Maritime Press Ltd., 1985) 12.
[690] Rohwer, *Chronology of the War At Sea*, 245

1943[692]	April 18 *MTB* *234, 241, 233*	The three MTBs attacked a German convoy off Scheveningen and registered many hits. **Builder:** Vosper; *241* Morgan Giles; *234, 233*-Berthon Boat Co. **Speed:** 40 knots maximum **Displacement:** 38 tons **Range:** 400 nm at 20 knots **Length:** 72'6" **Beam:** 17'0" **Engines:** 3 x Packard 3,540 bhp, petrol **Armament:** 2 x 21" torpedo tubes, 1-1x2,.5 inch MG, 2x2,.303 inch MG **Crew:** 13	English Channel
1943[693]	April 25 *MTB* *635, 639*	On their journey from Bone to Sousse, as they crossed Bizerta Bay, they attacked two Axis troop ships and sank them both with torpedoes and gunfire. They rescued many survivors. **Builder:** *635* Boat Construction Co. Ltd.; *639* Brooke Marine Ltd, Fairmile "D," 115'	Mediterranean
1943[694]	April 27/28 MTB *633, 637, 639*	Near Cap Bon on Kelibia Point in Tunisia, the boats discovered German evacuation facilities being built. They attacked the ships in the small harbor, damaging or sinking two Italian motor minesweepers, two 300 ton coasters and a German R boat. In the fighting, the boats turned their guns on several dozen Ju 52 transports on the nearby airfield, destroying most of them. **Builder:** *633* Dorset Yacht Co. Ltd; *637* Alex Robertson; *639* Brookes Marine; Fairmile "D", 115'	Mediterranean
1943[695]	April 28 *MTB* *633, 637* *639*	After their successful raid on Kelibia Point, these boats attacked the merchant ship SS *Teranio*, which was escorted by the Italian destroyer *Sagittario* in the Gulf of Tunis, off Pantelleria Island between Sicily and Tunisia. The *Sagittario* hit the *639* with crippling gunfire and German fighters cannon fire killed most of its crew. The *633* boat rescued most of the survivors and then sank the hulk with her gunfire.	Strait of Sicily Mediterranean

[691] Rohwer, *Chronology of the War At Sea*, 246
[692] Rohwer, *Chronology of the War At Sea*, 245
[693] Reynolds, *Dog Boats At War*, 51, 53.
[694] *Ibid.*
[695] *Ibid.* at 53.

1943[696]	April 30 *MAS-552*	This was the final battle for Tunisia. The Italians were desperate to get troops and munitions to Tunisia. RAF fighter and fighter-bomber attacked all convoys from Sicily to Tunisia. The *MAS-552* was sunk while escorting the *Hermes* (formerly *ZG3*, German, formerly the Greek destroyer *Vasilefs Georgioue*), which was under tow.	Strait of Sicily Mediterranean

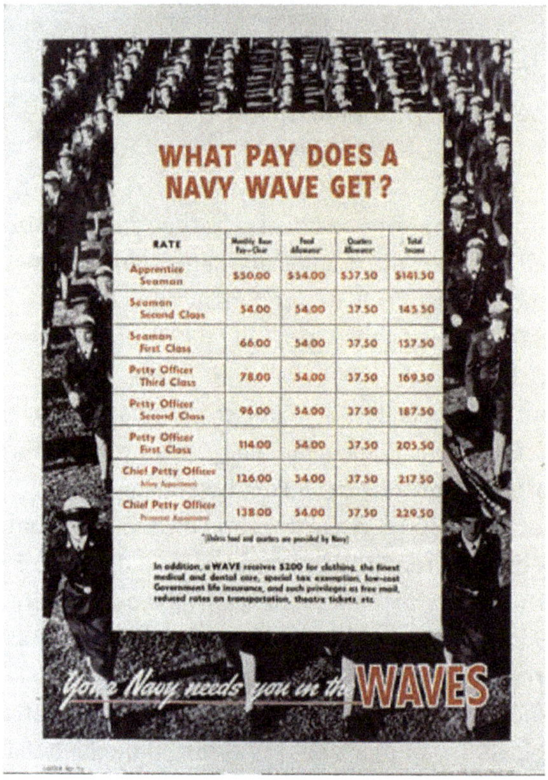

Figure 262. World War II saw dramatic changes in American culture, including increased participation by women in all phases of the war. (National Archives)

696 Rohwer, *Chronology of the War At Sea*, 248 and Worth, Richard, *Fleets of World War II*, De Capo Press, 2001.

Figure 263. The Corsair was a heavy duty fighter with tremendous fire power. On April 28, 1943, the PT boats experienced the worst attack ever-by their own air cover.[697]

[697] http://commons.wikimedia.org/wiki/File:F4U_Corsair.jpg.

Date	Craft, Name, Event or Country	Event, Builder and Boat Specifications	City, River, Lake, Sea, Ocean
1943[698]	April 30 *RA 10* (*MTB-314)*	This MTB fell into German hands earlier in the war. The Germans commissioned the boat into the German Navy as *RA 10.* She was sunk by the RAF on this date. **Type:** Elco, 77'	Off La Goulete Tunisia Mediterranean
1943[699]	April 30 31st MTB Flotilla	5 MTBs and 2 MGBs attacked 4 large flak ships off the Dutch coast. One flak ship burned and sank and two others burned to the waterline and most likely sank. The fourth ship escaped. The *MTB-630* was badly shot up with 4 dead and 8 wounded.	North Sea
1943[700]	May 1 31st MTB Flotilla, 17th MGB Flotilla	The two flotillas attacked four German patrol boats of the 12th Patrol Boat Flotilla off Terschelling, sinking the *V-1241.*	North Sea
1943[701]	May 2 MTB-31 (ex-*PT-53)*	Mined and sunk off the Cani Rocks, between Bone and Bizerte, Tunisia. **Builder:** Elco, 77'	Off North Africa Mediterranean
1943[702]	May 7 "Spanish Armada"	A group of Spanish Naval personnel went from Berlin to Swinemünde for training assignments. Some went to S-Boat training. A second group went to the submarine chaser *UJ-1205,* Bay of Kronstadt and the third went to Submarine Flotilla 25 for U-boat training.	Baltic Sea
1943[703]	May 9 *MTB-61*	Grounded in Kelibia Harbor after combat **Builder:** Vosper, 72'-6"	Tunisia Mediterranean
1943[704]	May 10 MTB 264	Struck a mine and sank off Sousse, Tunisia. **Builder:** Elco, 81'	Mediterranean
1943[705]	May 13/14 *MTB-234, 244, 241*	Near the Hook of Holland, the MTBs ran into the 1st & 7th M Flotilla and the 9Th Motor Minesweeper Flotilla. The MTBs attacked and the German M-8 was hit by two torpedoes and sunk. **Builder:** Vosper, 72'-6"	North Sea

---

698 Rohwer, *Chronology of the War At Sea*, 248.
699 Reynolds, *Dog Boats At War*, 26-27.
700 Rohwer, *Chronology of the War At Sea*, 248.
701 Lambert, *Allied Coastal Forces*, Vol. 2, 10.
702 *Spanish in the Kriegsmarine*, http://www.libreopinion.com/members/ch1492/english/blue/kriegsm..., 3.
703 *Warship Supplement* 65, 6.
704 *Warship Supplement* 65, 10.

Figure 264. An element of the "Spanish Armada".[706]

Figure 265. It was a rare moment when S-boat crewmen could sit down and take a short break.[707]

---

[705] Rohwer, *Chronology of the War At Sea*, 251.
[706] http://www.prinzeugen.com/Armada.htm.
[707] Busch, *Unsere Schnellboote im Kanal,* 55

Date	Craft, Name, Event or Country	Event, Builder and Boat Specifications	City, River, Lake, Sea, Ocean
1943[708]	May 13 *TKA-115, 125*	The two Soviet boats attacked German shipping, firing torpedoes into ships packed into the pier in Anapa Harbor. The full extent of the damage is not known.	Kuban Black Sea
1943[709]	May 14 *PT-150, 152*	The Japanese submarine *RO-102 (Kanemoto)* was sunk 5 nm west of Lae Island by the two PT boats. Both boats fired two torpedoes each at 6,000 yards. The sub stopped and evaded the torpedoes, but then the boats fired two more torpedoes at 4,000 yards and these split the sub in two. **Type:** Elco *PT-103* 80'	South-West Pacific
1943[710]	May 20 *S-72, 49*	The S-boats sank two small craft off Sochi. **Type:** S-38	Black Sea
1943[711]	May 23 *PT-173, 165*	Both boats were lost in transit when the Japanese submarine I 17 sank their transport 100 miles south of Noumea, New Caledonia. The PT-boat tender Niagara (AGP-1) was sunk by IJN bomber in the Solomon Islands area. **Type:** Elco PT 103 class 80'	South Pacific
1943[712]	May 28 "Spanish Armada"	Spanish Naval Captain Corbetta Alvaro Urzaiz with 38 Spanish marines reported to Swinemünde base for Training with the *Schnellbootsschulfotille* under *Freg. Kap* Rüdolf Peterson. At this time they received five *S-boats, S-126, 134, 125, 126, 73.* The was one of many Spanish groups to receive training under actual combat conditions in the Baltic War Zone since 1942.	Baltic Sea

---

708 Rohwer, *Chronology of the War At Sea*, 251.

709 Rohwer, *Chronology of the War At Sea*, 251 and *United States Submarine Losses World War II*, 175, 217 and Polmar and Morison, *PT-Boats At War*, 60.

710 Rohwer, *Chronology of the War At Sea,* 251.

711 U.S. Naval Chronology of W.W. II, 1943, 7.

712 *Kriegsmarine: Spanish in the Kriegsmarine* http://www.libreopinion.com/members/ch1492/english/blue/kriegsmarine.html ("a brief resume of the book by Alfonso Escuadra, *Bajo las banderas de la Kriegsmarine*").

Figure 266. The British *MTB 49-1*.[713]

Figure 267. Soviet *G-5* boats in formation.[714]

[713] http://military-photo.blogspot.com/2008_12_01_archive.html.

Date	Craft, Name, Event or Country	Event, Builder and Boat Specifications	City, River, Lake, Sea, Ocean
1943[715]	May 28 *MTB 219, 221*	The 50th and 52nd ML Flotillas with *MTB-219* and *221* worked in a joint mine laying operation off West Dyck Bank. German patrol boats from the 13th and 14th Flotillas engaged the British boats in a vicious firefight. *MGB-110* and *108* were sunk. The *V-1181* was badly damaged and attacks by *MTB 629, 628, 697* were unsuccessful. **Builder:** Vosper,72'-6" Portsmouth & Portchester	Off Isle of Wight English Channel
1943[716]	June *MTB-654*	The *654* had suffered structural problems and was awaiting repairs when, due to an unseen fuel leak, she exploded into a fire-ball and was declared as a "Constructive Total Loss" or CTL. **Builder:** Lady Bee, Southwick Fairmile "D", 115'	Bone, Tunisia Mediterranean
1943[717]	June 4 *MTB 620, 626* (RNorN)	The two boats sank the German freighter *Altenfels* in Korsfjord. They then fired on a German M class minesweeper (*M468*) and came under fire from five shore batteries. *MTB-626* sustained two dead and two wounded sailors and returned to Lerwick on one engine. **Builder:** *620*, Dickie, Bangor & Tarbert; *621*, Silver, Roseneath Fairmile "D", 115'	Norwegian Sea Norway
June8[718]	June 8 *MTB 73, 77, 84*	The three boats were a part of the massive naval and army force sent to capture Pantelleria Island in Operation Corkscrew. After great effort by all forces, the fortress fell on June 11th. **Type:** Vosper, 72'-6"	Strait of Sicily Mediterranean
1943[719]	June 11 *PT-22*	Lost to severe storm damage at Williwaw, Adak Island. **Type:** Elco, *PT-20* class **Speed:** 39 knots **Displacement:** 35 tons **Length:** 77' **Beam:** 19' 11" **Draft:** 5' 6" **Engines:** 3x 1,200 hp Packard liquid-cooled, petrol **Armament:** 2 x 21" torpedo tubes, 2 x twin .50 cal.MG; **Crew:** 17	Aleutian Islands Bering Sea

---

714 Central Naval Museum, St. Petersburg.
715 Rohwer, *Chronology of the War At Sea,* 251.
716 Reynolds, *Dog Boats At War,* 48, 60, 273.
717 Reynolds, *Dog Boats At War,* 39, 273.
718 Rohwer, *Chronology of the War At Sea* , 254.
719 HyperWar: *US Navy Torpedo Boats* (PT-Boats), 2.

1943[720]	June 14 *MTB-656*	The *656* and *MGB-648* were escorting a senior officer to Pantelleria. On reaching the island the two boats were attacked by FW-190s and a bomb from the German aircraft blew the bottom out of the *MGB-648.* The crew was rescued by the *656* boat. **Builder:** Wallasea Bay Yacht Station, Fairmile "D", 115'	Pantelleria Island Strait of Sicily (Sicilian Narrows) Mediterranean
1943[721]	June 19 British MTB	A British MTB's torpedoes sank a German motor minesweeper in the Seine Bight.	English Channel
1943[722]	June 20 *PT-166, 168, 164*	Coming from the PT-base at Rendova, in Fergusson Passage, the three boats were attacked by three special purpose B-25 bombers. The bombers have eight 50 cal. machine guns in the nose for strafing Japanese warships. *PT-166* was set on fire and blew up, the *168* was bullet-ridden and sunk. The PT crews shot down one B-25 with three dead, one wounded and two saved. The PTs had eleven men dead and wounded. **Type:** Elco, *PT-103* class, 80'	New Georgia Islands Solomon Sea
1943[723]	June 22/25 1st SFltl.	The flotilla, attacked by Soviet Patrol Cutters while laying mines off Gelendzhik, South of Kerch. The S-102 sank from a mine blast. **Type:** *S*-38	Black Sea
1943[724]	June 26 *MTB-651*	The *651* was dispatched with MGBs *643 and 644* to attack the port of Marsala on the southwestern coats of Sicily. Before the trio reached the port, they encountered floating mines. The *MGB-644* struck a mine and incurred many dead, missing, and wounded. The *MTB-651* took all the sailors that could be found on board and then proceeded to sink the *MGB-644,* along with the *MTB-643*, with gunfire. **Builder:** Tough Brothers, Teddington Fairmile "D", 115'	Off the coast of Marsala, Sicily Mediterranean
1943[725]	June 27 TKA	The German ferry lighter MFP *F-325A* was damaged by a blind torpedo near Chauda Cape, Ukraine. **Type***:* *G-5*, 62.5'	Black Sea

---

720 Reynolds, *Dog Boats At War,* 57-58, and Lambert, John, *The Fairmile 'D' Motor Torpedo Boat*, 11.
721 Rohwer, *Chronology of the War At Sea*, 257.
722 Keating, *The Mosquito Fleet*, 73.
723 Rohwer, *Chronology of the War At Sea*, 257.
724 Reynolds, *Dog Boats At War*, 60,271 and Lambert, John *The Fairmile 'D' Motor Torpedo Boat.*
725 Internet correspondence with Richard Worth, July 12, 2007.

1943[726]	June 30 PTs	Three PTs in Blanche Channel, off Rendova Island, mistakenly sank the US attack transport *McCawley* while she was undertow for repairs.	New Georgia Is. Coral Sea
1943[727]	July *MTB-635*	In sailing for Gibraltar the boat hit rough weather and split her keel. She is listed as a CTL. **Builder:** Boat Construction Co. Ltd. Barracks Fairmile "D" 115'	Mediterranean
1943[728]	July 1 "Spanish Armada"	The Spanish contingent of Officer and enlisted trainees was split into groups to serve on S-boats of the 9th *S*-Boat Flotilla in the Baltic Sea war zone.	Baltic Sea
1943[729]	July 2 *PT-153*	The boat ran aground and was then towed to Tulagi for repairs. The damage was too extensive and she was scrapped off Munda, New Georgia. **Type:** Elco *PT-103* class, 80'	New Georgia Coral Sea
1943[730]	July 5 *PT-158*	Ran aground off Munda Point and destroyed to prevent capture. **Type:** Elco *PT-103* class 80'	New Georgia Is. Coral Sea

726 Keating, *The Mosquito Fleet*, 71, 72.
727 Reynolds, *Dog Boats At War*, 273, 261, 55, 51.
728 *Kriegsmarine, Spanish in the Kriegsmarine, supra.*
729 Keating, *The Mosquito Fleet*, 119; Lambert, *Allied Coastal Forces*, 140.
730 Lambert, *Allied Coastal Forces*, Vol.2, 141.

Figure 268. On June 20, 1943, friendly fire by USAAF B-25s killed 11 U.S. Navy crewmen. One B-25-was lost in the mistaken confrontation.[731]

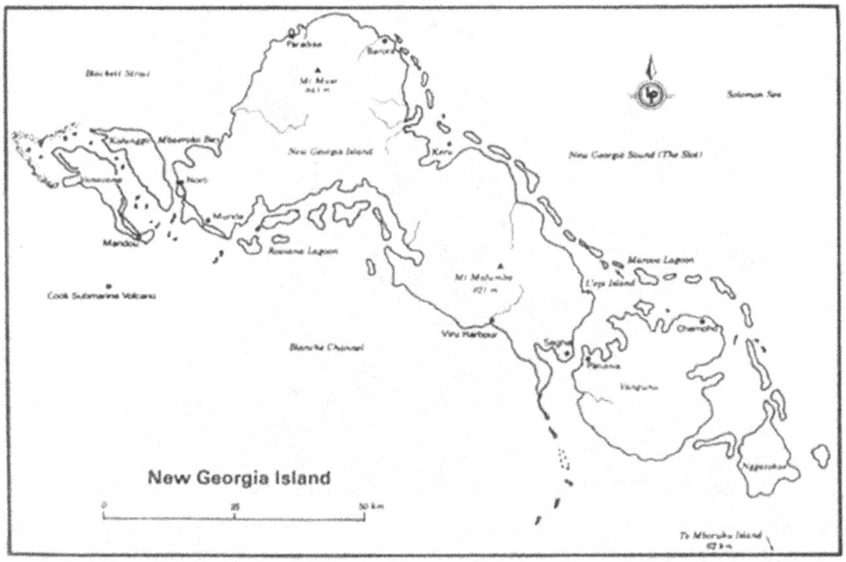

Figure 269. The New Georgia Islands.

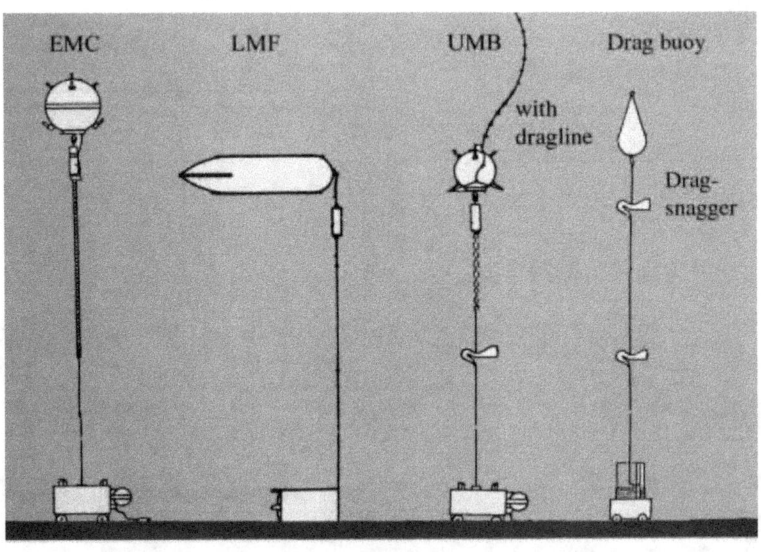

Figure 270. German mines reached new levels of sophistication during WWII and were a significant threat to torpedo boats.[732]

---

[731] http://en.wikipedia.org/wiki/File:B-25s_in_New_Guinea.jpg

[732] Frank, *German S-Boats At War,* 172.

Figure 271. The design of the Soviet G-5 meant that the entire crew was confined to a small space and they rode on top of the, ammunition, gasoline, and torpedo warheads. Their Duralumin skin was no protection against even the smallest bullet.[733]

Figure 272. The *McCawley* was a high speed attack transport that had just suffered an attack by Japanese aircraft. The ship fought back with its AA guns but the

[733] http://www.navweaps.com/Weapons/WNRussian_12.7mm-79_DShK.htm.

command abandoned ship as it settled in the water. Within a few hours the PTs sank her with torpedoes and they were then placed under direct command of Admiral Turner who assigned them a liaison officer to prevent further accidents. (U.S. Navy Historical Center)

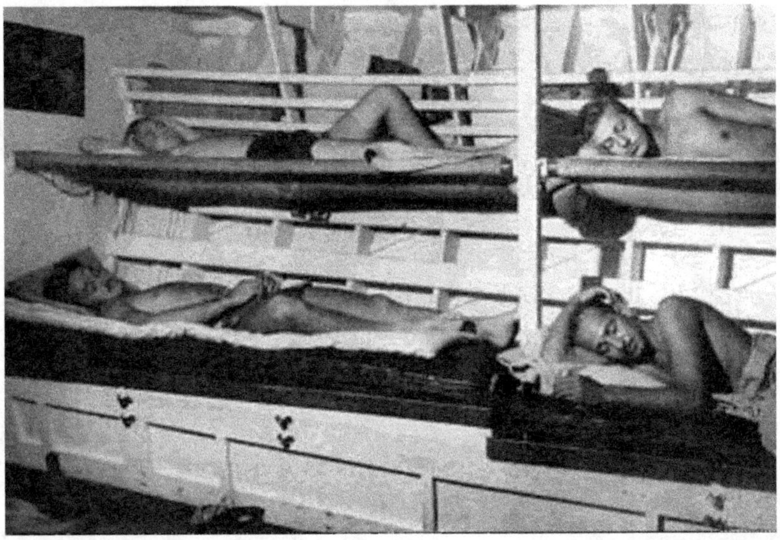

Figure 273. Sleeping quarters on the PTs were seldom used in the South Pacific. Most men preferred to lay out a mattress on deck and sleep in the night air. (U.S. Navy Historical Center)

Figure 274. An aerial view of Tulagi Island in 1942.[734]

---

[734] U.S. Navy Historical Center Photograph

Date	Craft, Name, Event or Country	Event, Builder and Boat Specifications	City, River, Lake, Sea, Ocean
1943[735]	July 5 *MTB-240, 235, 202*	The German torpedo ships *T-24* and *T-25* made their way from the North Sea through the English Channel and to the sea to the West. The Dover batteries could not stop them. The three MTBs attacked them off the coast of Dunkirk to no avail. Builders: *MTB-240, 235* Vosper, 72'-6; *MTB 202,* White, 72'-6"	North Sea English Channel
1943[736]	July 6 *S-9*	The *S*-boat sank during an Allied air attack on Porto Empedocle, Proviance, Agrigento, Sicily **Type:** *S-30*	Mediterranean
1943[737]	July 6/7 *S-102*	In the Black Sea, the S-102 struck a German laid mine and sank South of Kerch. **Type:** *S-38*	Off Kerch Black Sea
1943[738]	July 12 *MTB-81*	In the Strait of Messina, *Kapitanleutnant* Fritz Henning, on his second day of patrolling in his new command (*U-561*) was attacked and sunk by *MTB-81.* The sub suffers 42 dead and there are only five survivors. **Type:** Vosper, 72'-6"	Strait of Messina Mediterranean
1943[739]	July 13 *MTB-640, 651, 670*	The 3 MTBs encountered the Italian submarine *Flutto* running on the surface in the southern section of the Strait of Messina. A fire fight started up. The sub almost slipped away, but finally was destroyed with all hands.	Strait of Messina Mediterranean
1943[740]	July 15 *PT-133*	Sunk by Japanese shore batteries off Cape Pus, New Guinea. **Type:** Elco *PT-103* class 80'	Coral Sea
1943[741]	July 16/17 7th SFltl.	The flotilla was heavily engaged by roving British MTBs and in the resulting firefight; several S-boats suffered heavy damage. The Italian cruiser *Scipone* sinks the *MTB 316* and damages the *MTB-260.*	Mediterranean
1943[742]	July 18 *PT-107*	Destroyed in refueling accident on Emirau Island, St. Matthias (Mussau) Group. **Type:** Elco *PT-103* class 80'	Bismarck Sea

---

735 Rohwer, *Chronology of the War At Sea,* 259.
736 Rohwer, *Chronology of the War At Sea,*260.
737 Rohwer, *Chronology of the War At Sea,* 257.
738 *United States Submarine Losses World War II,* 164, 210, 235.
739 *United States Submarine Losses World War II,* 164, 210, 235.
740 Lambert, *Allied Coastal Forces,*Vol.2, 140.
741 Rohwer, *Chronology of the War At Sea,* 257

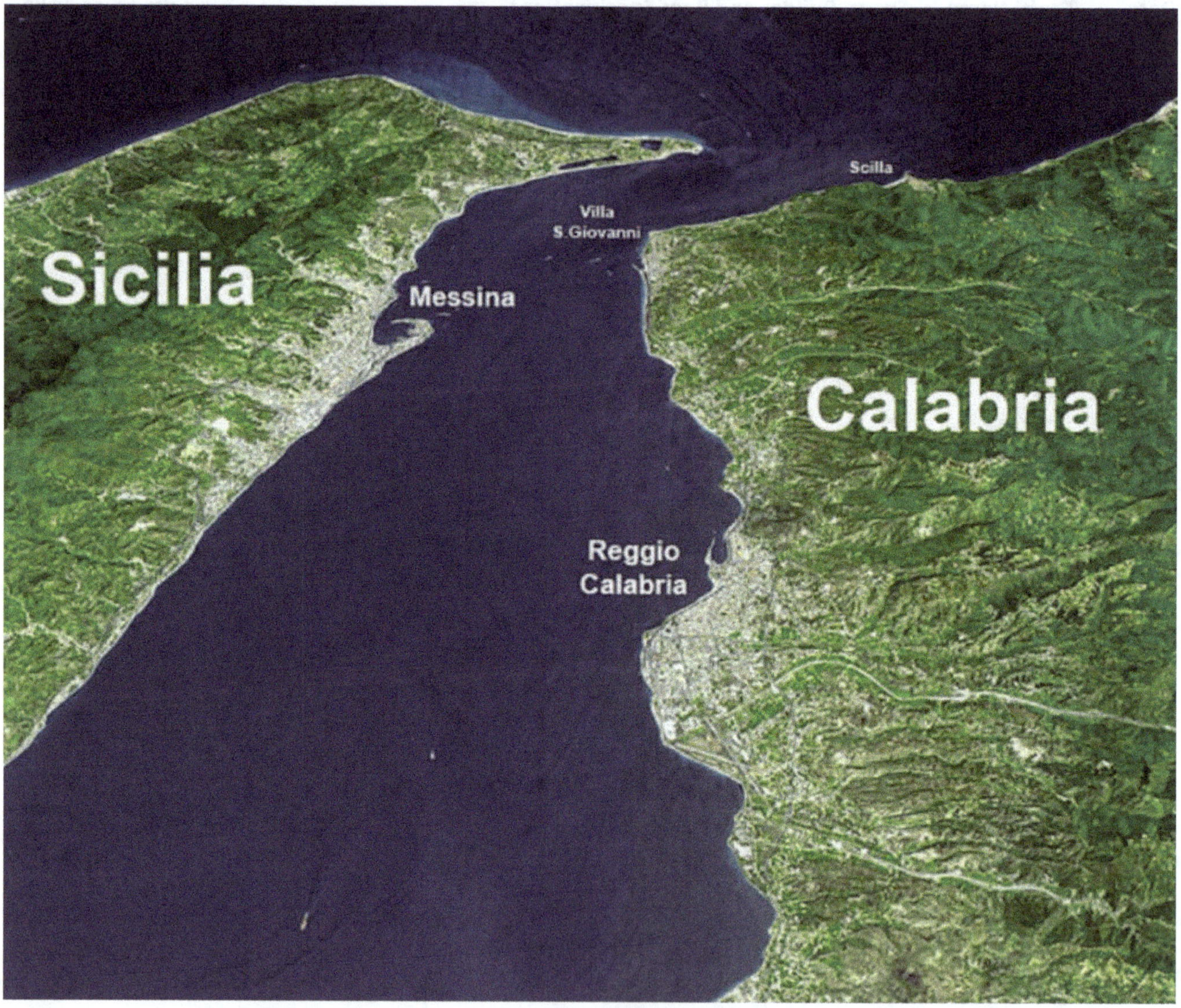

Figure 275. The Straits of Messina are only twenty miles long but thousands of ships over the years in many wars have been lost in the small body of water.[743]

---

[742] Lambert, *Allied Coastal Forces,* Vol. 2, 140.
[743] http://en.wikipedia.org/wiki/File:MessinaStrait.jpg.

Date	Craft, Name, Event or Country	Event, Builder and Boat Specifications	City, River, Lake, Sea, Ocean
1943[744]	July 20 *PT 166*	A sad case of over-anxious American pilots who strafed and bombed a PT boat. The bombers had been especially outfitted with forward firing 8-.50 cal machine guns to attack Japanese shipping. They were B-25's of the 69th Bomber Squadron.	Off Blanche Channel
1943[745]	July 22 *MTB-288*	Bombed and destroyed by enemy aircraft off Augusta, Sicily. **Builder:** Herreshoff,-Vosper 72'-6"	Mediterranean
1943[746]	July 22 RON 15	The city fell to the U.S. Army on this date and the US 15th PT Squadron (RON 15) immediately began operations from this port against German and Italian shipping.	Palermo, Sicily, Tyrrhenian Sea
1943[747]	July 24/25 *S-68, 77*	The two S-boats on patrol are attacked north of Dunkirk by British MTB's and MGB's. The enemy flotilla sinks the *S*-77 boat. The next night the *S*-68 struck a mine off Ostend and was towed back to port by the *S-110, 136* and *81.*	English Channel
1943[748]	July 27 *MTB-345* (RNoR)	The British MTB ran aground off Aspoy near Bergen and captured by Auxiliary patrol vessel V 5301. The entire crew of the MTB 345 were shot by the Gestapo on July 30th. The boat was re-commissioned into the Kriegsmarine as *SA 12.* **Type:** Thorneycroft 55'	Norwegian Sea Norway
1943[749]	August 1/2 *PT-109*	The Japanese destroyer *Amagiri* rammed and sank the *109.* Two members of the crew were killed and two badly injured. The boat's CO received the Navy and Marine Corps Medal for saving a badly burned sailor by swimming almost six miles to safety with the sailor in tow. **Type:** Elco *PT*-103, 80'	Rendova Is, New Georgia Is. Coral Sea
1943[750]	August 1 *PT-117, 164*	Both boats destroyed by Japanese aircraft, Rendova harbor. **Type:** Elco *PT-103* 80'	New Georgia Is. Coral Sea

744 HyperWar: U.S. Navy Torpedo Boats (PT-Boats). 1940-1945, 9
745 *Warships Supplement* 65, 10.
746 Rohwer, *Chronology of the War At Sea*, 264.
747 Rohwer, *Chronology of the War At Sea*, 264.
748 Rohwer, *Chronology of the War At Sea*, 264; *Warship Supplement* 65, 11.
749 Rohwer, *Chronology of the War At Sea*, 141.
750 Lambert, *Allied Coastal Forces,* 140-141.

Figure 276. An S-boat's senior petty officer manned his own gun and led the fighting for the boat on all occasions. As in almost all navies, the senior enlisted man of the boat was considered more of a father figure than anything else.[751]

Figure 277. As this photo shows, the B-25s that attacked *PT-166* on July 20 had been highly successful in strafing Japanese ships. [752]

[751] Busch, *Unsere Schnellboote im Kanal,* 90.

Figure 278. *PT 109* was hunted down and rammed by this Japanese destroyer, *Amagiri*, seen here in 1930. Japanese intelligence had plotted a way to identify and destroy American PTs.[753]

Figure 279. The crew of PT-109 included a future President of the United States.

[752] http://www.ibiblio.net/hyperwar/AAF/IV/AAF-IV-10.html.
[753] http://en.wifipedia.org/wiki/File:Amagiri_1930.jpg

Figure 280. Japanese airmen had developed great skill in destroying PT boats. By cruising at high altitudes, spotting the "V" wake of a PT, and then cutting the throttle, they were able to glide soundlessly down and start shooting before the PT crew could spot them. This is an Aichi E13A-5 "Jake" long-range reconnaissance seaplane.[754] (Library of Congress Print and Photographic Division)

[754] http://en.wikipedia.org/wiki/Aichi_E13A

Date	Craft, Name, Event or Country	Event, Builder and Boat Specifications	City, River, Lake, Sea, Ocean
1943[755]	August 5 2nd, 6Th Shoat Flotilla	The flotilla *(S-39, 74, 80, 83, 86, 89, and 94)* sank the armed trawler HMS *Red Gauntlet* off Harwich.	Hook-of-Holland North Sea
1943[756]	August 8 PT-113	Ran aground on Veale Reef, Tufi Island, New Guinea, scrapped. **Type:** Elco *PT-103* class 80'	New Guinea Coral Sea
1943[757]	August 9 *MTB-284, 285*	Both boats, destined for the Royal Indian Navy, were lost in transit while deck cargo on the freighter, *Larchbank*, which was sunk by submarines south of Italy. **Builder:** Herreshoff, Vosper 72'-6"	Mediterranean
1943[758]	August 11 4th, 5th SFltl.	The two flotillas were attacked by British RAF fighter Bombers while they were making a sortie into Plymouth Sound. The *S-112* and *S-117* were damaged, The other boat received light damage. The *S-110* was unscratched. **Type:** *S-38*	English Channel
1943[759]	August 15 *MTB-665*	Sunk by enemy shore batteries off Messina. **Builder:** Harris-Appledore, Fairmile "D" 115'	Strait of Messina Mediterranean
1943 760	August 17 *PT-136*	After making several successful attacks against Japanese barges the *147* ran hard aground on a reef in the Vitiaz Strait. There was no time to try and remove the craft and she was destroyed with gunfire by PT 142 and PT 136. Type; PT 103 Class, 80'	Strait of Vitiaz
1943	August 23 TKA Force	The Finnish minelayer *Riilahti* sank near Tisskeri Island after an attack by a Soviet TKA force.	

755 Hümmelchen,"*German Schnellboote* (E-Boats), 170; Rohwer, *Chronology of the War At Sea*, 266.
756 Lambert, *Allied Coastal Forces*, Vol.2, 140.
757 North, *Royal Naval Coastal Forces 1939-1945*, 35 and *Warship Supplement* 65, 10.
758 Rohwer, *Chronology of the War At Sea,* 266.
759 *Warships Supplemental* 65, 17.
760 Bulkley, *At Close Quarters,* 197.

Figure 281. *S-137* alongside a tender.

Figure 282. Now and then when aircraft were involved, S-boats took a real beating. This one lost its bow in a fight in the Channel.[761]

[761] Frank, *German S-Boats At War*, 57.

Date	Craft, Name, Event or Country	Event, Builder and Boat Specifications	City, River, Lake, Sea, Ocean
1943[762]	August 30 *MTB-255*	Lost to fire at Ostend harbor. **Builder:** White, Cowes 72-6" **Speed:** 36 knots **Displacement:** 41 tons **Length:** 72'6" **Beam:** 20' **Engines:** 3 x Sterling 3,360 bhp, petrol **Armament:** 2 x 21" torpedo tubes, 2 x .50-cal. MG, 2 x .303-cal. MG, 2 x depth charges	Netherlands North Sea
1943[763]	September 7 *PT-172, 118*	Both boats ran aground on Vella Lavella Island, Wilson Strait, Vella Gulf. **Type:** Elco *PT-103* 80'	New Georgia Is. Solomon Sea
1943[764]	September 8 *MTB-77*	Attacked and bombed by enemy aircraft off Vibo Valentia, Calabria, SW Italy. **Builder:** Vosper 72'-6"	Tyrrhenian Sea Mediterranean
1943[765]	September 9 *S-54, 61*	The last German S-boats left the Taranto harbor on the evening of the Italian surrender. *S-54* and *61* dropped mines in the harbor on their way out. The British cruiser-minelayer, *Abdiel* with 400 troops (1st Airborne) aboard, struck a mine and sinks. The S-boats sank the Italian gunboat, *Aurora* off Ancona and the Italian destroyer *Sella* off Venice, and then surrendered off Venice. **Type:** *S*-30	Adriatic Sea
1943[766]	September 10 TKA Division	The Soviet command ordered 129 small craft to attack the Germans at the Kuban bridgehead. Twenty-five TKAs forced open the approach and eliminated German resistance. A firefight in the harbor of Taman cost the Soviets the *TKAs 124, 125, SKA 025 ,032,* and *084.* **Type:** *G-5*, 62.5'	Kerch Strait, Sea of Azov, Black Sea
1943[767]	September 11 *S-46*	South East of Feodsiya the *S-46* was sunk by an Allied fighter-bomber. **Type:** *S-38*	Black Sea

[762] *Warship Supplemental* 65, 10.
[763] Lambert, *Allied Coastal Forces*, Vol. 2, 140.
[764] *Warship Supplemental* 65, 7.
[765] Rohwer, *Chronology of the War At Sea*, 272 and Coastal Forces and Special Units http://www.naval-history.net/WW2CampaignCoastal-SpecialWarfare.htm.
[766] Rohwer, *Chronology of the War At Sea*, 274.
[767] Rohwer, *Chronology of the War At Sea*, 274.

Figure 283. The commanders of the Soviet TKA forces tended to be more senior in age then the other services and navies.[768]

Figure 284. The Soviet developed the *D-3* torpedo cutter to replace the *G-5*. The new class needed much less maintenance and stood up better in heavy seas. (U.S. Navy Historical Department)

[768] http://en.wikipedia.org/wiki/Vasama_class_motor_torp...

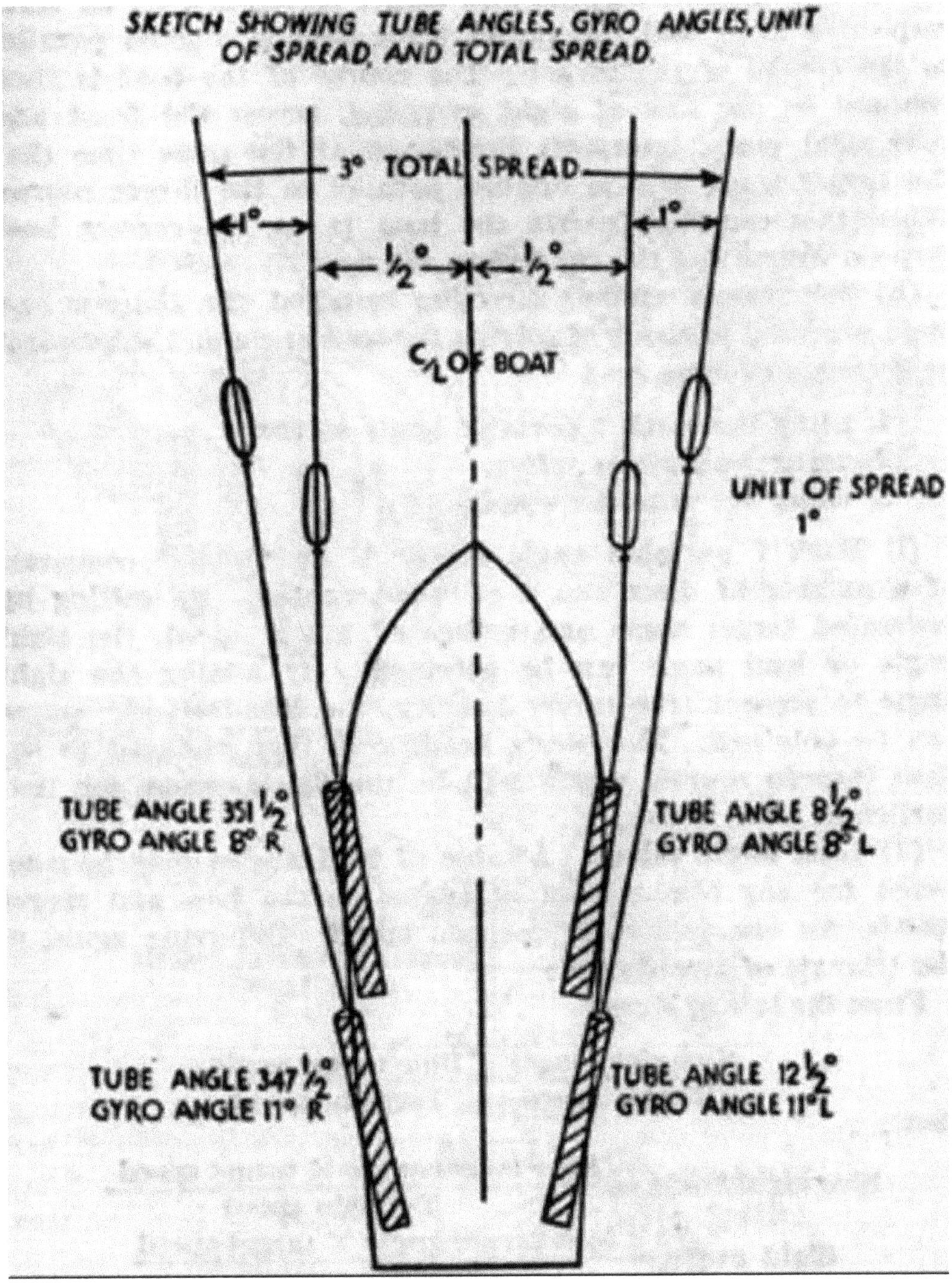

Figure 285. A page from the PT boat training manual showing the angles at which torpedoes were launched from the boats. (U.S. Navy Historical Department)

Figure 286. The Black Sea area of operations.

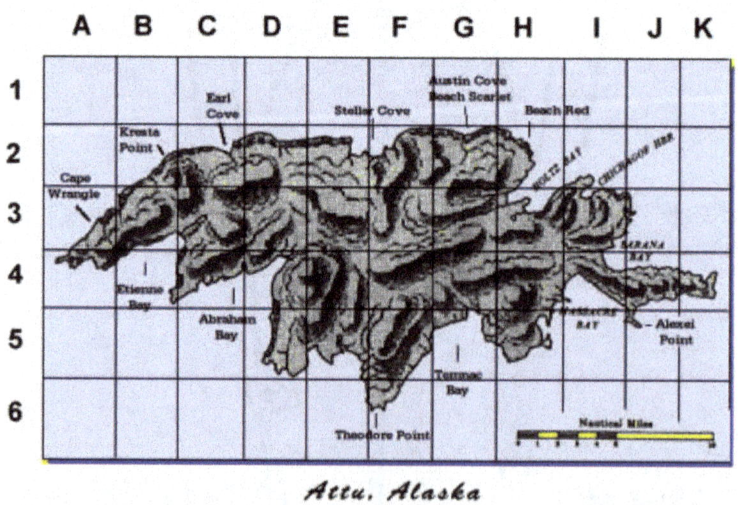

Figure 287. Attu, the westernmost piece of American territory and largest island in the Near Islands grouping of the Aleutian islands, is nearly 1,100 miles from the Alaskan mainland and 750 miles northeast of the northernmost of the Japanese Kurile Islands. Attu is about 20 by 35 miles in size.[769] PT boats suffered many weather-related casualties in the near-Arctic seas.

---

[769] http://www.hlswilliwaw.com/aleutians/attu-homepage.htm.

Date	Craft, Name, Event or Country	Event, Builder and Boat Specifications	City, River, Lake, Sea, Ocean
1943[770]	September 14 *PT-219*	The boat ran aground in high winds and rough seas and parted its mooring cable. She was badly damaged on the rocks and later scrapped as CTL. **Type:** Higgins *PT-20*0 class, 78'	Attu Island, Aleutian Islands, Bering Sea
`1943[771]	September 15 *TKA-13, 21*	Soviet torpedo cutters were assigned to operations against the German supply traffic off the Norwegian coast. On this date the *TKA-13* and *21* launched an attack on shipping near Kiberg Island, Norway. **Type:** *G-5,* 62.5'	Barents Sea Arctic
1943[772]	September 17 *PT-136*	Ran aground on Malai Island and destroyed to prevent capture. **Type:** Elco *PT-103* class 80'	New Guinea Coral Sea
1943[773]	September 17/20 1st SFltl.	During the German withdrawal of troops from the Kuban bridgehead, they enacted Operation *Brunhild.* The defense of the Kerch Strait fell to the 1st *S*-Boat Flotilla and the 3rd Motor Minesweeper Flotilla. They fought off several Soviet naval traffic attacks, and torpedoed several Soviet supply ships tied up in the piers of Anapa Harbor.	Off Anapa Harbor Black Sea
1943[774]	September 21 *TKA-15*	Working off the Norwegian coast near Kiberg Island, off Nurmensetty Cape, the *15* boat sank the steamship *Antje Fritzen.* **Type:** *G-5,* 62.5'	Barents Sea Arctic

770 Polmar and Morison, *PT-Boats At War,* 52.
771 Richard Worth, Internet correspondence with author July 24, 2007; Jordon, Roger W., *The World's Merchant Fleets 1939* (U.S. Naval Institute Press, 1999), 466.
772 Lambert, *Allied Coastal Forces,* Vol. 2, 140.
773 Rohwer, *Chronology of the Wart At Sea.*
774 Richard Worth with Kirill Klement'ev, Internet correspondence 5 May 2004, forwarded to author July 24, 2007; Rohwer, *Chronology of the War At Sea,* 267; Jordon, Roger W. *The World's Merchant Fleets 1939* (U.S. Naval Institute Press, 1999), 466.

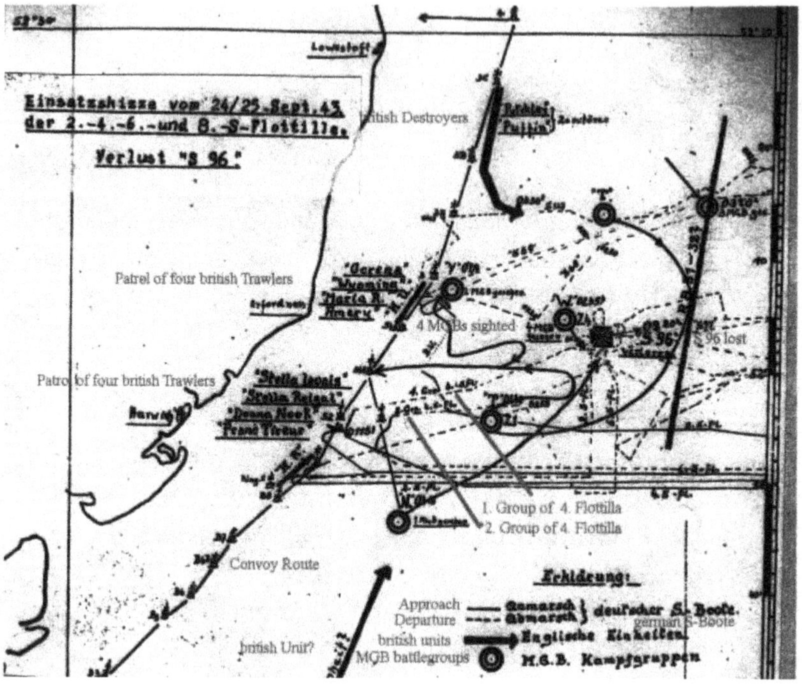

Figure 288. The after-action report and map show the layout of the contenders in the English Channel on September 24/25, 1943. The 25 S-boats were swarmed over by eight armed trawlers, two destroyers and six flotillas of MGB's. The *S-96* was sunk and the *S-69* badly damaged.[775]

Figure 289. The springbok logo of the 8th Schnelleboote Flotilla. Note the rivets in the background.

[775] Frank, *German S-Boats in Action*, 61-62. Translation notes by Hans Frank.

Date	Craft, Name, Event or Country	Event, Builder and Boat Specifications	City, River, Lake, Sea, Ocean
1943[776]	September 21/22 *MTB-226, 228*	The MTBs attacked the old German cruiser *Niobe* (ex-*Dalmacija*) north-west of Zara on the Croatian coast. The *Niobe* is not sunk.	NE Adriatic Sea
1943[777]	September 24/25 S Boat Operation *Probestuck* *4Th SFltl.*	German S-boat mining operations called *Probestuck* off Harwich and Orfordnes were completed, and then *S*-96 sank the armed trawler *Franc Tieur* and *Donna Nook.* The boat was rammed by a British ML-145 and then abandoned. The flotilla sank the armed trawlers *Franc Tieur* and *Donna Nook.* The *S-96* collided with ML 145 and sank off Harwich. **Type**: S-38	Harwich Hook-of-Holland North Sea
1943[778]	September 26/27 *MTB-202, 204, 231* (NRoN)	The German convoy running from Le Havre to Dunkirk was accompanied by 15th Patrol Boat and 2nd Motor Minesweeper Flotillas. Three British MGBs made an attack. The Dutch SGB force attacked from shoreward off Fe'camp and Berq-sur-mer, France. The MTBs, MGBs, and SGBs sank the German *V-1501*, the escort vessel *M-534* (*Jungingen)* and a freighter. **Builder:** *202, 204* White, 72'-6"**;** *231* Vosper,72'6".	English Channel
1943[779]	September 30 *PT-219*	Foundered and sank off Attu Island. **Type:** Higgins *PT-20*0 Class, 78'	Aleutians Bering Sea
1943[780]	October 1 PT-68	Ran aground on Vincke Point. *PT-191* sank her with gunfire to prevent capture. **Type:** Elco *PT-20* class 77'	Huon Peninsula E. New Guinea
1943[781]	October 7 4th, 6th, 8th SFltl.	The S-boats began Operation *Gesellenprifting* by sowing mines south of Smiths Knoll. The Flotillas mustered 29 boats for the operation. *S-62, 83, 93 and 197* returned to base due to engine trouble.	English Channel

---

776 Rohwer, *Chronology of the War At Sea*, 277.
777 Rohwer, *Chronology of the War At Sea*, 278 and Hümmelchen, *German Schnellboote*, 170.
778 Rohwer, *Chronology of the War At Sea*, 278.
779 HyperWar: US Navy Torpedo Boats (PT-Boats). 1940-1945 http://www.ibiblio.org/hyperwar/USN/ships/ships-pt.html.
780 Lambert, *Allied Coastal Forces*, Vol. 2, 123.
781 Rohwer, *Chronology of the War At Sea*, 280.

1943[782]	October 5/6 1st SFltl.	A force of Soviet destroyers was sent to stop the German evacuation from Sevastopol by attacking transport ships off the coast of Crimea. The 1st flotilla boats *S-45, 28, 42, 51* and *52* (all older and well-worn S-boat models) drove the Soviet destroyers away from the German ships. The next day the German Air Force sank the Soviet destroyers.	Black Sea
1943[783]	October 8/10 1st SFltl.	The flotilla was part of the last evacuation effort of 240 small craft from the Kuban bridgehead. The 30th Motor Minesweeper and 23rd Submarine-Chaser Flotillas joined in the effort, the Assault Boat 229 was lost to Soviet aircraft, and Ferry barge 474 was lost to a Soviet submarine.	Black Sea
1943[784]	October 11 *PT-21*	Suffered extreme storm damage, scuttled. Williwaw, Adak Island. **Type:** Elco *PT-20* class 77'	Aleutian Islands Bering Sea
1943[785]	October 13 *PT-183*	Japanese fighters attacked the PT base at Lambu, off Choiseul Island, and the *183* suffered damage. **Type:** Elco *PT-103* class, 80'	Choiseul Island Solomon Sea
1943[786]	October 15 *MTB-636* "D"	Sunk in error by *MGB-658* "D" in action off Elba Island. **Builder:** Silver, Rosneth Fairmile -D 115'	Ligurian Sea
1943[787]	October 16 *MTB-356* "D"	Scuttled after damage in combat off Dutch Coast. **Builder:** Harland & Wolfe, Vosper 72'-6	Dutch Coast North Sea"
1943[788]	October 23 MTB Force	British MTBs attacked the German mine layer *Juminda* (ex-Italian E. *Gasperi*) 2 nm west of San Stefano.	E. Bulgaria Black Sea

---

782 Rohwer, *Chronology of the War At Sea,* 280.
783 Rohwer, *Chronology of the War At Sea,* 280.
784 HyperWar: *US Navy Torpedo Boats (PT-Boats),* 2.
785 Rohwer, *Chronology of the War At Sea,* 281.
786 *Warship Supplement* 65, 18.
787 *Warship Supplement* 65, 12.
788 Rohwer, *Chronology of the War At Sea,* 282.

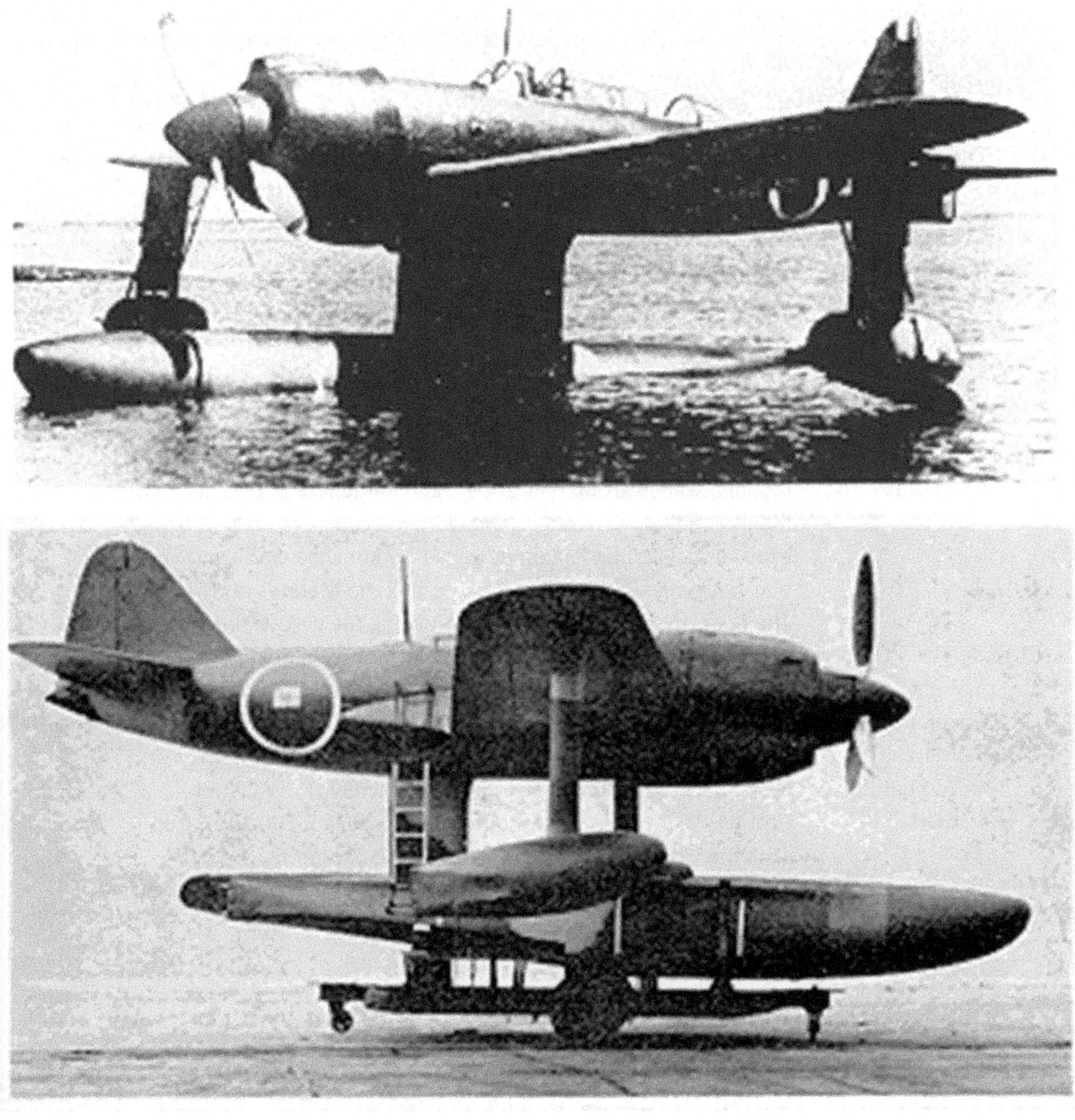

Figure 290. Japanese float planes became the number one enemy of the PT crews. Here, the Kawanishi E15K Shiun "Norm" (top) and the Kawanishi N1K Kyofu "Rex" (bottom).[789]

[789] http://www.combinedfleet.com/ijna/e15k.htm.

Date	Craft, Name, Event or Country	Event, Builder and Boat Specifications	City, River, Lake, Sea, Ocean
1943[790]	October 24/25 2nd, 4th, 6th, 8th SFltl.	This strong force attacked convoy FN 1160 off Cromer, North Sea. They failed to reach the convoy due to five British destroyers and several MGBs. *S-74* sank the armed trawler *William Stephen.* The destroyer HMS *Mackay* rammed and sank the *S-63. S-88* (4th SFltl.) was sunk by MGBs and the *MTB-439* sustained damage in the fight. Sixteen survivors were saved by the British and twenty-four by German S-boats. The *S-74* (6th SFltl.) sank the armed trawler *William Stephen.*	North Sea
1943[791]	October 26 *MTB-669* "D"	Sunk in combat off Norway. **Builder:** Lady Bee, Fairmile "D" 115'	North Sea
1943[792],[793]	October 28 PT 59, 236	The two boats were on patrol when they received word that a reconnaissance group of Marines, 2nd Parachute Battalion (Reinforced), on Choiseul Island were in desperate need of being evacuated. CO John F. Kennedy in *PT-59* knew he was low on fuel, but proceeded with the 236 boat to pull the Marines off the island. The boats found the Marines by following the vicious fire fight on the beaches and rescued about 50 Marines from being overrun by Japanese troops.	Choiseul Island
1943[794]	November 2 5Th SFltl.	The flotilla boats attacked convoy 221 southwest of Dungeness and sank three freighters.	North Sea
1943[795]	November 2/3 5th SFltl.	Nine boats of the flotilla attacked British convoy CW 221 of Hastings. *S-146* sank the freighter *Dona Isabel.* S-100 and *S-138* sank the freighters SS *Foam Queen* and SS *Storaa.* *S-100* and *146*: **Type:** *S-38,* with "Skull Cap"(the crew slang for fully armored bridge for all new S-boats). *S*-138: **Type** *S-38.*	English Channel

790Hümmelchen, *German Schnellboote, 170* and Rohwer, *Chronology of the War At Sea,* 83.
791 *Warship Supplement 65,* 19.
792 "JFK Rescue." http://www.orwelltoday.com/jfkpt59rescue.shtml
793 James F. Christ. "Mission Raise Hell: The U.S. Marines on Choiseul, October-November, 1943." Annapolis, Maryland: Naval Institute Press, 2006
794 Hümmelchen,*German Schnellboote,* 170.
795 Rohwer, *Chronology of the War At Sea,* 286.

Figure 291. The image of a Royal Navy trawler is based on the original artwork by Chris Mayger for the Arrow paperback edition of Proud Waters, a novel about minesweeping by Royal Naval Patrol Service member Ewart Brookes. Trawlers became an image of great pride to the British people. The crews were all civilian volunteers.[796]

---

[796] http://www.rnps.lowestoft.org.uk/history.htm.

Figure 292. The *59* boat had been converted to a gun boat before JFK took over its command. The boat was fitted with 40mm Cannon and three .50 cal MG stations on each side of the deckhouse. The forward station was a twin .50 cal mount and the other two were single .50 cal MG. This meant twelve machine guns where there were previously four. The change in these boats had dramatic effects in the devastating firepower they could bring to the fight. JFK saved 50 Marines with his *59* boat. Here, a firing exercise is conducted. (Frank Andruss, Sr., The Mosquito Fleet Exhibit).

Figure 293. PT-59, CO John F. Kennedy.

Date	Craft, Name, Event or Country	Event, Builder and Boat Specifications	City, River, Lake, Sea, Ocean
1943[797]	November 4 *MTB-606* "D"	Two British flotillas, the 55th and 50th, sailed from Yarmouth to land commandos on the Dutch coast. The flotillas ran into four German S-boats laying in wait. After a short firefight the *MTB-617* took fire and returned to base. The *MTB-606* became the lead boat and ran into the German buoy-laying ship *Main* and escorts: *Vp-1401, Vp-1419* and flak vessel *FJ23.* The *606* was immediately hit by 88 mm fire and suffered heavy damage with many dead and wounded. The surviving crews were taken off by *MGB-630.* The *606* lost six dead and 21 wounded. **Builder:** Wallasea Bay Yacht Station, Fairmile "D", 115'	Smiths Knoll North Sea
1943[798]	November 4 5Th SFltl.	Attacked Allied convoy FN 1170 between Cromer and Great Yarmouth off English Coast. The *S-74* was attacked and sunk by British Mosquito fighter bombers.	North Sea
1943[799]	November 4/5 8th, 4th, 2nd SFltl.	18 boats of the flotillas went on a mine laying operation off Smith's Knoll, Humber estuary. *S-116* and *S-114* returned due to engine trouble. The 8th flotilla attacks convoy FN 1170 East of Cromer. They torpedoed two ships. The 2nd flotilla was attacked by the RAF; *S-116* is damaged and *S-74* must be abandoned. The *S-135* sank the *S-74.* **Type***: S-38*	English Channel
1943[800]	November 9 *MTB-230*	Rammed and sunk by *MTB-222* during combat action, North Sea. **Builder:** McGruer, Vosper 72'-6"	North Sea
1943[801]	November 10 *MTB-222* "Sperwer"	Sunk because of the collision with *MTB 230.* **Builder:** Vosper 72'-6" Royal Netherlands Navy.	North Sra
1943[802]	November 15/16 *MTB-266, 315*	The boats sank one Axis landing craft and one barge off Leros Island by gunfire. **Type:** *266* Elco, 81'; *315 (ex-PT-57)* Elco, 77'	Dodecanese Islands. Sea of Crete

797 Reynolds, *Dog Boats At War*, 88-90 and Lambert, John, *The Fairmile 'D' Motor Torpedo Boat*, 13.
798 Hümmelchen, *German Schnellboote (E-Boats)*, 170.
799 Rohwer, *Chronology of the War At Sea*, 286.
800 *Warship Supplement* 65, 9.
801 *Warship Supplement* 65, 9.
802 *Warship Supplement* 65, 10, 11.

Figure 294. The RAF Mosquito fighter-bomber was a nightmare to the men of the S-boats and actually all enemy ships. It carried heavy machine guns, 20mm cannons a, rockets and bombs. It also had a long range which meant it could stay in the air for as long as it took to sink S-boats.[803]

Figure 295. The White-built boats were equally successful as MGBs. (Illustration by author)

[803] http://en.wikipedia.org/wiki/File:De_Havilland_Mosquito-DK338-1942.jpg

Date	Craft, Name, Event or Country	Event, Builder and Boat Specifications	City, River, Lake, Sea, Ocean
1943[804]	November 18 *PT-311*	The 311 boat struck a mine off the coast of Corsica and sank. **Type:** Higgins PT-200 Class, 78'	Italian Rivera Ligurian Sea
1943[805]	November 24 *S-56*	A USAAF raid on Toulon, France, caught and sank the *56* boat, along with several large ships and motor minesweepers.	Off Toulon Bay NW Med.
1943[806]	December 1 5[Th] SFltl.	Off Beachy Head, England, three-armed trawlers and the HMS *Avanturine* were sunk by boats of the 5th *S*-Flotilla.	Beachy Head, English Channel
1943[807]	December 4 *S-511*	British aircraft attacked and sank the S-511 (ex-Italian *MAS-522*) at Makronisos. **Type:** *MAS-533,* 55.8'	Aegean Sea
1943[808]	November 21 TKA	TKAs sank the German ferry lighter MFP *F-472* near Eltigen, Crimea. **Type***:* *G-5*, 62.3	Kerch Strait Black Sea
1943[809]	November 22 *MTB-626, 686 "D"*	Both boats were lost when petrol cans on the open deck of *686* caught fire. The explosions that followed set the *626* on fire. Ammunition explosions destroyed both boats. Scottish Quay, Anglo, North Ness. **Builder:** Robertson, Fairmile "D" 115'	Shetland Islands North Sea
1943[810], [811]	November 23 PT-322	While on a mission the 322 ran aground off Hardenberg Point, Sansapor Island, New Guinea and the wreck was destroyed by U.S. forces to prevent capture. **Type:** Elco *PT-103* class, 80'	New Guinea
1943[812]	November 24 *MTB-73*	Attacked by enemy aircraft off Sardinia and sunk by bombs. **Builder:** Vosper 72'-6"	Sardinia Island Tyrrhenian Sea

---

804 *U.S. Naval Chronology Of W.W. II*, 1943, 17
805 Rohwer, *Chronology of the War At Sea*, 290.
806 Hümmelchen, *German Schnellboote (E-Boats)*, 170.
807 Rohwer, *Chronology of the War At Sea*, 291.
808 Internet correspondence with Richard Worth, July 12, 2007.
809 Reynolds, *Dog Boats At War*, 106-107.
810 *U.S. Naval Chronology Of W.W. II*, 23
811 Lambert, *Allied Coastal Forces* Vol 2, 141.

Figure 296. The *G-5* light Duralumin boats operated in the ice choked sea of the Artic Ocean and the frigid waters off Norway as well as the Baltic Sea. Surviving a tour of duty in one of these craft was a miracle. (Central Naval Museum, St. Petersburg)

---

[812] *Warship Supplement* 65, 7.

Figure 297. "Know Your PT Boat", an introductory pamphlet for the PT training crews.

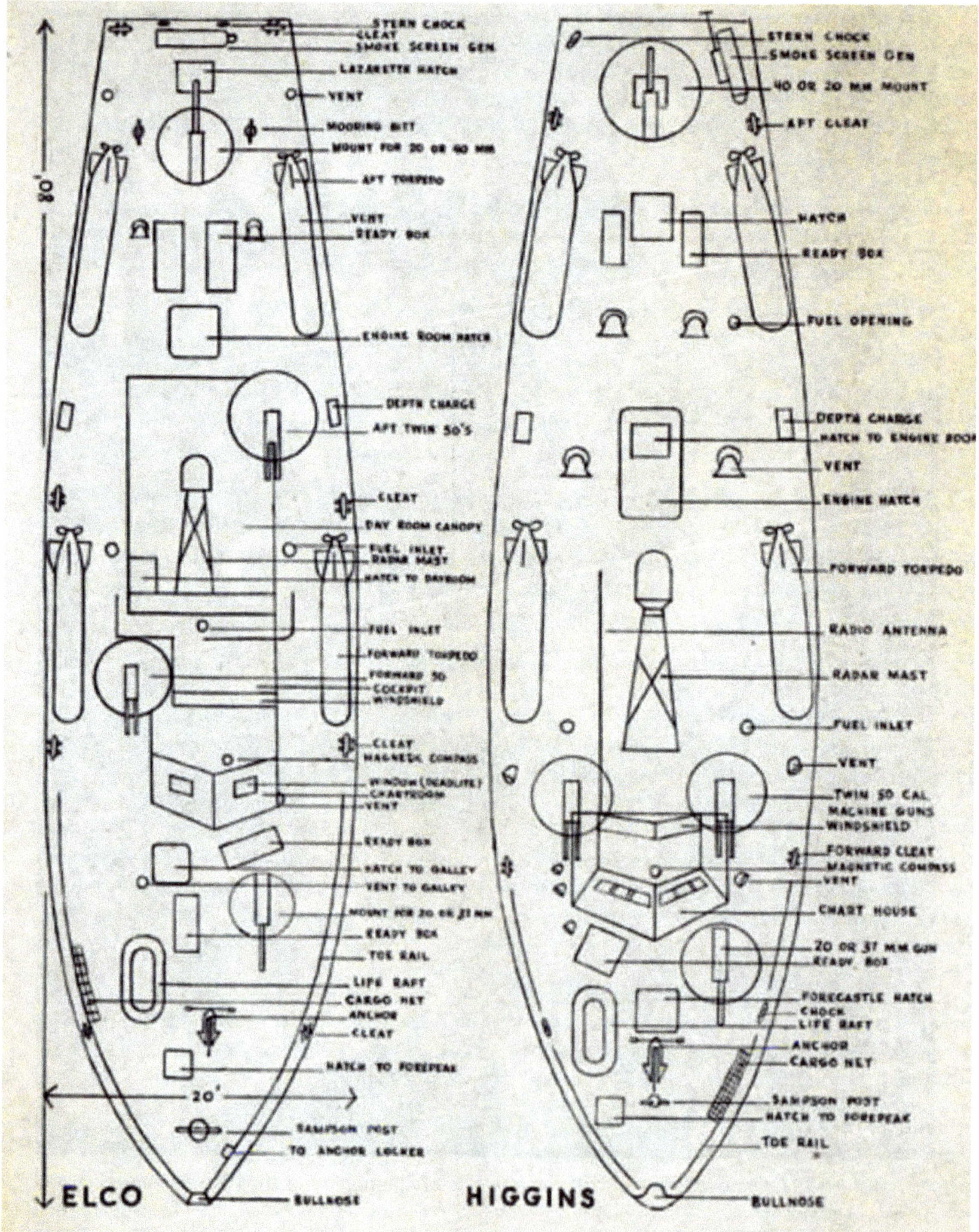

Figure 298. An interior page from "Know Your PT Boat."

Date	Craft, Name, Event or Country	Event, Builder and Boat Specifications	City, River, Lake, Sea, Ocean
1943[813]	December 11 *TKA-101*	The 1st *S*-boat Flotilla sank the Soviet *TKA-101* and *SKA 01012{010-012?}* near Eltgen. **Type:** *G-5*, 62.5'	Black Sea
1943[814]	December 12 MTB-649 "D"	The *649* boat was engaged in a shore bombardment on the Peljesac peninsula when she sighted the two-funnel, ex-Austrian, and steam torpedo boat leaving the Neretva Channel. The *649* pressed an attack and the enemy ship broke off and made it to Lagosta Island. The RAF later bombed her.	Croatia Coast Adriatic Sea
1943[815]	December 12 *MTB-637, 297 "D"*	The two boats were operating in support of a Partisan raid on the north coasts of Brac, Croatia, whose goal was to put ashore several hundred fighters near the garrison of Omis, south of Split. They became engaged in an attack on several vessels, and sank the heavily laden *Siebel* Ferry (*SF-193*), full of cans of gasoline, and two smaller craft that were acting as escorts.	Croatia Coast Adriatic Sea
1943[816]	December 14 *PT-239*	Destroyed in a dockside fire. **Type:** Higgins *PT-200* class, 78'	Solomon Islands

---

[813] Rohwer, *Chronology of the War At Sea,* 295.
[814] Reynolds, *Dog Boats At War*, 130.
[815] Reynolds, *Dog Boats At War*, 130.
[816] Polmar and Morison *PT-Boats At War,* 122.

1943[817]	December 22 *MTB-276, 298 "D"*	Once again, MTBs were after the German cruiser *Niobe* (ex-Italian *Cattro,* ex-Yugoslav *Dalmacija,* ex-German *Niobe*). The ship was grounded on Silba Island and this time the MTBs destroyed her. **Builder** *(276):* Elco, USN (Vosper) **Speed:** 38 knots **Displacement:** 36/40 tons **Length:** 72'6" **Beam:** 19' **Crew:** 13 **Engines:** 3 x 1,350 hp Packard, petrol (aviation gas) **Armament:** 2 x 21" torpedo tubes, 2 x -.50cal MG, 3 x .303-cal MG	Off Croatian coast North Adriatic
1943[818]	December 22 *TKA-13*	The *13* boat hit the German D/S *Marie* with two torpedo hits off Varangerfjord and sank her.	Arctic Sea
1943[819]	December 22/23 TKA Div.	TKAs attacked a German convoy and sank the ships *Vardo, Kirkenes,* and *Petsamo,* the freighter *Antje Fritzen* and patrol trawler *V-6106.* The Soviets lose six TKAs. **Type:** *G-5,* 62.5'.	Barents Sea
1943[820]	December 24 *MTB-357*	Suffered after action damage in Dover Straits **Builder:** Vosper 72'-6"	Dover Straits
1943[821]	December 25 *PT-190, 191*	Off Arawe, New Britain, 30 to 38 Japanese dive-bombers attacked the two boats. Two of the PT boat's crew members were seriously wounded and four Japanese planes shot down. **Type:** Elco *PT-103* class 80'	New Britain Solomon Sea

[817] Rohwer, *Chronology of the War At Sea*, 294 (see September 21/22).
[818] Internet correspondence with Richard Worth, 2007.
[819] Internet correspondence with Richard Worth, July 12, 2007.
[820] *Warship Supplement* 65, 12.
[821] *Keating, The Mosquito Fleet*, 85.

Figure 299. PTs carried only a few depth charges, but they were deadly. They were mostly used to discourage enemy surface ships from chasing the PTs. (U.S. Navy Historical Center)

Figure 300. Refueling at sea is always a shaky situation. While the boat keeps pace with the ship, its engines are running. This means hot engines are surrounded by highly volatile aviation gas fumes. (National Archives)

THE SHIP KILLERS

THE DEFINITIVE ILLUSTRATED HISTORY OF THE TORPEDO BOAT

VOLUME VIII: TIMELINE 1944

Photo # NH 44308 PT boats crossing the English Channel, with B-17s overhead, 6 June 1944

BY JOE HINDS

NIMBLE BOOKS LLC

CONTENTS (INCLUDING FIGURES)

DEDICATION

To Fred Zimmerman

It is not common for publishers to get noticed in a book dedication but in this special case it is fully warranted. Fred never understood that dealing with someone with two disabilities (Emphysema and Traumatic Brain injury) would be next to impossible to maintain his sanity.

Because of my condition in suffering from the effects of a massive brain injury he has had to shoulder to results of the load that comes from that on a near daily basis. My memory has about 72 hours duration and then the, thought, event, phone call, or instructions, is gone for good. I could be accused of stealing the Lindbergh baby and I could not honestly tell you I did not do the deed. I simply don't know. I do know what 'You' tell me, for about 72 hours. I could confess to anything because of this one defect.

This defect has led to some hot exchanges and smashed feelings, but we somehow keep muddling on. Fred is as much responsible for this work as me. He has cleaned up the format, edited everything so well, kept tabs on over a thousand entries, delivered spectacular layouts that add so much to the readability of these volumes and made it a success.

While it can take me days to do a few hours work, I eventually get the job done, but only because Fred is right there to keep me going.

TIMELINE 1944

Date	Craft, Name, Event or Country	Event, Builder and Boat Specifications	City, River, Lake, Sea, Ocean
1944[822]	January 1-15 S boat testing	The Germans had made significant advances in anti-reflection paints for ships when the war started. The *Schnellbootewise* color proved to help the boat melt or blend into the seascape at night and in the day. The Germans held trials to attempt to reduce the reflective surfaces reduced or eliminated by an absorbent coating. The project was codenamed *Netzemd.* The *S-147* was selected for the trials held off Pelzerhakeken in Lubeck Bay. Uncoated S-boats and coated U-boats were also tested. While the trials were impressive, the results were never applied.	Baltic Sea
1944[823]	January 4 *PT-145*	The boat ran aground on the coast near Mindiri, Madang Province, destroyed to prevent capture. **Type:** *PT-103*-class 80'	Papua New Guinea Bismarck Sea
1944[824]	January 5/6 5[Th] SFltl.	Seven of 5th flotilla's S-boats (*S-84, 100, 112, 136, 138, 139, 140, 141, 142, 143*) attacked the Allied convoy WP. 45 off Lands End. Four armed trawlers and the destroyers HMS *Talybont* and *Melbreak* acted as escorts. The S-boats fired 23 torpedoes, sinking the HMS *Polperro, Underwood* , *Solstad,* and the armed trawler HMS *Wallasea.* The *S-142* and *S-138* were credited with sinking the Allied ships.	Atlantic Ocean
1944[825]	January 12/13 *MTB-260, 309*	Deep within the Greek Islands, the boats sank a caigue (fishing boat) off Piscopi (officially Tilos) Island. **Type:** Elco 70'	Dodecanese Islands Aegean Sea

[822] Hummelchen, G. Dr. *German Schnellboote,* 174.
[823] Lambert, *Allied Coastal Forces* Vol. 2, 140.
[824] Whitley, *German Coastal Forces,* 70.
[825] Lambert, *Allied Coastal Forces,* Vol. 2, 110.

Figure 301. Attempting to test the dispersion of poison gas from a fast moving S-Boat in the bay was a serious mistake. Luckily, no one died but it was close call and a foolish idea. The absurdity of even considering this kind of warfare would haunt the German people for decades to come.[826]

[826] http://en.wikipedia.org/wiki/Baltic_Sea

Figure 302. *PT-145* underway. The single largest reason for so many PT being put out of commission was running into shoals and reefs. This stemmed mainly from the maps that were well over fifty years old.[827]

[827] William Maloney, www.williammaloney.com

Date	Craft, Name, Event or Country	Event, Builder and Boat Specifications	City, River, Lake, Sea, Ocean
1944[828]	January 14 5th SFltl.	The flotilla boats, off Beachy Head, sank the freighters, SS *Emerald and SS Caleb Sprague.*	English Channel
1944[829]	January 16/17 4th, 5th SFltl.	The 5th *S*-Boat Flotilla attacked a British convoy off The Lizard and were driven off after firing at least eleven torpedoes. The 4th flotilla boats *laid Luftmine-B (LMB)* mines off Orfordness.	English Channel
1944[830]	January 26 *PT 110, 114*	Lost in collision with *PT-114* off Cape Gloucester Island. *PT-110* accidentally dropped a depth charge, which further damaged both boats. **Type:** Elco *PT-103* class, 80'	New Britain Is. Bismarck Sea
1944[831]	January 27-28 *MAS-541*	(SOE Op's "Ladbroke"). The *541* took along two American PT boats on this mission to land two agents on Cap-Camarat. The first attempt failed due to poor navigation. The second attempt failed because of enemy surface craft encountered off shore.	S.E. coast France
1944[832]	February 8 *MTB-625* "D"	The boat ran into a fierce gale and the rough beating she took split her keel. With the forward compartments flooded, ammunition tossed, torpedoes fired, and one-engine dead, the boat could barely make headway. She was beached on Shetland Island and later wrecked in a salvage attempt.	North Sea

828 Rohwer, *Chronology of the War At Sea,* 303
829 Rohwer, *Chronology of the War At Sea,* 301.
830 Lambert, Allied Coastal Forces Vol. 2, 110; Hyper War: *US Navy Torpedo Boats (PT-Boats),* 1940-1945 http://www.ibiblio.org/hyperwar.USN/ships/ships-pt.html, 8.
831 Richards, *Secret Flotilla,* 688.
832 Reynolds, *Dog Boats At War,* 107, 271.

Figure 303. Fighter bombers and fighters were the terror of the skies for all motor torpedo boats, no matter how large or small. The sight of the plane would often more than not spell death for the entire crew.[833]

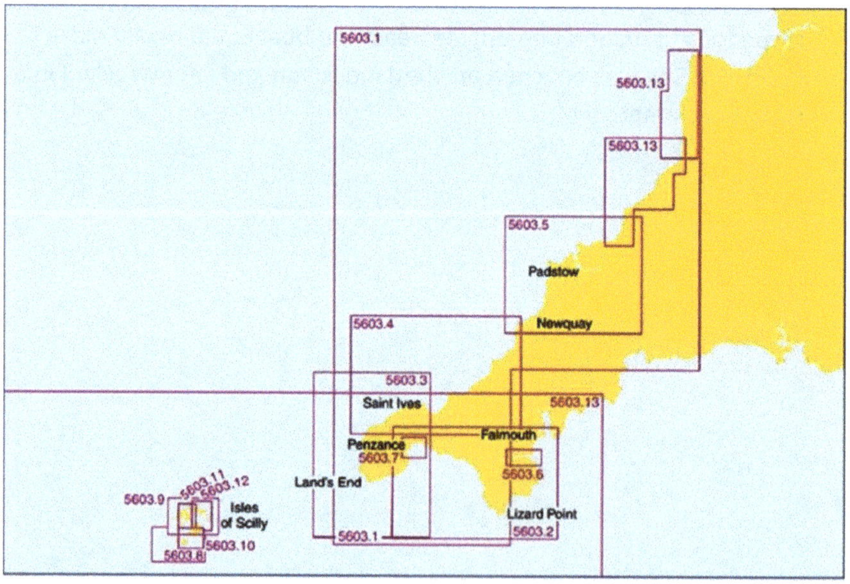

Figure 304. Action off The Lizard. [834]

---

[833] http://img241.imageshack.us/img241/9068/mosquito02rx2.jpg
[834] http://www.seachest.co.uk/acatalog/a137.html

Figure 305. American manufacturers took pride in being associated with the glamorous PT boats. (Library of Congress Prints and Photographic Division).

Date	Craft, Name, Event or Country	Event, Builder and Boat Specifications	City, River, Lake, Sea, Ocean
1944[835]	February 11 *PT-279, 282*	The two boats collided off Cape Torokina and the *279* sank as a result. **Type:** Higgins PT-200-class 78'.	Bougainville Is. Solomon Islands
1944[836]	February 12/13 2nd, 8th SFltl.	The flotillas were conducting mining operations South East of Grimsby and South East of Humber estuary. During operations the S-99 and S-65 sank the armed trawler *Cap d' Antifer* with torpedoes.	North Sea
1944[837]	February 12 *MTB-627, 653* "D" (RNorN)	The two Royal Norwegian Navy "Dog boats" sank two coasters, BOS's/D/S *Irma* and *Henry,* in a German convoy off Hustadvika south of Kristiansand. The loss of 100 civilian passengers both Norwegian and German caused controversy. **Builder:** *627*-Harris, Appledore; *653*-Robertson, Sandbank**. Type:** Fairmile "D" ,115'	Norwegian Sea
1944[838]	February 13/14 MTBs and S-boats	A huge firefight as 17 S-boats from 3 different flotillas plus three boats from the German 34th Minesweeping Flotilla engaged 5 British MTBs, along with the corvettes HMS *Mallard* and HMS *Shearwater*. The battle left the MTBs *444* and *455,* and the *S-89* and *133,* all heavily damaged.	North Sea English Channel
1944[839]	February 22 *PT 200*	The boat was in a collision off Newport, Rhode Island (Naval Torpedo Boat Testing facilities) and did not sink until the next day.	Atlantic Ocean USA
1944[840]	February 22/23 *MTB-690, 610* "D"	These two boats, along with the British destroyer HMS *Garth* and HMS *Southdown* repelled an attack by 15 S-boats near Smith's Knoll. The *Garth* sank the*S-94* and the S-128 was also lost. **Builder:** Boat Construction Co., Falmouth, Fairmile "D", 115'	English Channel

---

[835] Polmar, and Morison *PT Boats At War,* 123 and Hoyt, Edwin *P MacArthur's War The Seventh Fleet and the Battle for the Philippines* (Orion Books, 1989), 186-187.
[836] "The Loss of *HMT Cap d'Antifer*". http://www.capdantifer.org.uk/index.php, last accessed March 5, 2010.
[837] Rohwer, *Chronology of the War At Sea,* 301 and *Norwegian MTB Flotilla in Shetland* , 4.
[838] Rohwer, *Chronology of the War At Sea,* 306.
[839] *Casualties, Navy and Coast Guard Ships* http://www.history.mil/faqs/faq82-1.htm.
[840] Reynolds, *Dog Boats At War*, 93 and Hümmelchen, *German Schnellboote*, 170.

Figure 306. The 20-mm bow gun was the most advanced mechanically engineered deadly weapon on the S-boats. Its distinctive advantage was that the entire mount was raised and lowered by the gunner through hydraulics. As the bow would rise or fall, the gunner could to raise or lower his gun mount and maintain a level firing angle against enemy ships at all times and at all speeds. His mount was also surrounded by heavy armor plate.[841]

Figure 307. *MTB-449,* a sister of *MTB-627* and *653.*[842]

[841] http://www.panzergrenadier.net/forum/viewtopic.php?f=88&t=8293&p=80214
[842] http://www.yalumba.co.uk/Framesets/British%20Power%20Boat%20Co%20-%20page%202.htm

Figure 308. This Shell Oil poster showed PT boats and Navy "Black Cat" aircraft working together against a Japanese ship. (Library of Congress Prints and Photographic Division)

Date	Craft, Name, Event or Country	Event, Builder and Boat Specifications	City, River, Lake, Sea, Ocean
1944[843]	February 25 *MTB-415, 431*	The two MTBs, while attacking a German convoy, managed to register several hits on the German escort, *M-4618.* The damaged boat was capable of reaching the Jersey Isle for repairs. **Builder:** British Power Boat Company	English Channel
1944[844]	February 24/25 2nd SFltl.	The flotilla attacked convoy FS 1371 and only one ship was sunk, the freighter SS *Philip*. The British destroyers HMS *Vivien* and HMS *Eglinton* drove off most of the flotilla.	English Channel
1944[845]	February 26 *PT-251*	The *251* was sunk by Japanese shore batteries off Sio, New Guinea. **Type:** Elco *PT-103*-class,80'	Bismarck Sea
1944[846]	February 26/27 5th & 9th SFltl.	Operating out of Cherbourg, France, the flotilla sank the freighter SS *Modavia* and LCT-381	English Channel
1944[847]	February 27/28 5th & 9th SFltl.	While hunting for their convoy off the English coast, the *S-142* and *S-136* collided with each other and sustained severe damage.	English Channel
1944[848]	March *S-87*	An improved form of radar was tested aboard the vessel. With the improvements, there was still difficulty in showing both the range and bearing on the small screen.	English Channel
1944[849]	March 6/7 53rd MTB Flotilla	The 53rd and other MTB groups hunt German convoys off the Dutch Coast. Off the Hook of Holland, MTBs *225, 241, 244* and *234* sink the patrol vessel *V1304,* damage a gun ferry and another patrol boat.	English Channel

[843] Rohwer, *Chronology of the War At Sea*, 309.
[844] Hümmelchen, *German Schnellboote*, 69.
[845] *Casualties, Navy and Coast Guard Ships, WW II http;//*www.history.navy.mil/faqs/faq82-1.htm.
[846] Hümmelchen, *German Schnellboote*, 170.
[847] Rohwer, *Chronology of the War At Sea*, 308.
[848] Frank, *German S-Boats in Action in the Second World War*, 138.
[849] Rohwer, *Chronology of the War At Sea* ,309.

Figure 309. Higgins made more PTs than Elco. The British and Russians preferred the Higgins boats over the ELCO because of their more rugged construction and better sea keeping abilities. Both the British and Russians used the Higgins boat as motor gun boats (MGBs).[850]

Figure 310. MTBs ran in larger and larger packs while on the hunt for the deadly R- and S- boats. "The MGB Flotilla of coastal Forces legend Lt. Commander R. P. Hitchens RNVR. This photograph was censored—the boat numbers were erased from the bows—before being used in Gordon Holman's book The Little Ships printed by Hodder and Stoughton Ltd., London .October 1943."[851]

---

[850] Collection of Frank Andruss, Sr., The Mosquito Fleet, used with permission

[851] http://www.yalumba.co.uk/Framesets/British%20Power%20Boat%20Co%20-%20page%202.htm

Date	Craft, Name, Event or Country	Event, Builder and Boat Specifications	City, River, Lake, Sea, Ocean
1944[852]	March 6/8 2nd Torpedo Cutter Brigade	The Soviet Brigade of 10 TKAs moved from their base at Gelendzhik around the Crimea Peninsula to the port of Skadousk as part of the Soviet major offensive north of the Lower Dnieper, Romania.	Black Sea
1944[853]	March 7 *PT-337*	She was destroyed by Japanese shore batteries in Hansa Bay, New Guinea. **Type:** Elco *PT-103*-class 80'	Papua New Guinea Solomon Sea
1944[854]	March 14/15 *MTB-353, 417*	British MTB forces attacked the German 36th Minesweeping Flotilla off Gravelines and sink the lead boat, *M-3630.* During the firefight, *MTB-417* was sunk by gunfire. *MTB-353* sank the minesweeper *M-10* off the Dunkirk coast.	English Channel
1944[855]	March 15/16 5th SFltl.	The flotilla of ten boats was assigned to attack British convoy WP.492 north of Lands End. A strong force of British warships intercepted the flotilla and scattered the S-boats.	English Channel
1944[856]	March 17 *PT-283*	Lost off Point Moki, Bougainville Island to IJN machine gun fire and a 5" shell from the destroyer USS *Guest.*` **Type:** Higgins *PT 200*-class, 78'	Solomon Islands Solomon Sea
1944	March 17 *MTB-266* (ex-*PT-17)*[857]	Seriously damaged by Italian naval gunfire and declared "Constructive Total Loss" or CTL. **Type:** Elco 80'	Mediterranean
1944[858]	March 19 PTs	(SOE Ops "Cromer") The two boats were to land two agents and pick up secret mail from the coast of Margiou Bay. The operation failed three times on three dates due to bad weather.	Seacoast France Mediterranean

---

852 Rohwer, *Chronology of the War At Sea*, 298.
853 Lambert, *Allied Coastal Forces*, Vol. 2, 140.
854 Rohwer, *Chronology of the War At Sea*, 311.
855 Rohwer, *Chronology of the War At Sea* ,312
856 Polmar, *PT Boats At War*, 123.
857 *Warships Supplemental*, 10.
858 Richards, *Secret Flotilla*, 688.

Figure 311. "Sterling engines developed a solid reputation in marine power plants early in their initial use. Low maintaince, reliability and cost made this engine a must have for the navies of the world" (Library of Congress Prints and Photographic Division)

Date	Craft, Name, Event or Country	Event, Builder and Boat Specifications		City, River, Lake, Sea, Ocean
1944[859]	March 26 *MTB-352*	The *352* boat was lost in a collision with another surface craft in the North Sea. **Builder:** USA -Vosper72'		North Sea
		Speed: 38 knots	**Displacement:** 40 tons	
		Length: 72'-6"	**Beam:** 19'-3"	
		Engines: 3 x Packard 3,600-bhp	**Armament:** 2 x .50-cal MG, 2 x .303-cal. MG, 2 x 21-in torpedo tubes, 2 x depth charges	
1944[860]	March 26 *MTB-717* (RNoN)	In an intense surface action off Bjorn Holmen, Norway, the *717* suffers one dead, one wounded and the boat badly shot up. Fairmile "D" 115'		North Sea
1944[861]	March 26 *S-93, 129*	On this night, 358 Marauder bombers attacked Ijmuiden Harbor and destroyed much of its shipping facilities, along with the *S-93, 129* and patrol vessel *V 1416.* **Type:** ***S-38***		English Channel

Figure 312. A German assault boat (*Sturmpioniere*). (Hans Liska)

[859] *Warship Supplement 65*, 12 .
[860] *Norwegian MTB Flotilla in Shetland*, http://shetlopedia.com/The_MTB_Flotilla_in_Shetland, 4.
[861] Rohwer, *Chronology of the War At Sea*, 312.

Figure 313. The Norwegian sailors were glad for the use of MTB's from the British but they worked at keeping their national identity intact.[862]

Figure 314. Mistaken identity between air power and PTs would continue to be a serious problem that led to lost, shot-up boats ,and downed planes with dead pilots. This photo shows P-40s flying for Australia.[863]

[862] RNoN Pennant http://images.google.com/imgres?imgurl=http://www.smesh.co.uk/
[863] http://www.pacificaces.ocatch.com/sabre.htm Ausie P-40 image

Date	Craft, Name, Event or Country	Event, Builder and Boat Specifications	City, River, Lake, Sea, Ocean
1944[864]	March 27 *PT 353, 121*	Australian *P-40* aircraft, by a mistake in identification, attacked and destroyed both boats in Bangula Bay, New Britain. There was a huge loss of life when the firing finally stopped. **Type:** Elco *PT-103*-class 80'	Bismarck Archipelago Solomon Sea
1944[865]	March 27 *PT 202, 214, 218*	The *218* was outfitted with radar, and joined with two British destroyers and several MGBs to stop Axis supply ships from running in the shallow water off the island of Elba. The radar proved its value and the flotilla sank 5 fully loaded F-lighters carrying, fuel, ammunition, and supplies. The *PT-202* is credited with putting 2 torpedoes into an F-lighter.	Mediterranean Elba Island
1944[866]	March 27/28 2nd, 4th, 8th SFltl.	The flotillas laid an ambush near Smith's Knoll for MTBs and MGBs, but it failed.	English Channel
1944[867]	March 29 2nd,8th,4th SFltl.	These flotillas were returning to base after attacking a British convoy off the Dutch coast when MTBs *350, 244, 245, 241* jumped them. In the ensuing firefight, *MTB-241* went down off Den Helder. **Builder:** Morgan Giles, Vosper 72'-6" **Speed:** 40 knots — **Displacement:** 38 tons **Length:** 72'-6" — **Beam:** 17' **Engines:** 3 x 3,450-bhp, petrol — **Armament:** 2 x 18" tubes, 4 x twin .303-cal. MG	Dutch Coast North Sea

[864] Lambert, *Allied Coastal Forces,* Vol.2, 140.
[865] Keating, *The Mosquito Fleet*, 158-159.
[866] Rohwer, *Chronology of the War At Sea,* 312.
[867] Whitley, *German Coastal Forces,* 69; *Warship* Supplement 65, 4, 9.

Figure 315. Gunner Stanton loading a 20-mm cannon. (Collection of James Stanton, used with permission)

Figure 316. The awesome fire power carried by the *S*- and *R*- boats was never clearly understood by the American forces. This photo shows the twin 20-mm cannon with a Luftwaffe 37-mm machine gun cannon on the stern. About mid 1944, the Germans started to furnish all boats with quad 20-mm cannons.[868]

[868] Frank, *German S-Boats in Action,* 60.

Date	Craft, Name, Event or Country	Event, Builder and Boat Specifications	
1944[869]	April 8 TKA Div.	The *TKA-101* torpedoed and sank the German minesweeper *M-37* in Narva Bay. **Type:** *G-5,* 61.8’	Baltic Sea
1944[870]	April 12 *PT-135*	The *135* ran aground on Crater Point, east of Cape Gloucester, and was destroyed to prevent capture. **Type:** Elco *PT-103*-class, 80’	New Britain Bismarck Sea
1944[871]	April 17 *MTB-266* (ex *PT-17*)	The boat was damaged by Italian naval surface craft and listed as a constructive total loss. **Type:** Elco, McLean, 81’ **Speed:** 40 knots **Displacement:** 32 tons **Length:** 81’ **Engines:** 3 x Hall Scott 2,700-bhp **Armament:** 4 x 18” torpedo tubes, 2x.50-cal. MG, 6 x .303-cal. MG, 4 x depth charges	Mediterranean
1944[872]	April 18 *MTB-707* “D”	Lost in a collision with the French frigate *L’Escarmouche* north of Ireland. **Builder:** Boat Construction Co., Fairmile “D” 115’	North Sea

869 Rohwer, *Chronology of the War At Sea*, 317.
870 Lambert, *Allied Coastal Forces* Vol. 2, 140; Hyper War: *US Navy Torpedo Boats (PT-Boats),* 1940-1945, 8.
871 *Warship* Supplement 65, 10.
872 Reynolds, *Dog Boats At War,* 273; *Warship* Supplement, 19.

Figure 317. The flag of the Soviet Baltic Fleet.[873]

[873] http://upload.wikimedia.org/wikipedia/commons/a/a3/Flag_of_Armed_Forces

Figure 318. *PT-302* from Squadron 22, moves into Bastia in 1944. Crewmember (left) Arthur Frogello is laughing although very seasick. He is holding onto Wagner, the boats Radio Operator. The crew was making fun of Frongello for being sick, so he decided to laugh along with them. (Collection of Frank Andruss, Sr.)

Date	Craft, Name, Event or Country	Event, Builder and Boat Specifications	City, River, Lake, Sea, Ocean
1944[874]	April 21/22 British Navy	On this day, British naval forces carried out the first truly successful shore-controlled interception of an S-boat strike. Ten S-boats sailed to attack a convoy off Dungeness. The British were able to place Force 113 at the scene: two destroyers, HMS *Volunteer* and HMS *Middleton,* plus several MTBs. In the fight that took place the *S-167* and *MTB-416* were both damaged.	North Sea
1944[875]	April 23/24 *S 100*	The S-boat sank the tug *Roode Zee* as it was making its way to Portsmouth. **Type:** *S-100*	English Channel
1944[876]	April 23/25 MAS boats	The Germans ordered hundreds of vessels to be scuttled in harbors under German control in Italian waters. The *MAS 549, 561, 558, 557, 505, 544,* and *551* were among those scuttled.	Mediterranean
1944[877]	April 24 *MTB-671, 617, 632*	The *671* was raked by heavy gunfire from 2 German *Mowes* "torpedo boats" (ships, considerably larger than *Schnelleboote*) off Cherbourg. The *617* and *632* returned to base with damage. The *671* lost all but two of its 30-man crew. **Builder:** Dickie Fairmile "D" 115'	Baie de la Seine English Channel

[874] Whitley, *German Coastal Forces,* 70-71.
[875] Hümmelchen, *German Schnellboote (E-Boats),* 71.
[876] Rohwer, *Chronology of the War At Sea,* 411.
[877] Reynolds, *Dog Boats At War,* 95-96; *Warship* Supplemental 65, 19.

Figure 319. *The Roode Zee* was an old ship and not much of a target for an S-boote to take on. Still, orders called for sinking all enemy ships.[878]

Figure 320. *S-100* v. *Roode Zee* was quite a mismatch. (Collection of Hans Frank)

---

[878] http://www.zeesleepvaart.com/gesch8.1.htm

Date	Craft, Name, Event or Country	Event, Builder and Boat Specifications	City, River, Lake, Sea, Ocean
1944[879]	April 24/25 *PT 218*	The American PT boats' main contribution to the fight in the Mediterranean was their radar . The *218* boat was put in the forefront of "Operation Newt" as a scout. The force of British MGB and MTBs sank 2 F-Lighters and the tug *Tebessa* off San Vincenzo, Italy. **Type:** *PT-200* class, Higgins, 78' **Speed:** 41 knots **Displacement:** 43 tons **Length: 78'** **Beam:** 19' 11" **Draft:** 5'3" **Crew:** 17 **Engines:** 3 x 1,200-hp Packard, petrol **Armament:** 4 x 21" Mark VIII torpedo tubes, 2 x 20-mm MG, 1 x 40-mm cannon	Tyrrhenian Sea
1944[880]	April 25/26 *PT-202, 213, 218*	The three PT boats attacked two German torpedo boats, the *T-26* and *29*. Heavy gunfire was exchanged and the fight resulted in a draw. **Type:** *PT-200* class, Higgins	Mediterranean
1944[881]	April 25/26 5th & 9th SFltl.	The twelve boats of the flotillas met the British convoy in the convoy route off Selsey Bill near Nab Tower and were fired on by British shore batteries from Culver Cliffs. At the same time two-destroyer patrols, HMS *Stevenstone* and *Seymore* ,and the French destroyer *La Combattante* with HMS *Rowley* converged on the spot and gave chase to the S-boats.	West Sussex English Channel

879 Reynolds, *Dog Boats At War*, 146.
880 Rohwer, *Chronology of the War At Sea*, 312.
881 Whitley, *German Coastal Forces*, 70.

Figure 321. Henry Beazley was a crewman on the *PT-209*. He spent much of his time in the Mediterranean and wound up posted to Japan at the end of the after surviving the close-in fighting in the last days of the war. Mr. Beazley is still with us and in good health. (Collection of James Stanton)

Figure 322. *PT-209* with big brass on board during the second invasion of France, Operation Dragon. Lt. Gen. Lucian K. Truscott, Jr. is in the foreground and Lt. Gen Jacob Devers is in the cockpit. Truscott's son, Lucian K. Truscott III, and grandson, Lucian K. Truscott IV, would follow him to West Point. James Stanton is on the extreme left. (Collection of James Stanton)

Figure 323. The 15th Squadron emblem was designed by Mrs. Pat Kennedy of Dallas, Texas, for her husband, Mack Kennedy, an ensign in the squadron. (Collection of James Stanton)

Date	Craft, Name, Event or Country	Event, Builder and Boat Specifications	City, River, Lake, Sea, Ocean
1944[882]	April 26 *MTB-623, 717* (RNoN)	The two Norwegian boats were about to return to base when they spotted the German submarine *U-637* off Karmoy. Both boats loosed torpedoes and immediately drew heavy cannon fire from the sub. It shortly caught fire; the MTB crews heard internal explosions and the sub sank, in flames, stern first. **Builder:** Fairmile "D", 115'	North Sea
1944[883]	April 26/27/28 Operation Tiger	"T4" was the operational name for US Navy convoy that sailed on a dry run for the invasion of Normandy. The convoy's heart was eight LSTs packed with Army Engineers destined for Omaha Beach. The escort was to have been composed of the British corvette HMS *Azalea* and the destroyers HMS *Scimitar* and *Saladin*. However, *Saladin* never made it to the convoy. German S-boats easily penetrated the convoy and wrecked total havoc with the eight LSTs with cannon and heavy machine gun fire. *LST-531* and *507* sank with most of the men on board and all of their invasion equipment. The LST-289 was severely damaged and went back to port with many dead and wounded soldiers and sailors. The LST was in the yards until war's end. No S-boats were sunk or damaged in the attack.	Lyme Bay English Channel
1944[884]	April 29 *PT-346, 347*	US Naval aircraft off Cape Pomas destroyed both boats. It was another case of faulty intelligence and poor communication between the surface fleet and the air arm. **Type:** Elco Type *PT-103*-class 80'	New Britain Bismarck Archipelago

882 *The Norwegian MTB Flotilla in Shetland*, 4.
883 Stillwell, *Paul Assault on Normandy*, 40; Whitely, German Coastal Forces, 71-72; Operation Tiger http://www.history.navy.mil/faqs/faq20-1.htm, Stillwell, Paul, Ed. Assault on Normandy First Person Accounts from the Sea Services. Exercise Tiger, a disaster of WW2-what did happen? http://www.mikeekemble.com/ww2/tiger1.html Accessed by Author 03/18/2006,.32.
884 Lambert, *Allied Coastal Forces*, Vol. 2, 142.

Figure 324. The horrific damage to *LST-289* hints at the tremendous loss of life.. It is not widely known that if one more LST had been sunk or put out of action, D-Day would have been set back by several months.[885]

---

[885] http://www.lstmemorial.org/wayitwas/wayitwas.htm LST 289

Figure 325. One of the S-boats involved in Operation Tiger.[886]

Figure 326. The mainstay of U.S. naval operations in the Pacific, the Hellcat was a substantial airplane heavily armed with .50-calibre MGs. With its armor plate and self-sealing gas tanks, it was difficult to shoot down.[887]

[886] http://img231.imageshack.us/img231/3350/operationtiger3lz2.jpg
[887] http://en.wikipedia.org/wiki/File:F6F_Hellcat.jpg

Date	Craft, Name, Event or Country	Event, Builder and Boat Specifications	City, River, Lake, Sea, Ocean
1944[888]	May 3 *MTB-711* "D" (RNoN)	The 711 was vectored to a position in the North Sea by aircraft to examine an unidentified motor launch. 50 nm north off Muckle Fugga the *711* crew found six Luftwaffe men in the act of deserting from the German Air force. They were taken to Lerwick and interned as POW's. **Builder:** Fairmile "D", 115'	North Sea
1941[889]	May 5 *PT-247*	Sunk by Japanese shore batteries off Cape Torokina, Empress Augusta Bay, Bougainville Island. **Type:** Elco *PT-103*-class,80'	Solomon Islands Solomon Sea

Figure 327. The "Dog" boats were a favorite with the Norwegian sailors for direct combat against the German S-boats. (Collection of John Lambert)

[888] The Norwegian MTB Flotilla in Shetland, 7.
[889] HyperWar :*US Navy Torpedo Boats* (PT-Boats), 4.

Figure 328. One of the more absurd uses of the PT boat image. Sometimes Hollywood and Madison Avenue did not know when to shut up. (Library of Congress Print and Photographic Division)

Date	Craft, Name, Event or Country	Event, Builder and Boat Specifications	City, River, Lake, Sea, Ocean
1944[890]	May 5 New Model S-100 boats *S-151-158*	A radical design change occurred to account for new armament (below) developed and to be used in Dutch waters. Instead, the boats were sent to the Mediterranean to reinforce 7th SFltl. The boats were slower than the rest of the flotilla's S-boats and poor sea keepers. They also suffered frequent engine failures. Seven boats were built by Gusto Werft, Schiedam, Holland. **Type:** *S-1* **Speed:** 30 knots cruise, 35 knots maximum; **Displacement**: 54 tons standard, 68 fully loaded; **Range:** 350 nm **Length:** 91.86' (28 m); **Beam:** 14.11'; **Draft**: 8.01' **Armament:** 2 x torpedo tubes; 1 x 20 mm cannon (2 cm AA 38 socket mount); 1 x 15-mm AA MG 39, 2 MG 34	Mediterranean
1944[891]	May 7 *MTB-666* "D"	Sunk in combat with enemy surface craft off Ijmuiden. **Builder:** Dorset Yacht Co, Fairmile "D", 115'	North Sea
1944[892]	May 7 TKA *215, 218, 219*	The TKAs captured the M//K *Moder II*, set her on fire and sank her in the Varangerfjord. **Type:** *G-5*, 62.5'	Finmark, Barents Sea

Figure 329. The new class of S-boote turned out to be an embarrassment to the S-boote name. None of the new type boats gave satisfactory service.[893]

---

[890] Frank, *German S-Boats in Action,* 151; Fock, *Fast Fighting Boats,* 297-299.
[891] Lambert, *The Fairmile 'D' Motor Torpedo Boat,* 15; *Warship* Supplemental 65, 19.
[892] Internet correspondence with Richard Worth, July 12, 2007.
[893] http://panzergrenadier.net/forum/viewtopis:php?f=88&t=8293=0214

Figure 330. Soviet gunners on the TKA were at first very vulnerable to being fired on by enemy aircraft. Later, more armor plating was added to the gunners mounts and larger caliber guns made their life expectancy a bit longer, but not by much.[894]

[894] http://blog.kievukraine.info/2007_09_01_archive.html

Date	Craft, Name, Event or Country	Event, Builder and Boat Specifications	City, River, Lake, Sea, Ocean
1944[895]	May 8 TKA	Near Sevastopol, the German submarine hunter Uj-104 was torpedoed by TKAs. The ship broke in half and they towed the stern to Sevastopol where it was finally destroyed by Allied aircraft. The bow sank later. **Type:** *G-5*, 62.5	Black Sea
1944[896]	May 8 TKA	TKAs sank the German lighter *Elbe-5*, patrol boat *G-3102* (*Drache*) and the freighter SS *Cornelis Anni*, near Sevastopol. Type:*G-5*, 62.5'	Black Sea
1944[897]	May 12/13 *S-141*	While escorting the convoy WP. 526 the French destroyer *La Combattante* sank the S-*141* and damaged two other S-boats in the English Channel.	English Channel
1944[898]	May 14/15 *PT-558, 552, 559*	Two German torpedo boats, *TA 26* and 30, were laying mines when they were attacked by the three PT boats. Both German vessels were sunk. **Type:** Elco PT-103 class,80'	Western Mediterranean

Figure 331. Patriotism ran high on all sides.

---

[895] Internet correspondence with Richard Worth, July 12, 2007.
[896] Internet correspondence with Richard Worth, July 12, 2007.
[897] Hümmelchen, *German Schnellboote*, 71.
[898] Rohwer, *Chronology of the War At Sea*, 324.

Figure 332. The destroyer *Le Combattante* was originally built for the British Navy as a *Hunt* Class destroyer, HMS *Halden* (L 19). She was given to the Free French (FNFL) March 14, 1941. *Le Combattante* had an active and fruitful life fighting the German Navy in the English Channel and North Sea. She hit a mine on February 23, 1945, sank and lost 64 men.[899]

Figure 333. The UMB mines deployed by S-boats were an older model, but still a very effective mine.[900]

---

[899] Wikipedia http://en.wikipedia.org/wiki/French_destroyer_La_Combattante

[900] Frank, *German S-Boats in Action,* 154, 172.

Date	Craft, Name, Event or Country	Event, Builder and Boat Specifications	City, River, Lake, Sea, Ocean
1944[901]	May 18 *MTB-203* (RNorN)	The Norwegian MTB struck a mine off Boulogne, France and sank. **Builder: White, Cowes** **Speed:** 38 knots **Displacement**: 39 tons **Length:** 72′ 6″ **Beam:** 20′ **Engines:** 3 x Packard 3,600-bhp **Armament:** 4 x .50-cal. MG, 2 x 21″ torpedo tubes	English Channel North Sea
1944[902]	May 18/19 8th SFltl. *S-83, S-87*	The flotilla sowed mines, which resulted in the minesweeper USS *Wyoming* and *MMS-227* being sunk. British Swordfish aircraft of the No. 819 Squadron bombed the *S-87* while it was being towed by S-83. **Type:** S-38	English Channel
1944[903]	May 19 *MTB-709, 712*	The two boats set an ambush in Inner Leads, near Askevold and sank the German Vp 5531 and Vp 5525. Builder:709, Dickie, Bangor & Tarbert; 712: Wallasea Bay Yacht Station, Fairmile 'D', 115′	Norwegian Sea
1944[904]	May 19 *PT-71, 72*	(SIS Ops."Glover") Near Île-Grande, the boats were to retrieve one agent and secret correspondence. The first attempt failed due to agents delayed by travel difficulties in France. The second attempt was successful on May 24th. **Type:** Higgins *PT-71* class, 78′	Mediterranean
1944[905]	May 19 *PT-556*	(SOE Ops "George") The 556 boat was to land four SOE agents and retrieve five agents from Morgiou Bay. The mission failed due to faulty navigation. **Type:** Elco *PT-103* class, 80′	English Channel
1944[906]	May 20 *PT-559*	(SOE Ops "George"---repeat) The mission was to land and retrieve OSS agents at Morgiou Bay. This time, the mission was successful. Elco: *PT-103*, 80′	Mediterranean

---

901 Whitley, *German Coastal Forces*, 70; *Warship* Supplemental 65, 8.
902 Whitley, *German Coastal Forces*, 70.
903 Reynolds, *Dog Boats At War*, 111.
904 Richards, *Secret Flotillas*, 318.
905 Richards, *Secret Flotillas*, 689.

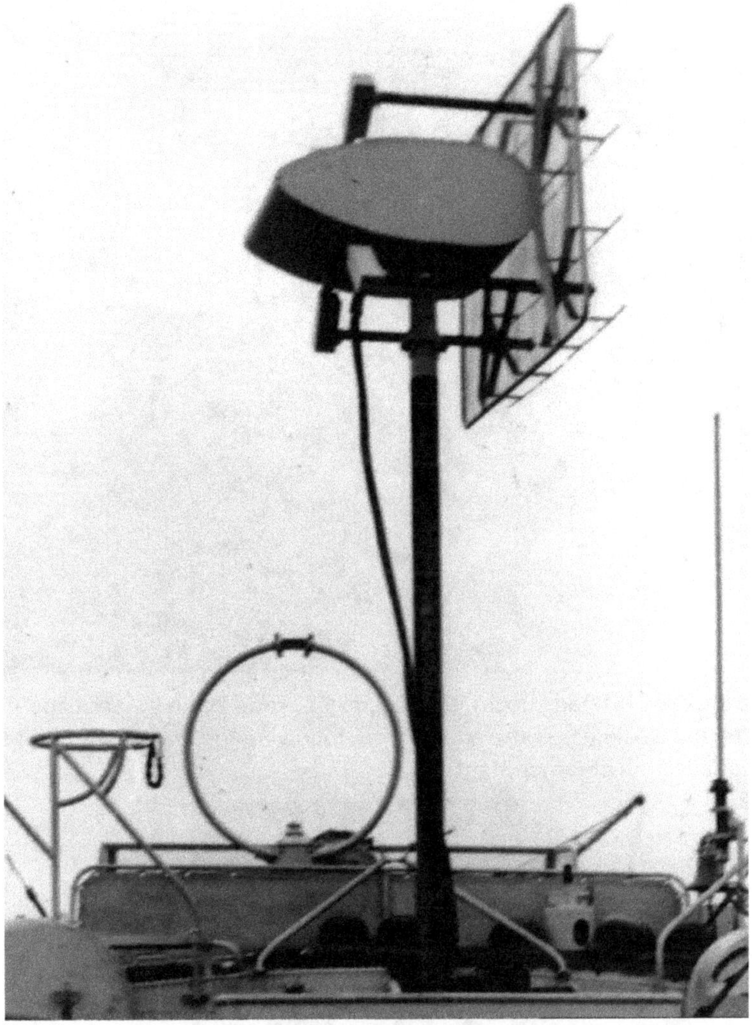

Figure 334. The *S-112* was selected for radar testing in 1942. The *S-112* was still in action in 1944 but did not carry radar. The early trials of Luftwaffe radar being adopted for S-boat use were not wholly successful. Because of its size, the S-boat crews jokingly called this device the "Enemy shell collector."[907]

---

906 Richards, *Secret Flotilla*, 689.
907 Frank, *German S-Boats in Action*, 137-139.

Figure 335. In action off Vada Rocks, the Higgins PT crews begin to learn about the real strength and weakness of the enemy from following the lead of the British MTB crews. (Collection of James Stanton)

Figure 336. A British wartime sentiment that has not lost its relevance today. (Library of Congress Print and Photographic Collection)

Date	Craft, Name, Event or Country	Event, Builder and Boat Specifications	City, River, Lake, Sea, Ocean
1944[908]	May 23/24 *S-112*	The destroyer HMS *Vanquisher* damaged the *S-112* in a surface firefight near convoy WP.526. The *112* had been chosen to test experimental radar sets (FuMG) in 1942.	North Sea
1944[909]	May 24 *PT-217*	(OSS Ops "Chatham") Off Cap-Lardier, France, the crew picked up airmen evading and escaping from German held territory. **Type:** Higgins *PT-200* class, 78'	English Channel
1944[910]	May 24 *PT-218*	(OSS Ops "Grapefruit") The crew of 218 were sent to retrieve agents Marius Chavant (Clement) and Lt. Jean Veyrat at Cap-Lardier and were successful. **Type:** Higgins *PT-200* class, 78'	English Channel
1944[911]	May 24 *PT-202, 213, 218*	The PTs attacked a strong force of German corvettes and sank the *Uj 2223* (ex Italian *Marangone*) and badly damaged the *Uj 2222* (ex-Italian *Tuffeto*) off Vada Rocks, Corsica. **Type:** Higgins, PT 200-class, 78'	Mediterranean
1944[912]	May 25 *PT-339*	The boat was run aground off the island of Pur Pur, New Guinea, and was destroyed to prevent capture. **Type:** Elco *PT-103*-class, 80'	Coral Sea
1944[913]	May 28 *MTB-732* (RNorN)	During an action against German surface craft the French destroyer *Le Combattante* accidentally sank the MTB. **Builder:** Dorset Yacht Co., Fairmile "D", 115'	English Channel
1944[914]	May 29 *MTB-672*	She struck a mine in the Schelde, Netherlands, and sank. Three men came up missing; the balance were rescued. **Builder:** Silver, Rosneth, Fairmile "D", 115'	North Sea

908 Whitley, *German Coastal Forces*, 72 and Frank Hans , *German S-Boats in Action in the Second World War*, 137.
909 Richards, *Secret Flotilla*, 689.
910 Richards, *Secret Flotilla*, 689.
911 HyperWar: US Navy Torpedo Boats (PT-Boats) 1940-1945 and Rohwer, *Chronology of the War At Sea*, 324.
912 Lambert, *Allied Coastal Forces* Vol. 2, 140.
913 Reynolds, *Dog Boats At War*, 273; Lambert, *The Fairmile D Motor Torpedo Boat*, 15.
914 Reynolds, *Dog Boats At War*, 186.

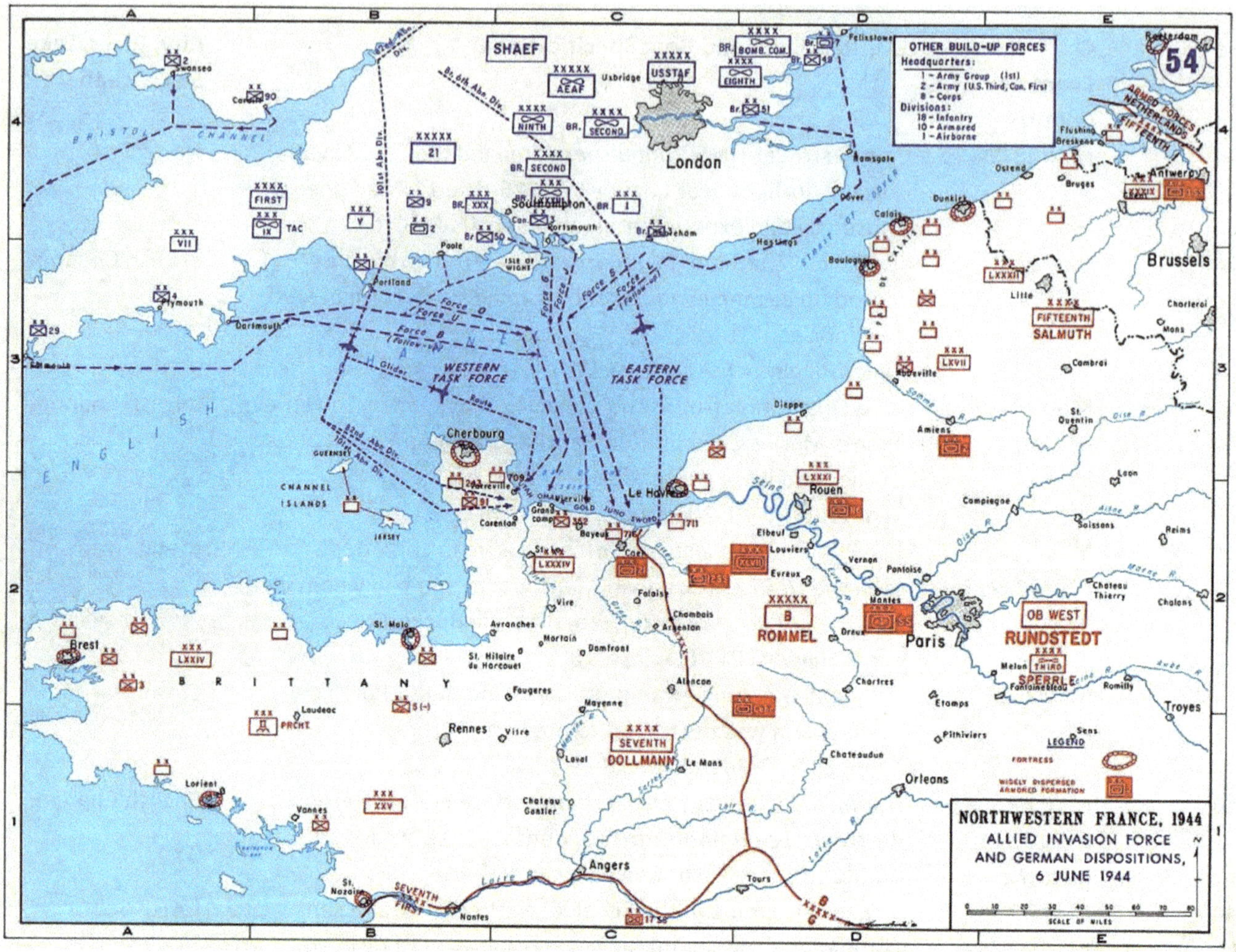

Figure 337. Neptune was the naval segment of the European invasion plan. The depth and breadth of the planning was far beyond the scope of any previous military operation.[915]

---

[915] http://upload.wikimedia.org/wikipedia/commons/1/1e/Allied_Invasion_Force.jpg

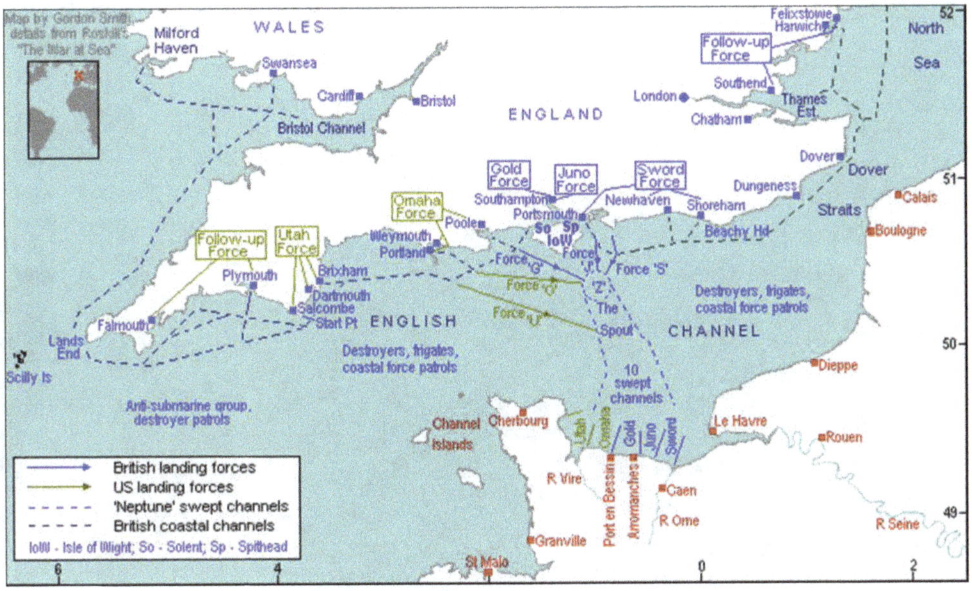

Figure 338. The delivery of the invasion forces required careful traffic planning.[916]

Figure 339. Operation Overlord was a "message from America" that Nazi domination of Europe would not stand.

[916] http://realityandsanity.wordpress.com /2007/06/06/d-day-june-6-1944-never-forget/

Photo # NH 44308 PT boats crossing the English Channel, with B-17s overhead, 6 June 1944

Figure 340. PT boats crossing the English Channel on D-Day, with B-17s overhead.

Figure 341. The German artillery crews on Juno Beach woke up to see the greatest Armada in human history on their doorstep. They caused heavy losses among the troops on the beachhead, but in the end the coastal defense system that had taken four years to construct was overrun in less than eight hours.[917]

[917] http://www.history.navy.mil/photos/events/wwii-eur/normandy/nor4u.htm

Date	Craft, Name, Event or Country	Event, Builder and Boat Specifications	
1944[918]	May 23/24 5th SFltl.	The flotilla is laying mines with an auxiliary minesweeper off the Isle of Wight when they encounter a British MTB force. The German M-39 is sunk by MTBs 354 and 361 off Quistreharn.	English Channel
1944[919]	May 31 PT *304, 306, 307*	The three PTs took on the German torpedo ships *TA 29* and *30* off the Ligurian coast. Mutual damage results and both forces retire. **Type:** Higgins *PT-200*-class, 78'	Ligurian Sea W. Mediterranean Sea
1944[920]	June 6 Operation Neptune	This began the amphibious phase of Operation Overlord, the invasion of Europe. MTBs and MGBs were in place to cover the Western approaches to the beaches of Normandy. The Germans had 36 *S*--boats in three bases within striking distance of the mass of Allied ships.	Normandy France

[918] Whitley, *German Coastal Forces*, 72; Rohwer, *Chronology of the War At Sea*, 324.
[919] Rohwer, *Chronology of the War At Sea*, 324.
[920] Rohwer, *Chronology of the War At Sea*, 331.

Date	Craft, Name, Event or Country	Event, Builder and Boat Specifications	City, River, Lake, Sea, Ocean
1944[921]	June 6 5[Th] SFltl.	The flotillas left their base at Cherbourg with 14 S-boats to make their first strike at the Allied invasion. Off Cap Barfleur, the *S-139* was sunk by a mine laid by the 64th MTB Squadron off St. Vaast. The 9th SFltl. came to the S-139's aid and lost the *S-140* to another British mine. *S-100, 142, 150,* and *168* broke through the defensive ring of ships and let loose their torpedoes. The Norwegian destroyer *HNoMS Svenner* (former HMS *Shark)* was sunk off Normandy Beach at dawn. An LST and LCT were sunk by torpedoes off St-Vaast. MTBs *448* and *478* and the frigate HMS *Stayner* drove off the S-boat force before more damage was done.	Off Normandy Baie de la Seine
1944[922]	June 6 MTB-624, 682	The German 4th Motor Minesweeper flotilla ran into the British 55th MTB and Canadian 29th MTB flotillas. During the following firefight, the *MTB-624, 682* and *R-49* were all damaged.	Off Normandy Baie de la Seine
1944[923]	June 6 5th, 9th SFltl	Boats from the 9th sank two small landing craft—*LC(1)105* and *LCT 875* off the Seine as they were escorted by ML 903 Torpedoes from the 5th sank the US destroyer *Meredith.*	Off Normandy Baie de la Seine
1944[924]	June 6 9th SFltl.	The Allies laid a mine barrage, "Greengage," off the invasion site and sank the *S--139* and *140.*	Off Normandy Baie de la Seine
1944[925]	June 7/8 4th, 5th SFltl.	The flotilla of eight boats attacked the invasion convoy being led by the destroyer HMS *Beagle*.. The 5th SFltl. attacked with the S-*100, 138* and *84*. All 3 boats were badly damaged in the firefight. The 4th SFltl. suffered damage to *S-145* and *168* and needed to be towed out of the invasion area.	Off Normandy Baie de la Seine

921 Whitley, *German Coastal Forces,* 72.
922 Rohwer, *Chronology of the War At Sea,* 331.
923 Whitley, *German Coastal Forces,* 72.
924 Frank, *German S-Boats in Action,* 110.
925 Rohwer, *Chronology of the War At Sea,* 332.

Figure 342. The engineering spaces of the S-boats were so vastly different from the PT's and MTB's that they cannot begin to compare. There was plenty of head room, adequate ventilation and they could be kept much cleaner while at sea.[926]

Figure 343. The S-boats broke through the protective shield of destroyers, MGB's and MTBs and reached the invasion fleet. (Collection of Hans Frank)

[926] Frank, *German S-boats in Action*, 157.

Date	Craft, Name, Event or Country	Event, Builder and Boat Specifications	City, River, Lake, Sea, Ocean
1944[927]	June 7 *PT 505*	Off Normandy Beach the *PT-505* gave to chase to what it thought was an enemy periscope. Before engaging the target the *505* ran over a mine and suffered two men injured, and extensive structural damage to the boat. The 505 had to be towed by *PT-507* to St. Marcouf Island. Temporary patches were applied to the boat and she was towed to Portland, England, for extensive re-building. **Type:** *PT-103* class, Elco, 80' **Speed:** 41 knots **Displacement:** 38 tons **Crew:** 17 **Length:** 80' **Beam:** 23' **Draft:** 5' **Engines:** 3 x 1,200 hp Packard 4M2500, 12-cylinder, petrol **Armament:** 4 x XIII 21" torpedo tubes, 1 x 20-mm MG,2 x twin .50-cal MG	Normandy Beach English Channel
1944[928]	June 7/8 9th SFltl.	The flotilla of six boats attacked landing craft being escorted by the British *ML-903*. *S-145* and *S-168* were damaged by the guns from landing craft.	Off Normandy Baie de la Seine
1944[929]	June 9 4th SFltl.	The flotilla was successful in breaking through Allied defenses at Normandy and sank a large section of the Mulberry harbor off Omaha and Gold beaches.	Off Normandy
1944[930]	June 9 5th, 9th SFltl.	German S-boats torpedoed and sank the *LST-376* and *LST-314* off Normandy Beach.	Off Normandy

927 Action Report- Into Action, The English Channel , 2-3 http://204.3.134.216/02-07-05-reports-006.html.
928 Rohwer, *Chronology of the War At Sea*, 332.
929 Frank, *German S-Boats in Action* , 111 and DANFS, 634.
930 *Dictionary of American Fighting Ships*, James L. Mooney, ED. (Naval Historical Center, Department of the Navy, 1981, Vol.VIII), 643.

Figure 344. The *505* boat had a close call when it struck a German mine off the invasion beach. Luckily, it was patched and towed back to England for repairs.

Figure 345. The Kriegsmarine was adept at laying some of the largest mine fields in all of WWII. The majority of the Allied ships sunk were by mines and not combat action.[931]

[931] Michael Emmerich, http://www.german- navy.de/kriegsmarine/ships/minehunter/index.html

Figure 346. Mines ready for deployment.[932]

Figure 347. The USS *Nelson* (DD-623) as she was before she was almost sunk by *S-138.*[933]

[932] *Ibid.*

[933] http://en.wikipedia.org/wiki/USS_Nelson_(DD-623)

Date	Craft, Name, Event or Country	Event, Builder and Boat Specifications	City, River, Lake, Sea, Ocean
1944[934]	June 10 2nd,4th ,5th, 8th, 9th SFltl.	Allied destroyers drove off the ten boats of the 5th and 9th flotillas from the invasion fleet on the "Dixie" line. The 4th flotilla laid 24 mines. *S-188* reported hits on an LST and *S-172* and *S-187* each scored hits on freighters. The 2nd flotilla attacked the invasion fleet with FaT torpedoes (pattern-running)[935] and claimed the sinking of one freighter. The 8th flotilla, on returning to Boulogne, lost the S-*180* and *190* to mines.	Off Normandy Baie de la Seine
1944[936]	June 10 58th MTB Flotilla	The British flotilla of six boats attacked German convoy 1253 off Den Helder. The British lost the *MTB-681* and sank the German Patrol vessels *V-1314, V-2020* and *V -2021*.	English Channel
1944[937]	June 10 *MTB-681* "D"	The MTB was sunk in surface combat off the Dutch coast. **Builder:** Brooke Marine, Fairmile "D", 115'	North Sea
1944[938]	June 10 RNorN 54th MTB flotilla	The Royal Norwegian Naval force of five boats attacked the German 11th minesweeping flotilla. In the fighting, *MTB-712* was damaged.	English Channel
1944[939]	June 10/11 2nd , 5th , 9th SFltl.	On this mission, the flotilla laid mines and then attacked Allied convoy WP 457 off the Isle of Wight. *S-177* and S-*178* sank the three freighters, SS *Dungrange,* SS *Ashanti* and SS *Brackenfield*. The 5th *S*-Flotilla attacked the two British frigates HMS *Stayner* and HMS *Halstead*. The *Halstead* had her bows blown off and was later declared a total constructive loss; torpedoes sank *MTB-448*, and the 9th SFltl. lost the S-*136*. The 9th flotilla went up against the two British steam gun boats, HMS *Grey Wolfe* and HMS *Grey Goose*, and their convoy of invasion craft. The flotilla sank the US Navy tug *Partridge* and LST-496 and damaged the *LST-538*..	English Channel

934 Rohwer, *Chronology of the War At Sea*, 332.

935 T IIIa FaT II (G7e) Pattern Running Torpedo, http://www.uboataces.com/torpedo-tiiia.shtml, last accessed 3/8/2010.

936 Rohwer, *Chronology of the War At Sea*,332.

937 *Warship* Supplement 65, 19

938 Rohwer, *Chronology of the War At Sea*, 332.

939 Rohwer, *Chronology of the War At Sea*, 332 and Whitley, *German Coastal Forces,* 73.

1944[940]	June 10/11 MTB 712 (RNorN)	After being hit by several large caliber guns off Texel, the boat grounded itself off Scapa Flow Builder: Wallasea Bay Yacht Station, Fairmile 'D', 115'	North Sea
1944[941]	June 12 *S-138*	On this day, the *S-138* blew the stern off the American destroyer USS *Nelson*. The *Nelson,* while damaged, was towed to Portsmouth for repairs. The destroyers *Nelson*, *Somers* and *Laffey* used radar to direct their big guns and kept the S-boats at bay. **Type:** *S-38*	"Dixie Line" Normandy Coast
1944[942]	June 12 4th SFltl.	The flotilla attacked the destroyers HMS *Onslow,* HMS *Offa,* HMS *Onslaught,* HMS *Oribi* and a MTB squadron off Normandy. The *S-171* sank the *MGB-17.* **Type:** *S-100*	Normandy Coast
1944[943]	June 12 2nd SFltl.	Off Boulogne, the flotilla was attacked by RAF fighter-bombers and lost the *S-178, 179, 189* and *R-97* to the attack. The flotilla is prevented from entering the invasion area by an Allied frigate and 29 Canadian MTBs The *S-181* and *179* are damaged by mines.	Normandy Coast
1944[944]	June 14/15 RAF raid	The RAF sent Lancaster bombers to bomb the S-boat facilities at Le Havre and sank 14 S-boats: *S-84, 100, 138, 142, 143, 144, 146, 150, 169, 171, 172, 173, 187,* and *188*.	Le Havre Baie de la Seine

940 Reynolds, *Dog Boats At War*, 273

941 Whitley, *German Coastal Forces*, 73.
942 Whitley, *German Coastal Forces*, 73.
943 Frank, *German S-Boats in Action*, 113.
944 Whitley, *German Coastal Forces*, 74.

Figure 348. The concrete boat pens built along the French coast became a serious problem to the RAF. Conventional weapons were not powerful enough to produce any real damage. The "Tall Boy" was the answer. The bomb was 21' in length, weighed 12,000 pounds and was dropped from 25,000 feet. The bomb was so aerodynamically perfect that it fell faster than the speed of sound. The fall of the bomb was heard *after* the explosion.[945]

---

[945] http://upload.wikimedia.org/wikipedia/commons/6/62/Tall_Boy_Bombe.jpg and http://www.grand-blockhaus.com/usauk.asp

Figure 349. As the war progressed, more and more American and British MTBs sported radar, which helped even the odds against the S-boats.[946]

[946] http://members.cox.net/squadron35/index.htm

Figure 350. Severe storms slashed the beaches of Normandy for two days. Small craft and large all took a terrible beating. Unloading was stopped and everybody hunkered down until the storm passed.[947]

[947] Artist, Mitchell Jamieson, U.S. Navy Art Collection http://www.history.navy.mil/ac/d-day/exdday/exdday.htm

Date	Craft, Name, Event or Country	Event, Builder and Boat Specifications	City, River, Lake, Sea, Ocean
1944[948]	June 15 Soviet TKA	The Soviet TKAs, under Captain 2nd class Osipov, sank the German Torpedo boat (ship) *T-31* in Viborg Bay, Koivisto Sound, Barents Sea. The drifter *Hugin* was sunk by *TKA-239* with gunfire and depth charges, off northern shore of Varangerfjorden. The *Hugin* was towed to Vadso and later sank from neglect.	Finmark, Norway
1944[949]	June 15/18 *S-133, S-83*	S-boats from the Cherbourg and Boulogne bases were driven from potential targets by the unrelenting attacks of British Mosquito fighter-bombers. *S-133* and *S-83* were damaged in these attacks. **Type:** *S*-38	English Channel
1944[950]	June 16/17 65th Flotilla	The Canadian 65th MTB Flotilla of four boats attacked a German convoy west of the Cotentin Peninsula. *MTB-748* torpedoed the German minesweeper *M-133*, which was then scuttled.	English Channel
1944[951]	June 17/18 *PT-207, MTB-633, 640, 655*	The PT and MTBs were joined by *MGB-658* to attack two German torpedo boats laying mines. There were no reports of any vessels sunk in the firefight. **Type:** PT 207, Higgins PT-200 class, 78'; 633,640,655, Fairmile "D", 115'	Western Mediterranean
1944[952]	June 18 *PT-107, 63*	The boats were lost to onboard fire at Emirau Island. *PT 107*, **Type:** Elco *PT-103*-class, 80'. *PT-63* **Type:** Elco *PT-20* class, 70'	New Ireland Is. Bismarck Sea
1944[953]	June 19 2nd SFltl.	British destroyers sank the *S-190.* The other flotilla boats continued to lay mines in the approaches of the invasion beach through August 1944.	Normandy Invasion Beach
1944[954]	June 20 TKA Div.	The Soviets made a major offensive against the Germans seeking control of the Islands of the Karelian Isthmus. **Type:** *G-5*, 61.8'	Karelian Islands Gulf of Finland

---

948 Rohwer, *Chronology of the War At Sea*, 343 and Internet correspondence from Richard Worth's communications with Kirill Klement'ev May 5, 2004.
949 Rohwer, *Chronology of the War At Sea*, 336.
950 Rohwer, *Chronology of the War At Sea*, 336.
951 Rohwer, *Chronology of the War At Sea*, 324.
952 HyperWar: *US Navy Torpedo Boats(PT-Boats)*,.8
953 Rohwer, *Chronology of the War At Sea*, 336; Frank, *German S-Boats in Action*, 115.

Figure 351. The RAF Mosquito fighter-bomber was a true killing machine. It carried a combination of cannon, rockets and heavy machine guns.[955]

Figure 352. The Royal Marines served on board major combatants for the RN.

---

[954] Internet correspondence with Richard Worth, July 12, 2007 and Rohwer, *Chronology of the War At Sea*, 336.
[955] ttp://img241.imageshack.us/img241/9068/mosquito02rx2.jpg.

Date	Craft, Name, Event or Country	Event, Builder and Boat Specifications	City, River, Lake, Sea, Ocean
1944[956]	June 21/25 3rd, 5th SFltl.	During their Baltic tour of duty, the flotillas laid mines off Libau (Now Liepa-a, Latvia) and Windau (Ventspils). The 5th flotilla mined the Irben Straits leading to the Gulf of Riga from the Baltic Sea. The *S-59* and *S-60* sink the Latvian steamer *f3ais-na.* The 1st flotilla captured the steamer *Estonia* and *S-39* and *S-103* sustained damaged from combat actions. **Type:** S-100	Gulf of Finland Gulf of Riga
1944[957]	June 22/23 65th MTB Flotilla	The Canadian flotilla of four boats and with four British MTBs attacked a German convoy near St. Heller (Jersey) and sank the German supply ship *Hydra.*	English Channel
1944[958]	June 23 *S-190*	This hard fighting boat sank after shelling by several American destroyers opened up on the boats. They blew the *190* to pieces in the exchange of gunfire. **Type:** S-100	Le Havre Baie de la Seine
1944[959]	June 23-24 *PT-557*	(SOE Ops "George II') In conjunction with other boats the 557 was tasked to land four agents and retrieve five agents from Morgiou Bay. The first two attempts failed from bad navigation. The third attempt on the 25th had been called off due to the arrest of agent "Octave", a W/T operator, and the capture of the Rayband family.	S.E French coast Mediterranean
1944[960]	June 24 *PT-19, 199*	(OSS Ops "Giraffe") Their mission was to deliver 4 OSS agents to the French coast, near Beg-an-Fry, which they accomplished. **Type:** Higgins *PT-71* class, 78'	English Channel

956 Whitley, *German Coastal Forces*, 80.
957 Rohwer, *Chronology of the War At Sea*, 336.
958 Whitley, *German Coastal Forces*, 74.
959 Richards, *Secret Flotillas*, 689.
960 Richards, *Secret Flotillas*, 319.

Figure 353. Captain Cyrus R. Taylor was the on board the "Bitchin Witch", PT 193 along with PT 196 when they were attacked by a Japanese float plane. The combined fire power of the two boats brought the plane down in flames. The 193 grounded off Bani Point. While tossing gasoline on the boats radar equipment, the captain received serious burns and died of his wounds later on a U.S. Naval hospital ship. He had been awarded a Silver Star for shooting down the float plane.[961]

[961] http://www.pacificwrecks.com/ships/ptboat/PT-193/cyrus-taylor.html

Date	Craft, Name, Event or Country	Event, Builder and Boat Specifications	City, River, Lake, Sea, Ocean
1944[962]	June 24 *MTB-670* "D"	The British MTB along with MGB's *659* and *652* chase the torpedo boat *TA-34* (ex-Yugoslav T-7), crewed by Croatians, and run it aground on the island of Murter. **Builder:** Wallasea Bay Yacht Station, Fairmile "D", 115'	Adriatic Sea
1944[963]	June 25 *PT-193* *"Bitchin' Witch"*	During operations, the *193* ran aground in enemy waters and destroyed to prevent capture, off Bani Point, Noemfoor Island. Before the mishap, the *193* sank three Japanese barges loaded with troops and supplies. **Type:** Elco *PT-103*-class, 80'	New Guinea Coral Sea
1944[964]	June 25/26 *MTB-734* "D"	The boat was attacked and sunk by RAF Swordfish torpedo planes, off Normandy by mistake. The *734* was badly damaged and *MTB-612* moved in to sink the boat with gunfire. **Builder:** Thomas & Balfour, Fairmile "D", 115'	Off Normandy
1944[965]	June 26 9th SFltl.	During the flight to evacuate Cherbourg the flotilla abandoned the *S-112* and *S-145* due to lack of materials and time to repair them.	English Channel

962 Reynolds, *Dog Boats At War*, 93.
963 HyperWar: *US Navy Torpedo Boats*, 9.
964 Reynolds, *Dog Boats At War*, 163 and Lambert, *The Fairmile 'D' Motor Torpedo Boat,* 12, 15; *Warship* Supplement 65, 20.
965 Frank, *German S-Boats in Action*, 115.

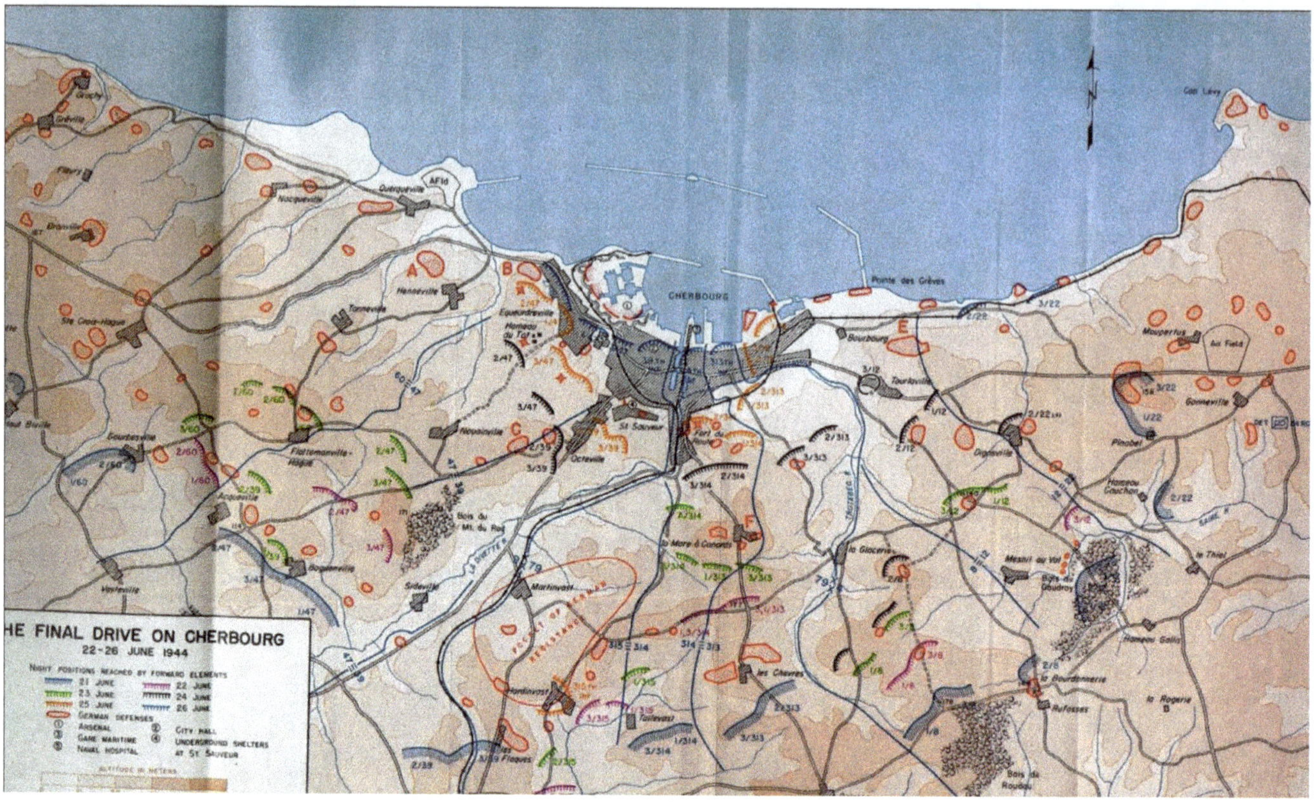

Figure 354. The German naval base at Cherbourg was pounded daily by Allied bombers, fighter bombers and fighter aircraft for days on end. Two precious S-boats had to be left during the withdrawal.[966]

[966] Frank, *German S-Boats in Action,* 168.

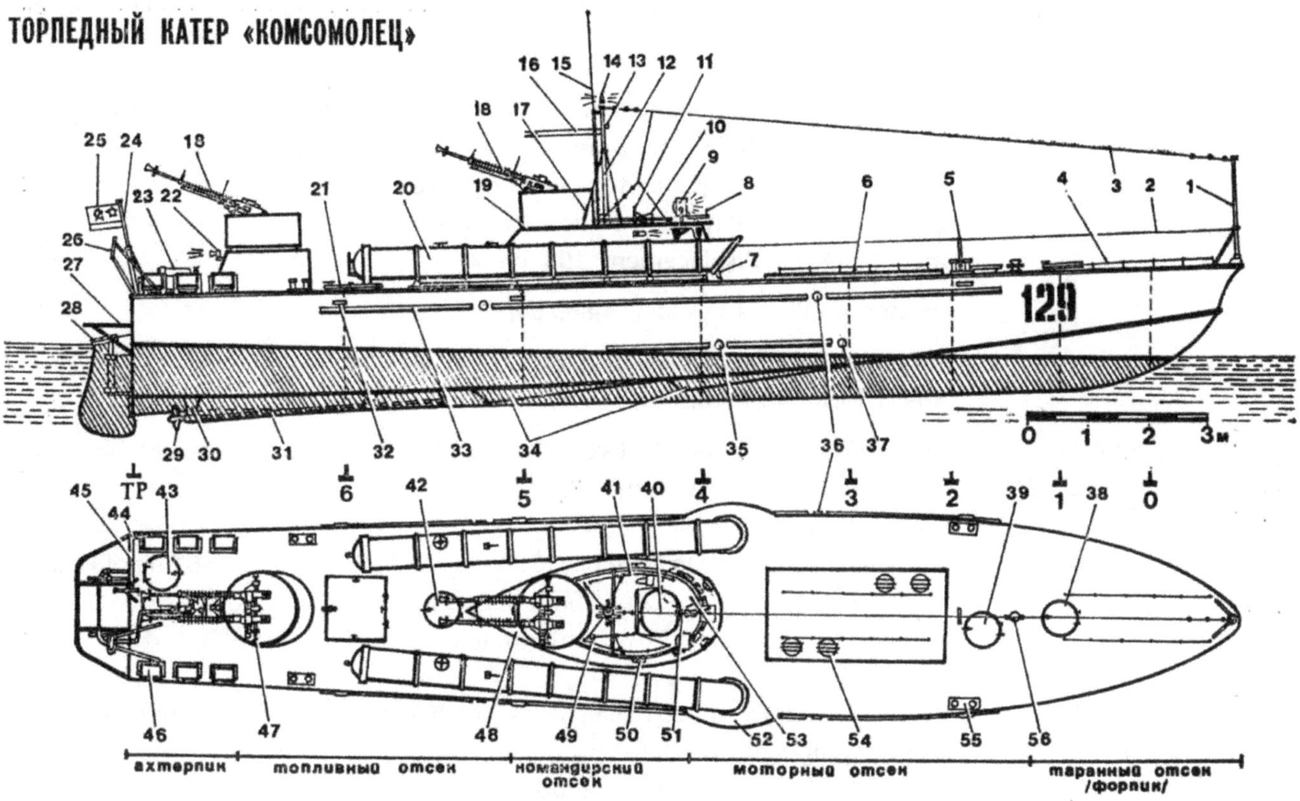

Figure 355. The Soviets pursued the testing of news MTB designs before WWI and through WWII. Russia had been the first country to adopt MTB into their navy in 1877. The Russian, and late, Soviet naval doctrine placed emphases on more powerful and better built MTBs. It was an endless pursuit for their engineers.[967]

[967] http://ru.wikipedia.org/wiki/%D0%A4%D0%B0%D0%B9%D0%BB:Komsomolets_torpedo_boat.JPG.

Date	Craft, Name, Event or Country	Event, Builder and Boat Specifications	City, River, Lake, Sea, Ocean
1944[968]	June 27 *MTB-640* "D"	Commanded by Lieutenant Campbell, RCNVR, the boat struck a mine north of Elba Island. 5 men were lost. The *MTB-658* sank the vessel with gunfire to prevent capture. **Builder:** Wallasea Bay Yacht Station, Fairmile "D" 115' **Speed:** 30 knots **Displacement:** 108 tons **Engines:** 4 x Packard yielding 4,800 bhp **Armament:** 2 x 6-pdr, 1 x twin 20-mm, 1 x single 20-mm, 4 x .50-cal. MG, 2 x 21" torpedo tubes	W .Italian Coast Ligurian Sea
1944[969]	June 28 TKA 239, 241	The two TKAs torpedoed and sank the freighter SS *Nerissa* on the Kirkenes-Petsamo convoy route. **Type:** *G-5*, 62.5	Barents Sea
1944[970]	June 30 TKA 43, 63, 102	German patrol vessels in Naru Bay, German DD *Z-39*, minesweepers *115, 29, 312* and ferry barges put up a fierce fight and sank the three TKAs with a great loss of life. **Type:** *G-5*, 62.5'	SE Gulf of Finland
1944[971]	June 30 *PT-308, 309*	The two PTs attacked two Italian MTBs, one being *MAS-562*, causing great damage and forcing the crew to abandon their boat. The *PT-306* towed the MAS-boat to the harbor of Basita.	Mediterranean
1944[972]	July *MAS-541*	Under the command of Tenente "Ginger" di Vascello Cosulich, the boat struck a mine off the Liberian coast and sank with all hands. This included their passengers, a French frigate captain and British Lieutenant Dow, RNR.	W. coast Africa
1944[973]	July 1 *MTB-460* (RCN)	During operations, the boat hit a mine and sank off the Normandy coast. **Type:** British Power Boat Co., 71'-9"	Normandy Coast English Channel

---

968 Reynolds, *Dog Boats At War*, 148; *Warship* Supplement 65, 18 and Lambert, *The Fairmile 'D' Motor Torpedo Boat*, 14.
969 Internet correspondence with Richard Worth, July 12, 2007.
970 Internet correspondence with Richard Worth, July 12, 2007.
971 Rohwer, *Chronology of the War At Sea*, 339.
972 Richards, *Secret Flotilla*, 650, 653.
973 *Warship* Supplement 65, 14.

Date	Craft, Name, Event or Country	Event, Builder and Boat Specifications	City, River, Lake, Sea, Ocean
1944[974]	July 1/2 MTB-632, 650	The two British MTBs attacked the German 10th Motor Minesweeping flotilla and sank the *R-189. MTB-632* withdrew on the arrival of additional R boats off Fecamp.	English Channel
1944[975]	July 2/3 55th MTB Flotilla	Five boats of the 55th attacked a German Naval Force off Cap d' Antifer. The resulting gunfight was a draw and no sinkings were recorded.	English Channel
1944[976]	July 3/4 Canadian 65th MTB Flotilla	The five boats attacked a German convoy and torpedoed the supply ship SS *Minotaure*. The MTBs went on to sink the German auxiliary patrol vessels *V -06* and *V-210*. Great damage was done to the *V-209* and *M-4622* in the fighting.	English Channel
1944[977]	July 3/4 S-boats	German S-boats made nightly sorties at both ends of the invasion area. In spite of the munitions shortages, they maintained a regular vigil of planting moored mines, drift mines and the deadly acoustic mines, all along the Normandy and English coasts and routes of transports. Hundreds of Allied vessels of all types were sunk or damaged by these mines.	English Channel
1944[978]	July 3/4 8th, 2nd, SFltl.	The 8th flotilla of three boats dropped 12 DM-1 mines then makes for Dieppe, France. The 2nd flotilla of three boats laid 12 DM-1 mines then was swept into a firefight with the British frigates HMS *Stayner* and HMS *Thornborough* off Cap d' Antifer. The frigates became leaders for all MTB flotillas for the duration of the war.	English Channel
1944[979]	July 4 British MTBs	British MTBs attack and sink the German Patrol vessels *V-208* and *V-210.*	English Channel
1944[980]	July 4/5 8th SFltl.	The flotilla was moving off Cap d' Antifer to the base at Le Havre when they ran into the frigate HMS *Trollope* with an accompanying force of MTBs. The ensuing firefight forced the S-boats to return to their base.	English Channel

974 Rohwer, *Chronology of the War At Sea*, 340.
975 Rohwer, *Chronology of the War At Sea*, 340.
976 Rohwer, *Chronology of the War At Sea*, 340.
977 Rohwer, *Chronology of the War At Sea*, 340.
978 Rohwer, *Chronology of the War At Sea*, 341.
979Rohwer, *Chronology of the War At Sea*, 340.
980 Rohwer, *Chronology of the War At Sea*, 341.

Figure 356. Kriegsmarine R Boats (Raumboot or Mine Hunter) carried a tremendous amount of fire power. They may not have been fast boats but they could lay down a devastating wall of lead when forced into a fight.[981]

Figure 357. These German stamps paid tribute to the *Schnellboote* flotillas scattered over the northern seas of Europe. The Germans paid tribute to their S-Boat force just as Americans honored their PT force.[982]

---

[981] Michael Emmerich http://www.german- navy.de/kriegsmarine/ships/minehunter/index.html.
[982] Michael Emmerich http://www.german-navy.de/kriegsmarine/ships/ minehunter/index.html.

Figure 358. HMS *Trollope* is other ship to the right. She was built in America for Lend-Lease and scrapped after suffering heavy damage at the hands of the S-soats.[983]

[983] http://www.navsource.org/archives/06/images/0651602.jpg .

Date	Craft, Name, Event or Country	Event, Builder and Boat Specifications	City, River, Lake, Sea, Ocean
1944[984]	July 4/5 2nd SFltl.	A British naval force of destroyers and MTBs from the Canadian 29th Flotilla met the German flotilla. A gunfight erupted but no sinkings occurred.	English Channel
1944[985]	July 4/5 *MTB-666, 723,729*	Near Ymuiden. The British MTBs engage a force of five armed boats, *Vps-1411, 1515* and *AF-41 and 47* (Schelde gun boats). The British *666* boat was shattered by heavy fire from shore batteries and captured by the Germans. She was intended to be displayed as a war trophy. Temporary patches were applied and the boat was set in an S-boat bunker, with three S-boats, to await transport. During the night, three scuttling charges that had malfunctioned now decided to work and blew up and sank the *666* along with an S-boat and damaging a second S-boat as well as the bunker. **Builder:** Dorset Yacht Co., Fairmile "D",115'	Ymuiden Harbor English Channel
1944[986]	July 5/6 2nd SFltl.	The flotilla attacked a force of British destroyers off Le Havre and the veteran destroyer HMS *Trollope* was badly mauled by the S-boats. She was beached and listed as a total constructive loss. After the S-boats returned to Le Havre, the repair depot blew up, destroying 41 of the flotilla's torpedoes and restricting subsequent operations. The 2nd Flotilla moved its-boats from Boulogne to Le Havre amidst heavy Allied fighter-bomber attack.	English Channel

[984] Rohwer, *Chronology of the War At Sea*, 341.
[985] Reynolds, *Dog Boats At War*, 170-172.
[986] Rohwer, *Chronology of the War At Sea*, 341.

Figure 359. What is not widely known is that in times of war, the Coast Guard is automatically a part of the U.S. Navy. This explains the huge number of Coast Guard sailors operating landing craft in the Pacific and European theatresThey brought their vast experience of small boat handling to the war of the "coastal forces."[987]

[987] http://nwcfoundation.org/zencart/index.php?main_page=index&cPath.

Date	Craft, Name, Event or Country	Event, Builder and Boat Specifications	City, River, Lake, Sea, Ocean
1944[988]	July 6 *MTB 248*	Sunk in a collision off the Normandy coast. **Builder:** White, Cowes, 72'-6" **Speed:** 36 knots **Displacement:** 41 tons **Engines:** 3 x Sterling yielding 3360 bhp **Armament:** 2 x 21" torpedo tubes, 2 x -.50 cal MG, 2 x .303-cal MG, two depth charges	English Channel
1944[989]	July 7/8 2nd, 9th SFltl.	The nine boats of the two flotillas targeted Allied destroyers and hits were scored on the destroyers *La Combattante* and *Cattistock* and on the frigate *Thornborough.*	English Channel
1944[990]	July 8 MTB Flotilla	Off Ijmuiden, three German patrol boats were attacked by seven British MTBs. The German *V 14 R* was sunk and three MTBs sustained much damage. *MTB 455, 457, 458, 476, 468, 469, 470* **Type:** British Power Boat Co 71'-9"	North Sea
1944[991]	July 8/9 2nd, 6th SFltl.	The twelve boats caught a reprieve from constant combat because the weather made sorties not worth the risk. Combat operations did not begin again until July 15th.	English Channel
1944[992]	July 14 TKA Div.	The TKA torpedo Brigade *(12, 13, 238, 239, 240, 245, 243)* attacked German convoys between Magnetos Sound and Porsangerfjord, near the Cape of Harbaken, Barents Sea. **Type:** *G-5*, 61.8'	Barents Sea

988 *Warship* Supplement 65, 9.
989 Rohwer, *Chronology of the War At Sea*, 343.
990 Rohwer, *Chronology of the War At Sea*, 343
991 Rohwer, *Chronology of the War At Sea*, 341.
992 Rohwer, *Chronology of the War At Sea*, 343.

Figure 360. The American destroyer *Rich* had already had her stern blown off by an S-Boat laid mine. As she was being prepared to be towed she drifted over another mine and split in half.

Date	Craft, Name, Event or Country	Event, Builder and Boat Specifications	City, River, Lake, Sea, Ocean
1944[993]	July 15/16 TKA Operations	Soviet combined operations against German convoys, off the Arctic coast, started with Operation RV6, from 13 to 28 July. Submarines, patrol craft and MTBs along with constant air cover by aircraft were involved. Eight TKAs missed the first convoy but were redirected to the target in Southern Busse Sound by YAK-9 recon aircraft. The TKA-12 torpedoed and sank the Norwegian ship *Storegga* during operations.	Arctic Coast
1944[994]	July 15/16 *TKA-239* (ex-*PT-412*)	The TKs attacked the German D/S *Hugin* with artillery and MG fire. The vessel was set ablaze and later towed to Vadso where she sank from neglect. The American PT-built, Vosper-designed, Soviet-operated torpedo cutter *TKA-239* was sunk by a German sub chaser, the *UJ-1211,* off North Cape, Norway later that same day.	North Cape
1944[995]	July 15/16 *PT-133*	Destroyed by Japanese shore batteries in Cinderawash Bay, Biak Island, Cap Pus. **Type:** Elco *PT-103*-class, 80'	Indonesia, North Pacific Ocean
1944[996]	July 15/16 6th, 9th, 2nd SFltl.	The three flotillas met an equally aggressive force of British MTBs. During the fighting, the *S-135, S-90* and *MTB-361* were badly damaged. *S-91, 79* and *114* move from Ostend base to Boulogne to join the 6th *S*-Boat Flotilla.	English Channel
1944[997]	July 18/19 *PT-107*	Lost in a refueling accident at Emirau, Hollanda (Jayapura) Island, St. Mathias (Mussau) Group **Type:** Elco, *PT-103* class, 80'	Pacific Ocean
1944[998]	July 19/20 *MTB-709, 722* (RNorN)	The two Royal Norwegian Navy *MTB-709* and *722* joined up with four British Fairmile "D" MTBs and attacked a German convoy off the Dutch coast. Both sides are shot up but no sinkings registered.	North Sea English Channel

993 Rohwer, *Chronology of the War At Sea,* 343.
994 Polmar, *PT Boats At War,* 127.
995 Lambert, *Allied Coastal Forces* Vol.2, 140 and HyperWar: US Navy Torpedo Boats (PT-Boats), 1940-1945, 8.
996 Rohwer, *Chronology of the War At Sea,* 341.
997 Lambert, *Allied Coastal; Forces* Vol. 2, 140 and HyperWar: US Navy Torpedo Boats (PT-Boats), 1940-1945, 8.
998 Rohwer, *Chronology of the War At Sea,* 340.

Figure 361.TKA as used in Operation RV6.[999]

Figure 362. *MTB-724*, Royal Norwegian Navy. The Norwegian sailors put their boats to hard use and fought with unbound determination against the men who held their country hostage.[1000]

---

[999] Central Naval Museum, St. Petersburg
[1000] Collection of John Lambert, Lambert Plans

Figure 363. The American war economy was enormous. (Library of Congress Print and Photographic Division)

Date	Craft, Name, Event or Country	Event, Builder and Boat Specifications	City, River, Lake, Sea, Ocean
1944[1001]	July 20/21 *PT-557*	(OSS Ops Sweet Pea I) On the 20th, the boat successfully landed two agents near Cap-Lardier. **Type:** Elco *PT-103*, 80'	English Channel
1944[1002]	July 21/22 British MTBs	Six British MTBs attacked the German 34th Motor Minesweeper flotilla and their escort vessel patrol vessel *V-1303*. In the firefight the *M-3413* is sunk.	English Channel
1944[1003]	July 21/22 MTB 242	Sunk while under tow to Malta. Builder: Vosper 72'-6"	Mediterranean
1944	July 24/25 PT 555	(OSS Ops Sweet Pea II) The made a successful pick up of three OSS agents. **Type:** Elco *PT-103* class, 80'	English Channel
1944[1004]	July 24/25 *MTB-372* (ex *BPT-62*)	Sunk in combat with Axis surface forces off Cape Loviste, Adriatic Sea **Builder:** Annapolis Yacht Yard (Nelson's) USA, Vosper **Speed:** 35.9 knots cruise, 38.94 maximum; **Displacement:** 40 tons; **Range:** 400 nm at 20 knots **Length:** 72' 6"; **Beam:** 19' 2½"; **Draft:** bow 2' 0", stern 6' 3" **Engines:** 3 x Packard 1,400-hp, petrol. Auxiliary light engines: 24 V & 110 V Ford 10 hp, petrol. **Armament:** 2 x 18" torpedo tubes, 1 x 2.5-inch MG, 2 x .303-cal. MG, two depth charges **Crew:** 13	Cape Loviste Adriatic Sea
1944[1005]	July 25/26 PT 193	She ran aground on Noemfour Island and was scuttled. **Type:** Elco *PT-103*-class 80'	New Guinea Solomon Sea

1001 Richards, Secret *Flotillas, 689.*

1002 Rohwer, *Chronology of the War At Sea*, 340.

1003 *Warship Supplement 65*, 9

1004 Lambert, *Allied Coastal Forces*, Vol. 2, 160.

Figure 364. PTs had been adding fire power since 1942. Originally, the boats were to be used for harbor defense and the navy did not think they needed to be heavily armed. After RON 3 fought the Japanese navy until all of its PTs were destroyed, the Navy decided to add more and larger weapons.[1006]

1005 Lambert, *Allied Coastal Forces* Vol. 2, 141 and Casualties, Navy and Coast Guard Ships, WW II http://www.history.mil/faqs/faq82-1.htm.
1006 U.S. Navy Historical Department.

Date	Craft, Name, Event or Country	Event, Builder and Boat Specifications	City, River, Lake, Sea, Ocean
1944[1007]	July 26/27 6th, 2nd, 9th SFltl.	Six boats from the 6th SFltl. attacked a British convoy off Dungeness and torpedoed two ships, sinking same. The *S-182* was sunk by two MTBs coming to the aid of the convoy. Four boats in the second group of the 6th SFltl. were attacked by three British MTBs and RAF Beaufighters. The 9th SFltl. engaged a British convoy with a strong naval escort force and withdrew.	English Channel
1944[1008]	July 27/28 6th SFltl. *S-97, 90, 91, 114*	Eight boats, in two groups, attacked a convoy off Dover and damaged two ships, SS *Fort Perrot* and SS *Empire Beatrice.* Allied destroyers drove off the second set of S-boats. **Type:** *S-38*	Strait of Dover
1944[1009]	July 27/28 9th SFltl.	The flotilla now operated out of Le Havre and encountered the heavily armed HMS *Retalick* (*K-555*, ex-*DE 90*) and several MTBs and MGBs. *MTB-430* was rammed and sunk by *S-182* (the 430 boat was later scuttled). *MTB*-412 struck a wreck and sank.	Strait of Dover
1944[1010]	July 30/31 9th SFltl. *S-97, 91, 114*	The three S-boats sailed to attack an east bound Allied convoy, FTM53, east of Eastbourne (near Beachy Head). The convoy consisted of eleven merchant ships, eight LST's, and an armed trawler. The escort vessels were seven destroyers, three groups of three MTBs, plus ML 59 and other ML's. Off Dungeness, five ships were torpedoed by six FaT torpedoes. The *Samwake* sank and the *Fort Dearborne, Ocean Volga, Fort Kaskaskia* and *Ocean Courier* sustained damage.	Off Dungeness English Channel

---

[1007] Frank, *German S-Boats in Action*, 117.
[1008] Whitley, *German Coastal Forces*, 74.
[1009] Whitley, *German Coastal Forces*, 74.
[1010] Whitley, German Coastal Forces, 74.

Figure 365. Collisions were quite frequent with all the MTB's of all the navies. Most engagements were night actions in light or heavy weather.

Figure 366. The Lancasters were heavily modified to carry the Tall Boy bombs used against S-boat bunkers. The bob-bay doors were re-worked to be two large doors to accommodate the bomb. Each plane could carry only one Tall Boy. The bunkers had been built to with stand the 1,000 pounds bombs in the Allies arsenal, but they could not withstand the Tall Boys.[1011]

[1011] http://en.wikipedia.org/wiki/File:Avro_Lancasters_flying_in_loose_formation.jpg and Frank, German S-Boats in Action, 51

Date	Craft, Name, Event or Country	Event, Builder and Boat Specifications	City, River, Lake, Sea, Ocean
1944[1012]	July 30/31 *S 39, S-114*	Both boats were destroyed by Allied aircraft that bombed the protective concrete pens at Le Havre. Numerous other small craft and the tender planet were also sunk. Two other S-boats, the S-*79* and *91*, were seriously damaged.	Le Havre English Channel;
1944[1013]	August 2/3 6th SFltl.	The flotilla attacked a convoy off the British coast and sank one freighter and damaged three LSTs.	English Channel
1944[1014]	August 3/4 British MTBs	Three British MTBs jumped two S-boats, *S-180* and *181,* of the 2nd flotilla and caused severe damage. The *S-167* and *MTB-608* "D" were damaged in a collision at sea.	English Channel
1944[1015]	August 5/6 *MTB-474, 475, 476*	3 British MTBs off Le Havre attacked a force of S-boats and the *S-91* was damaged. **Builder:** British Power Boat Co., 71'-9"	English Channel
1944[1016]	August 6/7 *PT-510, 512, 514*	The three PTs and HMS T*hornborough* get in a gunfight with the German B Group of the 15th Patrol Boat Flotilla off FéCamp. The fight was broken off before any boats are sunk. **Type:** Elco *PT-103* class, 80'	English Channel
1944[1017]	August 7/8 PT force	Four PT boats and the destroyer escort *Retalick* attacked the German 14th Motor Minesweeping Flotilla off FéCamp. The PTs *520* and *521* received a great deal of damage.	English Channel
1944[1018]	August 8 *MTB-93*	Off Harwich the *93* boat collided with the much larger *MTB*-729 (115' "Dog" Boat) and sank. **Builder:** Berthon Boat Co. Vosper, 72'-6"	English Channel

---

1012 Frank, *German S-Boats in Action*, 117.
1013 Frank, *German S-Boats in Action* 117.
1014 Rohwer, *Chronology of the War At Sea*, 346.
1015 Rohwer, *Chronology of the War At Sea*, 346.
1016 Rohwer, *Chronology of the War At Sea*, 346.
1017 Rohwer, *Chronology of the War At Sea*, 346.
1018 *Warship* Supplemental 65, 7.

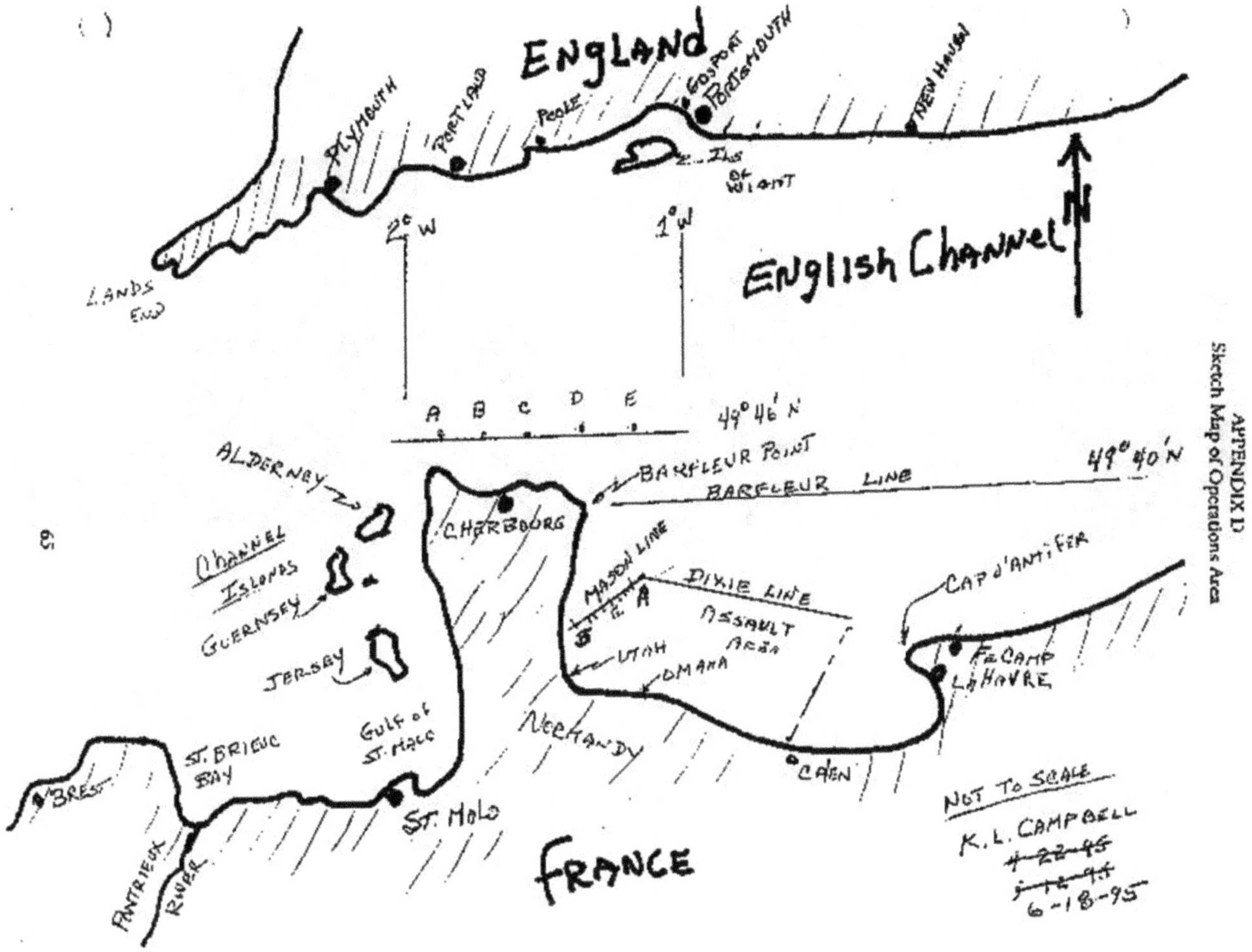

Figure 367. The map shows the changes that transpired over the course of a few days for RON 35. In the lower right hand corner you can see the dates crossed out as the situation developed.[1019]

[1019] http://members.cox.net/squadron35/picframe.htm

Figure 368. Ike on board PT-58 off Normandy.[1020]

[1020] *Ibid.*

Date	Craft, Name, Event or Country	Event, Builder and Boat Specifications	City, River, Lake, Sea, Ocean
1944[1021]	August 9/10 34th RON *PT-509, 508*	*PT-508* and *509* attacked six German minesweepers off La Corbeire, Isle of Jersey. In the ensuing battle; the helmsman of the *509* was killed; his death caused the boat to run out of control at high speed. The bow of the boat was stoved in and the PT boat became locked into the M boat. Eventually the M boat crew freed themselves from the wreck and at that moment, the *509* blew up, off Point La Corbeire. **Type:** Elco *PT-103*-class, 80'	Channel Islands Isle of Jersey Gulf de St-Malo
1944[1022]	August 11/12 10th SFltl.	The flotilla of six boats transferred from Ostend to Ijmuiden from whence it started mining operations off Orfordness.	English Channel
1944[1023]	August 14/15 8th SFltl.	Eight boats of the flotilla, including the newest S-boat model, the *S-701,* were transferred from Rotterdam to Boulogne to bolster the remaining S-boats.	English Channel
1944[1024]	August 15/16 Operation Dragon *PT-202, 218*	2nd invasion of France. American forces staged an invasion of France from the Mediterranean side, landing between Cannes and Toulon. The landing needed 5,000 men of the 1st Airborne Division, 396 troop transport gliders, 1,300 aircraft, 1,370 LST's, LSGs, LCVPs, LCCSs, LCMs,881 naval assault vessels and 24 PT boats. The PTs lost the *202* and *218* to German mines off Point Aygulf. **Type:** Higgins, PT-200 class, 78'	Western Mediterranean, France
1944[1025]	August 17/18 8th SFltl.	The eight boat flotilla attacked a British convoy off Dover and faced off with British destroyers, accompanied by five MTBs. The S-boats managed to torpedo one freighter.	English Channel

[1021] Hümmelchen, *German Schnellboote (E-Boats)*, 169.
[1022] Rohwer, *Chronology of the War At Sea*, 346.
[1023] Rohwer, *Chronology of the War At Sea*, 346.
[1024] Rohwer, *Chronology of the War At Sea*, 353.
[1025] Rohwer, *Chronology of the War At Sea*, 346.

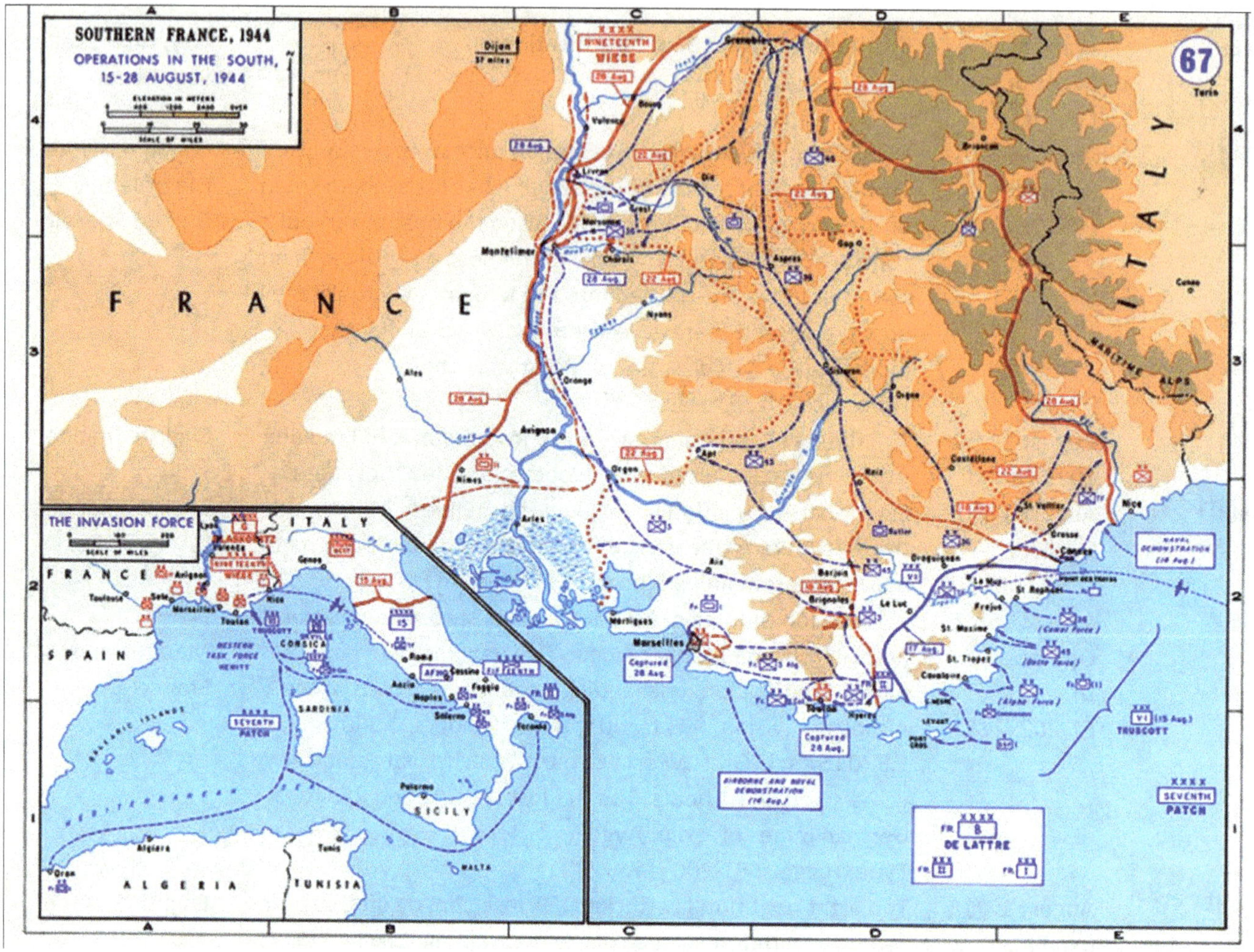

Figure 369. Operation Dragoon, the second invasion of France, did not garner the same headlines as the first did. Still, it was another major naval operation that was supported by PT boats. Two PT's were lost to mines, *PT-202* and *PT-218*.[1026]

[1026] http://en.wikipedia.org/wiki/File:Operation_Dragoon_-_map.jpg.

Figure 370. The emblem on the bow of the S-Boat is the 6fltl. The boat in front is the S 212 and the next to it is the S 213.The sailors in the foreground are heading for a transfer or leave.[1027]

[1027] http://www.panzergrenadier.net/forum/viewtopic.php?f=88&t=8293&p=80214

Date	Craft, Name, Event or Country	Event, Builder and Boat Specifications	City, River, Lake, Sea, Ocean
1944[1028]	August 4th, 5th, 9th SFltl.	The flotillas received a new "wonder" weapon, the T3d Dackel[1029] long- range pattern-running torpedo. The Dackel cruised at 9 knots with a 300-kg warhead. The torpedo could run for 3.5 hours in set pre-assigned zigzag course or a random zigzag pattern. On its first live use, the cruiser HMS *Frobisher* and two other vessels were damaged by this new torpedo.	English Channel
1944[1030]	August 19 *S-26, 40, 72*	Soviet aircraft sank the *S-26, S-40,* and *S-72* off the Sulina Estuary . A fourth boat was badly damaged. *S-26,* **Type**: *S-14; S-40* **Type***: S-30; S-72,* **Type:** *S-38*	Black Sea
1944[1031]	August 19 *S-57*	The *S-57* was sunk by British MTBs off Dubrovnik (Korcula). **Type:** *S-30*	Croatian coast Adriatic Sea
1944[1032]	August 17/19 TKA Div.	During Arctic operations, three TKAs laid mines south of Busse Sound. Four TKAs laid down heavy smoke screen and two TKAs reconnoiter Persjord. *TKA-205* and *206* sank the German patrol ship *V 6102* (Koln) and the steamship D/S *Colmar* (ex-*Volgoles*). South of Busse Sound, German *M-202* and *M-203* each sank one TKA, one was the *TKA-203.* **Type:** *G-5*, 61.8'	Kibergness Cape Barents Sea

---

1028 Hümmelchen, *German Schnellboote* , 172 and Frank, *German S-Boats in Action,* 170.
1029 German torpedoes of World War II, http://www.navweaps.com/Weapons/WTGER_WWII.htm. Last accessed March 9, 2010.
1030 Frank, *German S-Boats in Action,* 104.
1031 Rohwer, *Chronology of the War At Sea,* 351.
1032 Internet correspondence with Richard Worth and Rohwer, *Chronology of the War At Sea,* 351.

Figure 371. Soviet aircraft such as this Tupolev Tu-2 were being built faster than the Germans could shoot them down. With huge factories working 24/7 they would over run the German forces with sheer numbers.[1033]

Figure 372. Soviet aircraft were flown by dedicated pilots who would stop at nothing to destroy German targets where ever they could found. S-boats were vulnerable to the heavy machinegun and cannon from this Ilyushin 10 Shturmovik aircraft.[1034]

1033 http://www.xs4all.nl/~fbonne/warbirds/ww2htmls/tupotu2.htm
1034 http://www.vectorsite.net/avil2.html

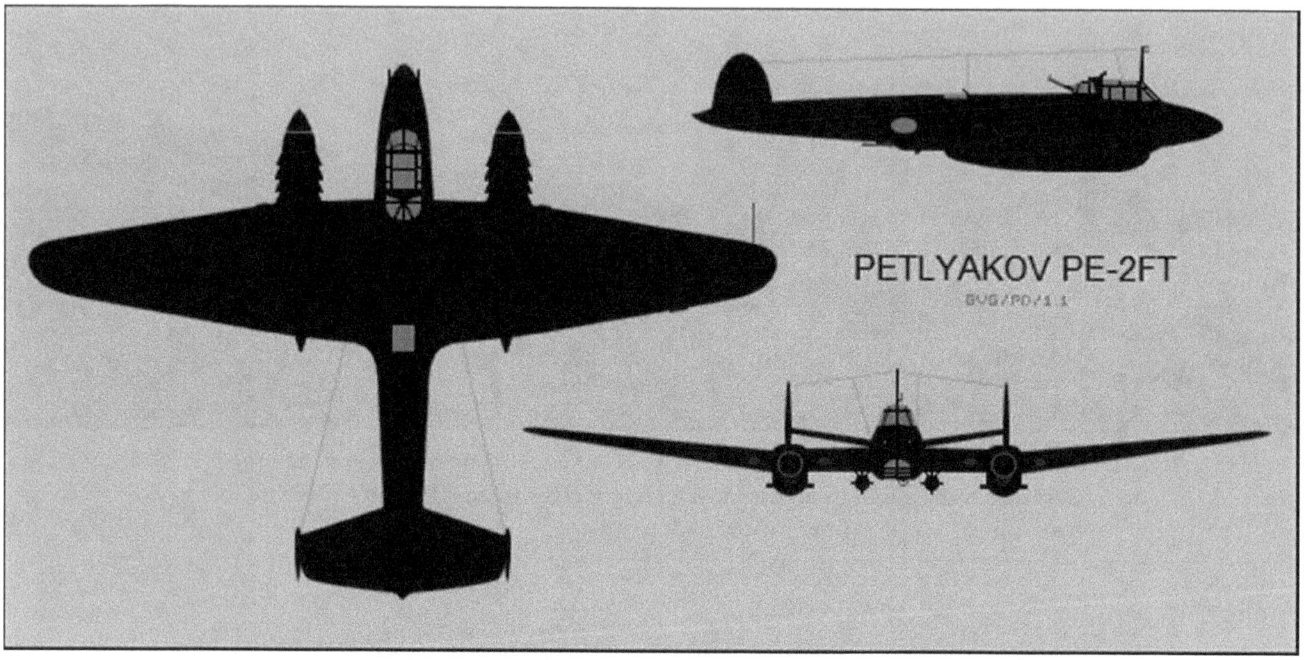

Figure 373. Petlaykov PE-2FT.[1035]

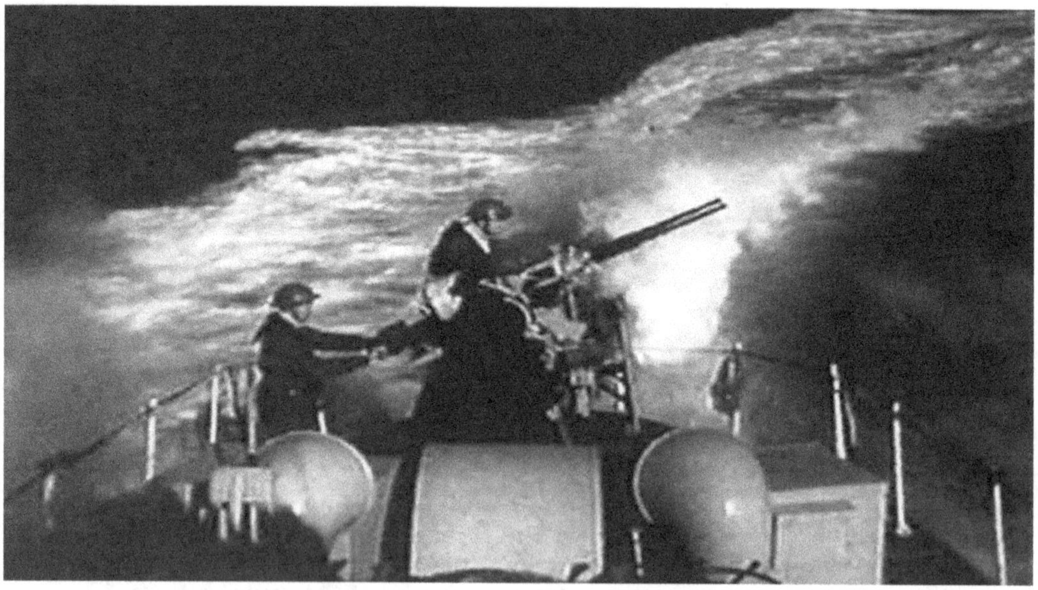

Figure 374. British MTBs at night.

[1035] http://www.vectorsite.net/avil2.html

Date	Craft, Name, Event or Country	Event, Builder and Boat Specifications	City, River, Lake, Sea, Ocean
1944[1036]	August 20 1st SFltl.	In Constantan Harbor a force of 182 Soviet aircraft bombs and strafes the harbor, sinking the *S-42, 52, 131* and *R 13* and *R 37* along with dozens of other small coastal craft. The *S-28, 45, 47, 49* and *S-51* sustained light damage. All surviving boats were later scuttled outside the harbor. *S-28* **Type**:*S-14; S-42 and 131 ,* ***Type*** *S:38; S-148,* ***Type:*** *S100*	Black Sea
1944[1037]	August 19/20 *MTB-473, 474, 479*	Off Cap de la Heve, the three MTBs with two MGBs attacked the A Group of the 38th Motor Minesweeper flotilla.	English Channel
1944[1038]	August 20/21 *MTB-471, 476, 477*	Off Cap d' Antifer the three MTBs attacked a force of German coastal vessels and sank two MS-boats of the 14th MMS (Motor Minesweeper Flotilla) and a submarine-chaser.	English Channel
1944[1039]	August 20/22 Black Sea SFltl. *S-148*	The *148* struck a mine off Sulina and sank with all hands. **Type:** *S-100*	Black Sea
1944[1040]	August 21 Black Sea SFltl. *S-28, 149*	After surviving the terrible attack at Sulina (August 19) the two boats were finally scuttled off Varna Harbor. Those crewmen that can walk, make their way back to Germany. *S-28,* **Type***: S-14; S-149* **Type:** *S-100*	Black Sea
1944[1041]	August 22 TKA Div.	Light craft of the of the Soviet Danube Flotilla supported the 46th Army in its crossing of the Dniester/Liman. The force contained six submarines and four TKAs, *221, 223, 227* and *233*. The torpedo cutters had a short fierce firefight with two German submarine chasers. Thirty TKAs and six patrol cutters took part in the capture of Constanta on the 30th of August. **Type:** *G-5*, 61.8'	Black Sea

[1036] Rohwer, *Chronology of the War At Sea*, 348.
[1037] Rohwer, *Chronology of the War At Sea*, 346.
[1038] Rohwer, *Chronology of the War At Sea*, 347.
[1039] Frank, *German S-Boats in Action*, 104.
[1040] Frank, *German S-Boats in Action*, 104.
[1041] Rohwer, *Chronology of the War At Sea*, 352.

1944[1042]	August 23 *PT-555*	Struck and sank by a German mine off Cape Couronne on the SE coast of France. **Type:** Elco *PT-103*-class, 80'	Gulf of Lyon Mediterranean
1944[1043]	August 24 *MTB-372*	Lost during combat in the Adriatic Sea. **Builder:** Annapolis Yacht Yard, Nelson's, Vosper 71'-6"	Adriatic Sea
1944[1044]	August 23/24 MTB 15th Patrol Flotilla	The German Naval evacuation of Le Havre harbor began this night with a military convoy of two armed trawlers, a submarine-chaser, two motor minesweepers towing one motor minesweeper each and 16 KFK's. Off Cap d'Antifer and Fecamp, they met a British patrol consisting of the frigate HMS *Thornbough,* the frigate *Retalick*, the destroyers HMS *Talybont* and HMS *Melbreak*, and MTBs *695, 694, 692, 212, 208,* and *205.* The convoy was scattered in the firefight with the *V-716* and *R-229* heavily damaged. The R-219 was sunk.	English Channel
1944[1045]	August 24/25 8th SFltl.	The flotilla attempted to attack a convoy off Beachy Head and was driven off by the steam gunboats HMS *Gray Wolfe* and HMS *Gray Goose.*	English Channel
1944[1046]	August 24/25 Black Sea SFltl.	When the Germans evacuated the Constanta Harbor, they received orders to scuttle all ships under German command.	Black Sea

[1042] HyperWar: *US Navy Torpedo Boats* (PT-Boats), 12.
[1043] *Warship Supplement 65*, 12.
[1044] Rohwer, *Chronology of the War At Sea*, 352.
[1045] Rohwer, *Chronology of the War At Sea*, 353.
[1046] Rohwer, *Chronology of the War At Sea*, 352.

Figure 375. The British SGB (Steam Gun Boat) gave valuable service and withstood all the rigors of combat at sea. These were built of steel and at first achieved a speed of 35 knots. After adding more armor plating around the engines the speed dropped to 30 knots. Seven were built with only one, *SGB-7,* sunk. The *Gray Goose* served until 1957. The ships were: *Gray Seal, Gray Fox, Gray Owl, Gray Shark, Gray Wolfe, Gray Goose* and *SGB-7.*[1047]

Figure 376. A Soviet T-series gun boat.

1047 http://www.naval-history.net/WXMemoirGS.htm and http://en.wikipedia.org/wiki/Steam_Gun_Boat

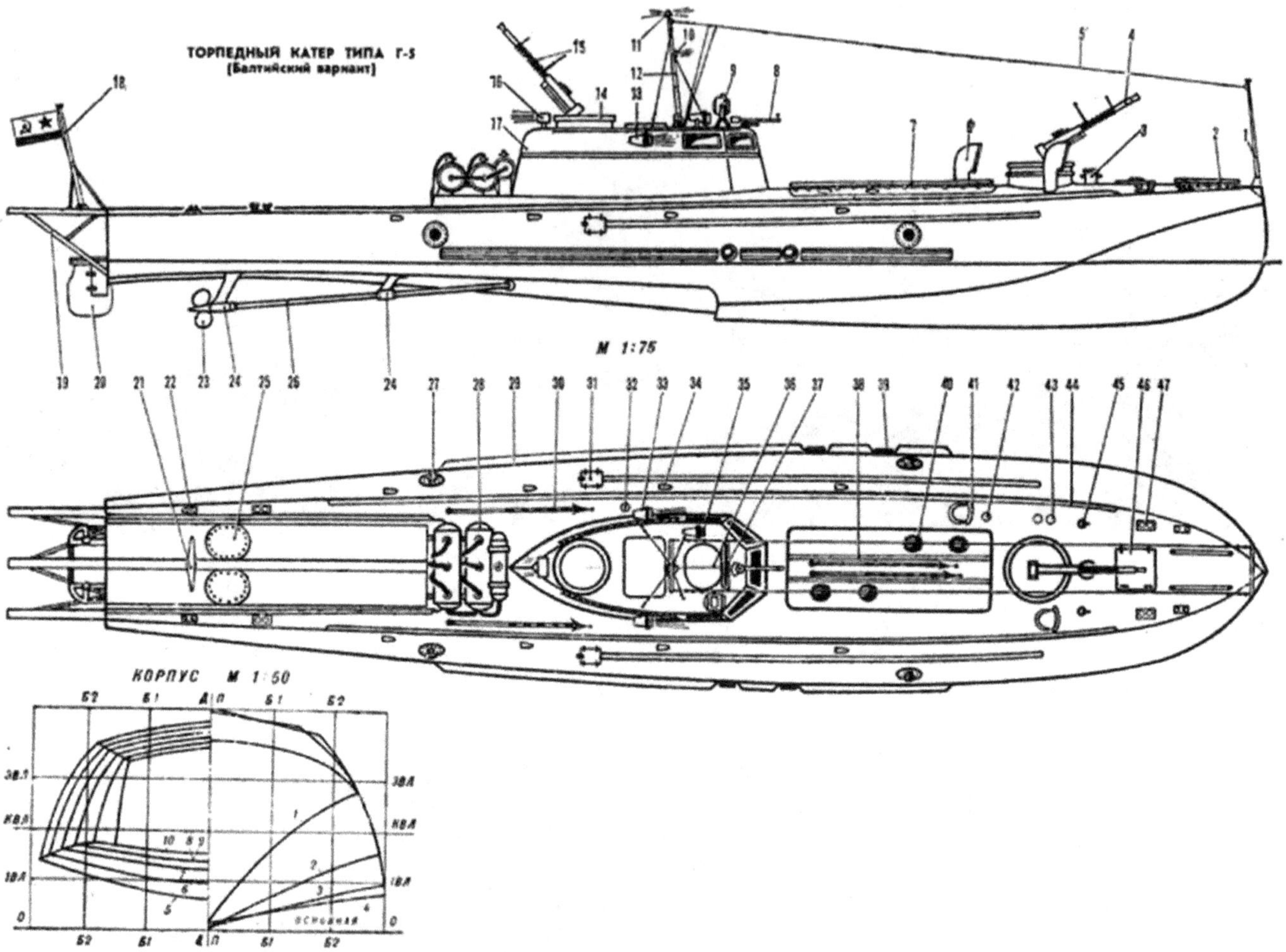

Figure 377. Side, above, and hull views of a Soviet G-5 boat.[1048]

[1048] http://ru.wikipedia.org/wiki/%D0%A4%D0%B0%D0%B9%D0%BB:Komsomolets_torpedo_boat.JPG

Date	Craft, Name, Event or Country	Event, Builder and Boat Specifications	City, River, Lake, Sea, Ocean
1944[1049]	August 24/25 8th Gun Carrier Flotilla	The Germans attempted to make a large move of naval vessels from Dieppe to Boulogne. They met with disastrous results. The 8th Gun Carrier Flotilla was running supplies to the *S*-Boat, M and V Flotillas still in Le Havre. The flotilla was attacked by *PT-250, 511,* and *514,* destroyer escort HMS *Seymour, MTB-257, 256, 254, 252, 452, 447, 453, 205, 209,* and *210,* the destroyers HMS *Talybon*t and frigate HMS *Retalick*. These 17 Allied vessels sank the *M 3857* and *S-91* and damaged the *V 243*. The AF-103 struck a mine and sank.	English Channel
1944[1050]	August 25/26 RON 3 PT-511, 514, 520	Another firefight with four S-boats. One S-boat was heavily damaged, one was less so. The *520* lost its radar to gunfire and had several 40-mm holes amidships. *511* received hits from small caliber MGs that started a fire. **Type:** Elco *PT-103* class,80'	English Channel
1944[1051]	August 25/26 *PT-513, 516, 519, 520, 514*	The PTs *513, 516* and *519* started a firefight with a German coastal convoy off Le Havre. *PT-511, 514,* and *516* sank the gun ferries, *AF 98* and *108*. PTs *511* and *514* sustained damage in the gunfight. **Type:** Elco *PT-103*-class, 80'	English Channel
1944[1052]	August 25/26 8th Gun Carrier Flotilla	The 8th was attacked again by a swarm of nine PT and MTB boats. The German flotilla comprised gun ferries *AF-97, 105, 110,* and *111.*	English Channel
194[1053]	August 27/28 *PT-511, 514,* 520	Just after the HMS *Middleton* finished a gunnery run the three PT's launched six torpedoes towards the *Middleton* target area. Second's later the PT's heard a gigantic explosion and then a series of small explosions. No PT received damage. The Middleton was undamaged. **Type:** *PT-103* class, 80'	Normandy Beach English Channel

1049 Rohwer, *Chronology of the War At Sea*, 353.
1050 Members Cox. net
1051 Rohwer, *Chronology of the War At Sea* , 353.
1052 Rohwer, *Chronology of the War At Sea*, 353.
1053 Members Cox. net

1944[1054]	August 27/28 14th MMS Flotilla	The flotilla's seven boats sowed mines in the Seine estuary near Le Havre to slow down the attacks of the Allies. Off Cap d'Antifer they ran into the destroyer escort HMS *Thornborough*, destroyer *La Combattante* and eight PT's and MTB's. The *UJ-1433* and KUJ-9 were sunk by four British MTB's. *R-231* was towed to shore badly damaged.	English Channel
1944[1055]	August 29/30 Black Sea SFltl. *S-45, 47, 49,* and *51*	The boats had already suffered damage in the air attacks of the 20th and 21st. Bulgaria declared its neutrality, and in order to not be interned under the 72 hour rule, the boats left harbor and the crews scuttled their boats off Varna harbor.	Black Sea

Figure 378. The group photos on this and the following page show the extreme youth of the PT crews that operated in dangerous waters.[1056]

[1054] Rohwer, *Chronology of the War At Sea*, 353.
[1055] Frank, *German S-Boats in Action*, 104.

[1056] http://members.cox.net/squadron35/picframe.htm

Date	Craft, Name, Event or Country	Event, Builder and Boat Specifications	City, River, Lake, Sea, Ocean
1944[1057]	August 29/30 8th SFltl.	The flotilla acted as escort for the German naval convoy going from Le Havre to Fécamp. The destroyers HMS *Retalick* and HMS *Cattistock* were badly damaged in their attempts to reach the convoy.	English Channel
1944[1058]	August 30 TKAs	In the final assault by the Soviets to push the Germans out of the Black Sea, 30 TKAs and six patrol cutters worked to help capture Constanta Harbor.	Black Sea
1944[1059]	September 1-9 *MTB-718* "D" *MGB-318*	(SIS Ops Korda I & II) The boats were assigned to photograph the coast of France, SW of Île-de-Batz (Clogourouan) The first sortie was not successful due to bad weather. The second attempt (September 12-17) succeeded. **Builder:** (*718*) Fairmile "D", 115'	English Channel
1944[1060]	September 1 *S-80*	The veteran S-boat *S-80* was sunk by a Soviet mine off Viborg.	Baltic Sea
1944[1061]	September 2 British MTBs	The British MTBs sank the German gun ferry, *AF 70*, off Cape Graz Net.	English Channel
1944[1062]	September 3/4 *S-184*	The last 13 surviving S-boats set sail from their Channel base for the last time as they made their way to new bases in Rotterdam and Ijmuiden. The *S-184* is sunk by gunfire from Dover shore batteries. **Type:** *S-100*	Strait of Dover
1944[1063]	September 11/12 *MTB-657* "D"	Near the port of Ancon, Lenice, Adriatic Sea, the boat struck a mine and was declared a constructive total loss. **Builder:** Woodnut & Co. Ltd., St. Helens, Isle of Wight, Fairmile "D", 115'	Adriatic Sea

---

1057 Rohwer, *Chronology of the War At Sea*, 353.
1058 Rohwer, *Chronology of the War At Sea*, 352.
1059 Richards, *Secret Flotillas*, 320.
1060 Rohwer, *Chronology of the War At Sea*, 354.
1061 Rohwer, *Chronology of the War At Sea*, 355.
1062 Rohwer, *Chronology of the War At Sea*, 353.
1063 Reynolds, *Dog Boats At War*, 218.

Figure 379. This illustration shows the Baltic color schemes that the S-boats adopted for winter waters. There were several variations of this arrangement. Each boat CO had their preferences for the adornment of camouflage. (Illustration by author)

Figure 380. In spite of all their effort the British Dover batteries only sank one ship and that was the *S-184*. It was truly, a very lucky shot.[1064]

[1064] http://www.panoramio.com/photo/2790423

Date	Craft, Name, Event or Country	Event, Builder and Boat Specifications	City, River, Lake, Sea, Ocean
1944[1065]	September 14 *TKA-211, 213, 242, 243*	The Soviet flotilla attacked and damaged the German minesweeper *M-252* off Vardo.	Barents Sea
1944[1066]	September 15 *TKA-214, 13*	Four TKAs, with 15 IL-2 ground attack aircraft, attacked the German convoy between Petsamo and Kirkenes. German escorts beat off the attack. *TKA 214* claimed one hit, but the *TKA-13* was sunk by one of the 61st Patrol Boats Flotilla. **Type:** *G-5*, 61.8'	Barents Sea
1944[1067]	September 18 *S-145*	In the Bay of Biscay, *S-145* was scuttled at Brest, France.	Bay of Biscay
1944[1068]	September 18/19 10th SFltl.	The four boats *S-185 , 186, 191* and *192* were part of the naval force bring supplies to the surrounded garrison at Dunkirk. Off Ostend, the convoy's escort group, *S-183, 200* and *702* ran into the British patrol consisting of the frigate HMS *Stayner* and "Dog Boat" MTBs *724* and *728*, of the Royal Norwegian Navy. The MTBs crippled one S-boat and the *Stayner* sank. The two remaining S-boats collided and both also sank. 57 German crewmen were recovered from the freezing waters, many more died of exposure.	North Sea
1944[1069]	September 19 *PT-371*	Grounded and scuttled, off Tagalasa, Halmahera Island. **Type:** Canadian Power Boat Co., PT 368-class (Vosper)	Indonesia Malucca Sea
1944[1070]	September 25 TKA Battalion	Soviet torpedo cutters attack and sink the German patrol ship *V-6101 (Gauleiter Bohle)* in Varangerfjord.	

1065 Internet correspondence with Richard Worth and Kirill Klement'ev; May 5, 2004. Last accessed by author, July 12, 2007.

1066 Rohwer, *Chronology of the War At Sea*, 357.

1067 Internet correspondence with Hans Frank, June 6, 2009.

1068 Hümmelchen, *German Schnellboote*, 172 and Frank,*German S-Boats in Action in the Second World War*, 121.

1069 HyperWar: *US Navy Torpedo Boats* (PT-Boats), 13.

1070 Internet correspondence with Richard Worth and Kirill Klement'ev; May 5, 2004, accessed by Author, July 12, 2007.

Figure 381. Smoke and damage control gear on an S-boat.

Figure 382. These pens were originally built for U-Boats. It became obvious that the S-boats could use the same shelters. An advantage of the pens was that water could be drained from any pen and leave an instant dry dock in its place. The British were slow in realizing the importance of the pens to the Kriegsmarine.[1071]

[1071] http://ahoy.tk-jk.net/macslog/TheU-BoatspensintheBayOfB.html.

Figure 383. This 1944 stamp design reflects the updated Schnellbootes now in the fleet. These boats were a great source of pride for the German people.[1072]

[1072] http://images.google.com/images?gbv=2&hl=en&q=S-Boats+WWII

Date	Craft, Name, Event or Country	Event, Builder and Boat Specifications	City, River, Lake, Sea, Ocean
1944[1073]	October 1 11th MTB Flotilla	Five flotilla boats, *351, 360, 349, 347* and *350*, attacked a German convoy running from Rotterdam to Bunkum. Eleven German coastal craft repulsed the attackers and sank MTBs *360* and *347* off Ijmuiden. **Builder**: Vosper, 72'-6"	Off Ijmuiden North Sea
1944[1074]	October 4-9 *PT-199* MGB-503	(SOE Ops Scrubby II) The mission to photograph the French coast, near Lezardrieux, was not successful due to continued bad weather. **Type:** Higgins PT-71 class.	W. coast France English Channel
1944[1075]	October 8/9 4th MTB Flotilla *MTB-473, 476, 480, 472, 475*	The flotilla attacked German armed trawlers, *V-1306, V-2007, and V-1303,* off the Hook of Holland, sinking the *V 1303.* **Builder:** British Power Boat Co, 71'-9"	Dutch Coast, near The Hague, North Sea
1944[1076]	October 8 54th MTB Flotilla (RNorN)	The three "D" boats, *712,711 and 722* attacked German ships north of Floor and drove the coaster *FreeCell* on shore. **Builders:** *712,* Wallasea Bay Yacht Station; *711,* Brooke Marine; *722,* Thompson & Balfour	Coast of Norway Norwegian Sea
1944[1077]	October 9/10 *MTB-663* "D"	Returning from a patrol off Venice, Italy, the *663* struck a mine and sank. 3-dead, 8-wounded. **Builder:** Osborne, Littlehampton.	North Adriatic Sea
1944[1078]	October 12 *TKA-230, 238, 241*	In attacks against convoy W.135 off Syltefjord, the Soviet boats sank the German minesweeper *M-303* with torpedoes near Kibergness Cape. **Type:** *G-5*, 61.8'	Barents Sea

---

[1073] Rohwer, *Chronology of the War At Sea* , 362.
[1074] Richards, *Secret Flotillas*, 321.
[1075] Rohwer, *Chronology of the War At Sea*, 352.
[1076] Rohwer, *Chronology of the War At Sea*, 352.
[1077] Reynolds, *Dog Boats At War*, 219.
[1078] Internet correspondence with Richard Worth, July 12, 2007.

Figure 384. These Vosper-type MTBs are showing large White stars on a dark background on the fore deck. Mistaken identity by Allied aircraft had become too common an occurrence. Evidently in this theater it was considered the priority to provide a symbol that American aviators would recognize immediately.[1079]

[1079] http://www.mikekemble.com/ww2/mgb.html.

Figure 385. The Soviet were fonder of their boats than they were of the Higgins or Elcos they were given.. What the Soviets did was to strip the boats of all hardware, guns, engines and electronics. Notice the Radar mast on the boat. It was beginning to be standard issue on all PT's in the last years of the war.[1080]

[1080] http://blog.kievukraine.info/2007_09_01_archive.htm

Figure 386. Soviet MTBs carried some of the heaviest machine guns in the world during WWII. They began to put more armor protection on the gunner's stations and plating over the decks of the engine space to stave off the effects of attacks by aircraft.[1081]

[1081] http://www.shiphistory.navy.ru/

Date	Craft, Name, Event or Country	Event, Builder and Boat Specifications	City, River, Lake, Sea, Ocean
1944[1082]	October 11 *PT-368*	Grounded and scuttled off Cape Salimoeli, Halmahera Island. **Builder:** Canadian Power Boat Co. **Type:** *PT-368*-class (Vosper) **Speed:** 41 knots **Displacement:** 33 tons **Crew:** 12 **Length:** 70′ **Beam:** 19′ **Draft:** 4′ 9″ **Engines:** 3 x4M2500 12-cylinder Packard, petrol (AV gas) **Armament:** 2 x 21″ torpedo tubes, 1 x 20-mm MG, 2 x 2 .50-cal MG	Indonesia Malucca Sea
1944[1083]	October 19 *PT-137*	During the Battle of Surigao Strait, the *137* boat sank the IJN light cruiser *Abukuma.* **Type:** Elco *PT-103*-class, 80′	Philippine Sea
1944[1084]	October 21 *TKA-237, 244*	The Soviets attacked six German convoys with ten torpedo cutters. The TKAs sank the German minesweeper *M-31* in Per fjord, Finmark, Norway. *R-311* was badly damaged by *TKAs-205* and *230* off Bussesund. **Type:** *G-5*, 62.5′	Barents Sea
1944[1085]	October 22 TKAs	TKAs of the Soviet Danube flotilla*TKA-221, 223, 227* and *233* had a short vicious gunfight with German submarine chasers East of Constanta.	Black Sea
1944[1086]	October 23 *MTB-688, 711, 653 (RNN)*	The three MTBs attacked a small German convoy near Stavfjord and are driven off by armed trawlers *V-5101* and *V 5113.* **Builder:** *(699)* Dorset Yacht Co., *(711)* Brooke Marine, *(653)* Robertson-Fairmile "D", 115′	Norwegian Sea

1082 HyperWar: *US Navy Torpedo Boats* (PT-boats), 13.
1083 Hoyt, *MacArthur's Navy*, 137.
1084 Rohwer, *Chronology of the War At Sea*, 365 and Internet correspondence with Richard Worth, July 12, 2007.
1085 Rohwer, *Chronology of the War At Sea*, 352.
1086 *Warship* Supplement 65, 5 and Lambert, *The Fairmile 'D' Motor Torpedo Boat*, 11, 13.

1944[1087]	October 24 *PT-131, 130, 152 "Lakacookie"*	The three boats were among fifteen PT has sent out to kill Japanese ships in the Surigao Strait. *PT-131,* near the island of Bohol, was the first to pick up radar signals of large ships. Shortly, the Japanese destroyer *Shigure* fired rounds to straddle the *131* boat. The *152* boat caught a 5" round in its 37-mm gun, which resulted in four dead or wounded. The *130* boat had a Japanese shell 5" pass through the mid section of the boat without exploding, but it did knock out all radio communication. **Type:** Elco *PT-103*-class, 80'	Bohol Island, Leyte Gulf, Philippine Sea
1945[1088]	October 24/25 *PT 321, PT 493 "Caroline Baby"*	These boats were positioned in the middle of the Strait between the islands of Panaon and Dinegat. These were two of the boats that were arranged into 13 sections of 3 boats each for the Battle of Surigao Strait. The *321* boat was hit by a shell from the IJN DD *Shigure.* The boat lost 2 dead and 9 wounded. The *PT-493 "Caroline Baby"* was the only American boat lost in the Battle of Surigao Strait. **Type** PT 103 Class, Elco 80'	Mindanao, Sulu Sea Philippines
1944[1089]	October 24 *PT-137, 146, 151, 190, 493*	The IJN cruiser *Mogami* fought off several attacks by *PT-146, 151* and *190,* and was hampered from fulfilling her mission. *PT-493* was sunk by cruiser gunfire in the Strait. **Type:** *(137, 146, 151, 49, 190)* Elco *PT-103* class, 80'	Surigao Strait Philippine Sea
1944[1090]	October 24 *PT-490, 493*	Section six of the PT boats worked off Panaon Island and when the Japanese fleet showed up, the boats fired all their torpedoes—and missed. The *493* sank from Japanese destroyer gunfire; the *490* had two killed and five wounded and was beached. **Type:** Elco *PT-103*-class, 80'	Panaon Island, Leyte Gulf, Surigao Strait Philippine Sea

[1087] Hoyt, *MacArthur's Navy*, 129, 130, 131 and Keating, *The Mosquito Fleet,* 216.
[1088] Keating, *The Mosquito Fleet*, 216-217

[1089] Rohwer, *Chronology of the War At Sea*, 367.
[1090] Hoyt, *MacArthur's Navy*, 132.

Figure 387. The sinking of the Japanese destroyer, *Abukuma,* was a major event for the PT crews. The IJN destroyers were a constant threat to the PTs of the Pacific.[1091]

[1091] http://en.wikipedia.org/wiki/Japanese_cruiser_Abukuma.

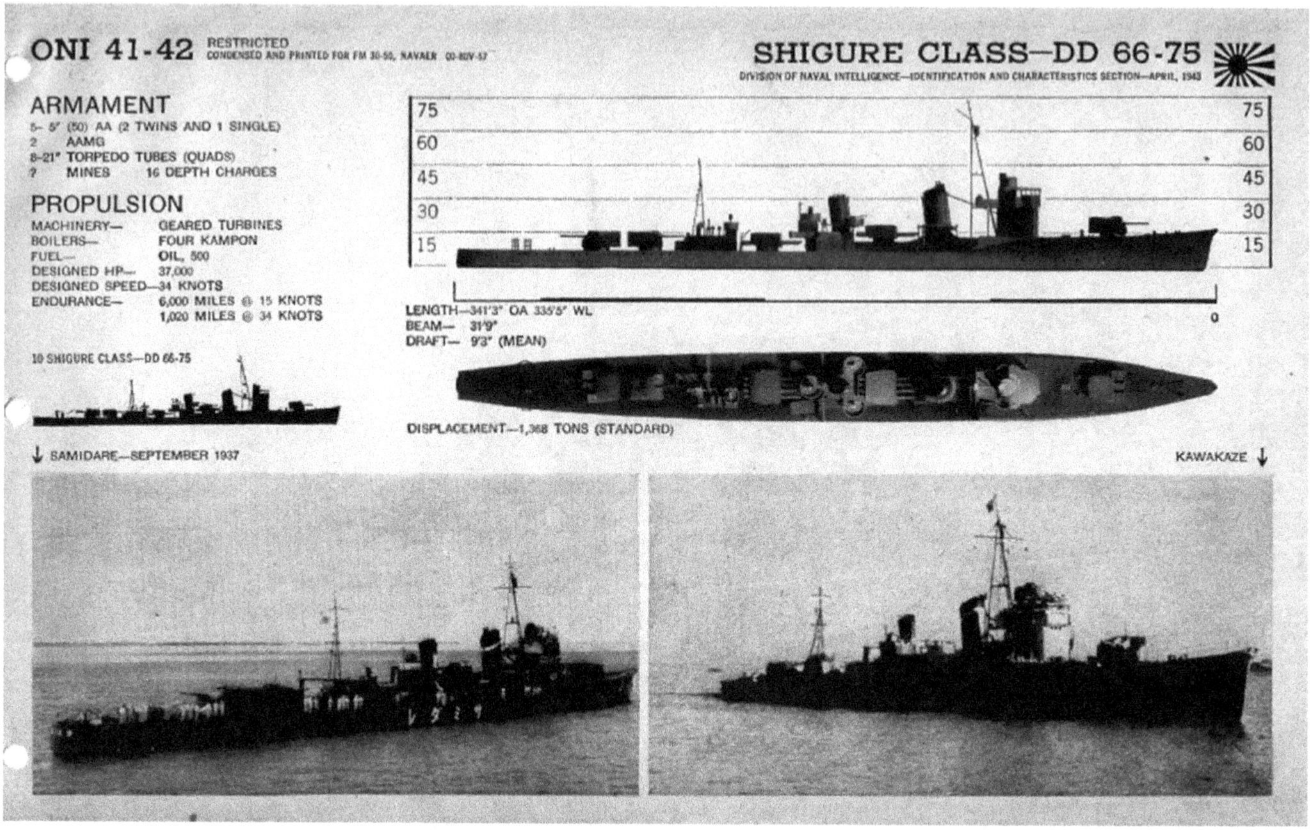

Figure 388. The *Shigure* was a modern Japanese destroyer that could easily outrun the PTs with its top speed of 42 knots. This along with her full armament made her a most dangerous adversary.[1092]

1092 http://www.acepilots.com/ships/ijn0051.jpg

Figure 389. Bow gunners on the later PT had fire power that had never been available before. Initially the PT boats were expected to be harbor defense craft and not the "Go Get 'Em" boats they turned out to be.[1093]

[1093] Library of Congress Print and Photographic Division

Date	Craft, Name, Event or Country	Event, Builder and Boat Specifications	City, River, Lake, Sea, Ocean
1944[1094]	October 24 *PT-194*	The *194* boat was hit by a shell from the IJN cruiser *Nachi* in the Battle of Surigao Strait. 3 men were wounded. **Type:** Elco *PT-103*-class, 80'	Philippine Sea
1944[1095]	November 1 *MTB-712, 709* (RNorN)	The Norwegian MTBs sank two German patrol boats, the V-5525 and V-5531. **Builders:** *(709)* Dickie, Bangor & Tarbert, *(712)* Wallasea Bay Yacht Station, Fairmile "D"115'	Sogenfjord, Norwegian Sea
1944[1096]	November 2 9th SFltl.	*S-175, 207* and *167* sank the British tanker *Rio Bravo* off Ostend and the armed trawler *Colsay* off Nieiport. **Type:** *S-100*	North Sea
1944[1097]	November 5 *PT-320*	Japanese Kamikaze aircraft sank the *320* boat off shore. **Type:** Elco *PT-103*-class, 80'	Leyte, Philippine Sea
1944[1098]	November 7 *PT-301*	Lost to fire and internal explosion, Mois Woendi, Biak Island, Cinderawash Bay. **Type:** Higgins *PT-200*-class, 78'	Indonesia North Pacific
1944[1099]	November 9 *PT-492, 497, 524, 525*	The PTs were sent to intercept IJN ships, loaded with troops, that were trying to reinforce the garrison at Ormac. The PTs torpedoed and damaged the frigates *Okinawa* and *Shimushu*, which were carrying 2,000 troops each. USAAF B-25 bombers sank the two ships the next day but not before all troops had been unloaded. **Type:** Elco *PT-103* class, 80'	Leyte Gulf Philippine Sea
1944[1100]	November 9 *S-203*	The S-boat collided with a minesweeper, R220, and sank. The craft was later salvaged and put back into action. **Type:** *S-100*	Norwegian Sea
1944[1101]	November 11 *PT-321*	Ran aground in San Isidro Bay, Ormac Island, and destroyed to prevent capture. **Type:** Elco *PT-103*-class,80'	Leyte Philippine Sea

---

1094 Hoyt, *MacArthur's Navy*, 140.
1095 Rohwer, *Chronology of The War At Sea*, 370 and Lambert, *The Fairmile 'D' Motor Torpedo Boat*, 13, 11.
1096 Hoyt, *MacArthur's Navy*, 142.
1097 HyperWar: *US Navy Torpedo Boats*, 9.
1098 HyperWar: *US Navy Torpedo Boats*, 9.
1099 Rohwer, *Chronology of the War At Sea*, 369.
1100 Internet communication with Hans Frank, August 10, 2009.
1101 Lambert, *Allied Coastal Forces*, Vol.2, 141.

Figure 390. The North Sea became a shooting gallery for the S-boats, but small victories could never add up to a total victory in this war. The Germans had no chance against the manufacturing facilities of the United States, the United Kingdom and the Soviet Union.[1102]

[1102] Collection of Hans Frank.

Figure 391. The *S-203's* collision with the *R220* cost the Germans precious resources and time in the war for Europe. They had already lost more S-boats than could be replaced. The Germans could build 100 hulls a month but only three engines in the same time frame.[1103]

[1103] http://www.panzergrenadier.net/forum/viewtopic.php?f=88&t=8293&p=80214.

Figure 392. The Navy continued to glamorize its PT boats. Many young men volunteered for the PTs based on the emotional effect of these posters.(Library of Congress Print and Photographic Division)

Date	Craft, Name, Event or Country	Event, Builder and Boat Specifications	City, River, Lake, Sea, Ocean
1944[1104]	November 13 MTB *688, 627* (RNoN)	The *688* and *627* boats attack the German convoy in the Southern exit of Krakhelle Sound (Sogenfjord). Two freighters avoid the four MTB torpedos and the escort vessel W *1430* is badly damaged in the fight. Builder: (*688*) A.M. Dickie & Sons, (*627*) P.K. Harris & Sons, Fairmile "D", 115'	Norwegian Sea
1944[1105]	November 18 PT 311	Hit a mine and sank in the Ligurian Sea, between Italian Rivera and Islands of Corsica and Elba. Type" Higgins PT 200-class, 78'	Northern Mediterranean
1944[1106]	November 24 MTB 287,371 (ex-BPT 29 & 61)	Wrecked on Levrera Island, North Dalmatian Islands in the Adriatic Sea. **Type:** Vosper, Herreshoff 70'	Adriatic Sea
1944[1107]	November 25 PT 363	Destroyed by Japanese shore batteries in Knoe Bay, Halmahera Island, **Type:** Elco *PT-103*-class, 80'	Maluku Islands, New Guinea
1944[1108]	November 26 Spanish Armada	The Spanish Navy bought the technology to build type *S-38* S-boats. The work began at the Arsenal de la Cardaca, Cadiz at the Carraca Cadiz. The first boats were numbered *LT-27, LT-28, LT-29, LT-30, LT-31,* and *LT-32.*	Spain

---

1104 Rohwer, *Chronology of the War At Sea*, 370 and Lambert, *The Fairmile 'D' Motor Torpedo Boat*, 11, 12.
1105 HyperWar: *US Navy Torpedo Boats*, 5.
1106 *Warship* Supplemental 65, 10.
1107 Lambert, *Allied Coastal Forces, Vol. 2*, 142.
1108 Spanish in the Kriegsmarine, http://www.libreopinion.com/members/ch1492/english/blue/kriegsmarine.html

Figure 393. As the war wore on, Japanese luggers, barges and coastal general supply ships took a huge beating from the PT boats and air power.[1109]

[1109] http://upload.wikimedia.org/wikipedia/commons/f/fc/GuadJapaneseSubandShip.jpg.

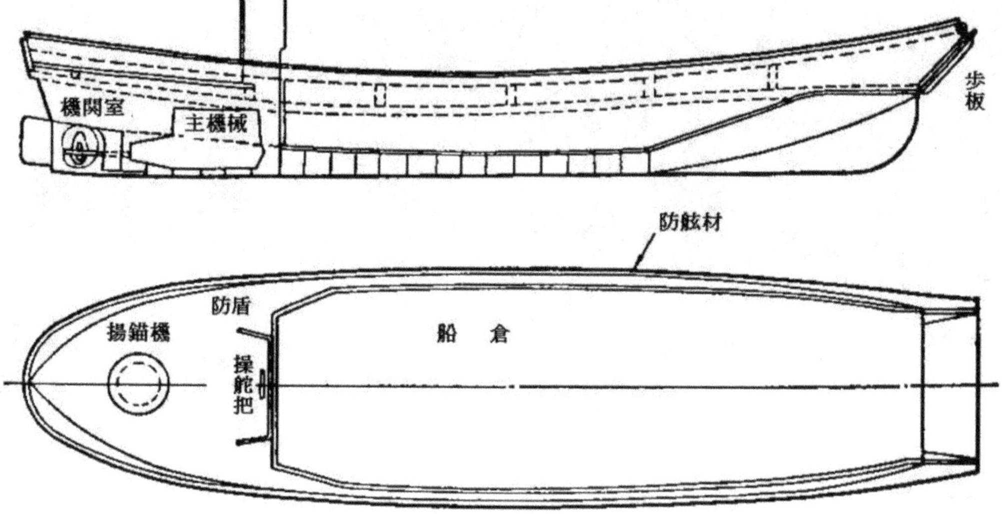

大発動艇一般配置図

Figure 394. The Japanese built barges and luggers that were no match for the fire power of the PT's. Poorly made, grossly under powered and without armor of any kind, they were a pitiful group of craft.[1110]

[1110] http://upload.wikimedia.org/wikipedia/commons/b/bd/Daihatsu_landing_craft.JPG

Date	Craft, Name, Event or Country	Event, Builder and Boat Specifications	City, River, Lake, Sea, Ocean
1944[1111]	November 27 *MTB-723* (RNoN)	The Norwegian MTB, off the coast of Norway near Hegdosen, exchanged heavy gunfire with German surface forces. 3 men were wounded and some damage to the boat sustained. **Builder:** P.K. Harris & Sons Ltd, Fairmile "D", 115'	North Sea
1944[1112]	November 27/28 *MTB-715, 623, 717, 627* (RNorN)	The Royal Norwegian Navy's MTBs made an unsuccessful attack on German patrol boats off Sognefjord. The MTBs 717 and 627 attacked a German convoy and the 717 torpedoed the freighter *Weilheim*. Then the German escort vessels *V 5303, 5312* and *R 312* damaged all MTBs with gunfire. **Builder:** Fairmile "D", 115'; *(715)* Woodnut, *(717)* Dickie, *(623)* Wallasea Bay Yacht Co., *(627)* Harris & Sons, Ltd.	Norwegian Sea
1944[1113]	November 28/29 *PT-127, 128, 331, 191*	The four PT boats began hunting Japanese small surface craft running supplies into Leyte through Ormac Bay and the Camotes Sea. On the first night, and the first attack, *127* and *331* sank the Japanese sub-chaser *SC-53.* The second attack by the 128 and 191 boats sank the Japanese *PC-105* (ex-Philippine *Arayat*). **Type:** Elco *PT-103*, 80'	Ormoc Bay, Camotes Sea, Philippines Is.
1944[1114]	December 6 *MTB-653, 717* (RNoN)	The two MTBs attacked and sank the 3,000-ton German supply ship *Dietmar Koel* in the port of Korsfjorden. The *653* began seriously leaking until her engines stopped. The *717* started towing the *653* and later transferred the towing to the whaler *Molde.* **Builder:** (653) Robertson, (717) Boat Construction Co., Fairmile "D", 115'	North Sea
1944[1115]	December 10 *PT-323*	Destroyed by two Kamikaze aircraft in Leyte Gulf. **Type:** Elco *PT-103*-class, 80'	Surigao Strait Philippine Sea

[1111] The Norwegian MTB Flotilla in Shetland, 7.
[1112] Rohwer, *Chronology of the War At Sea.*
[1113] Keating, *The Mosquito Fleet*, 228-229.
[1114] The Norwegian MTB Flotilla in Shetland supra. at 7 and Lambert, *The Fairmile 'D' Motor Torpedo Boat,* 11, 13.
[1115] Lambert, *Allied Coastal Forces,* Vol. 2, 141 and Keating, *The Mosquito Fleet*, 230.

Figure 395. PT 323, cut almost in half by a Kamikaze plane on December 10, 1944.

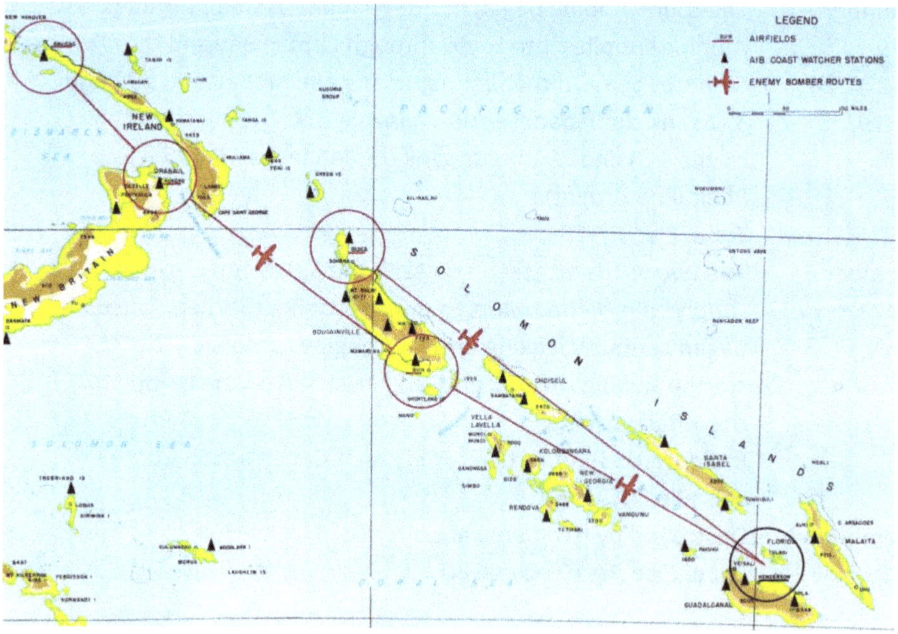

Figure 396. The Australian coast watchers could not prevent the enemy from firing on the PTs but when the Americans were in trouble the coats watchers could radio for help. They were also useful in identifying routes of the Japanese barges and luggers. The information passed on to the PTs led to many successful night attacks.[1116]

[1116] http://www.history.army.mil/books/wwii/macarthur%20reports/macarthur%20v1/Images/p_025.jpg

Date	Craft, Name, Event or Country	Event, Builder and Boat Specifications	City, River, Lake, Sea, Ocean
1944[1117]	December 11/12 *TKA-301, 353*	The remains of the wreck *Romania* were sunk by the two TKAs off Chersones (Kherson). **Type:** *G-5*,62.5'	Gallipoli Peninsula Black Sea
1944[1118]	December 12 PT 492, 490	The two boats attacked a Japanese convoy in Ormac Bay and sank the destroyer *Uzuk*i. **Type:** Higgins PT 200-class, 78' **Speed:** 41 knots **Displacement:** 43 tons **Length:** 78' **Beam:** 19' **Draft:** 5' 3" **Engines:** 3 x Packard 1,200-hp, petrol **Armament:**4 x 21" torpedo tubes, 1 x 40-mm, 2 x 20-mm **Crew:** 17	Ormac Bay Camotes Sea
1944[1119]	December 16 *PT-230, 77, 223*	The 3 PTs were attacked by Japanese kamikaze airplanes in Mangarin Bay, Mindoro. **Type:** *(230, 223)* Higgins *PT-200* class, 78'; *(77)* Higgins *PT-71* class, 78'	Mindoro Island Philippines Is.
1944[1120]	December 17 *PT-300*	A kamikaze crashed into the *300* boat cutting it in two. Four-dead, four missing, five-wounded. **Type:** Higgins *PT-200*-class, 78'	Mindoro Island Philippines Is.

1117 Internet correspondence with Richard Worth, July 12, 2007.
1118 Keating, *The Mosquito Fleet*, 230.
1119 Keating, *The Mosquito Fleet*, 233.
1120 Keating, *The Mosquito Fleet*, 230.

Figure 397. The Imjuiden S-boat bunkers were the next target for the RAF Lancasters and their Tall Boys. The damage was great but the casualties were light.[1121]

[1121] http://upload.wikimedia.org/wikipedia/commons/1/1c/IJmuiden-4.59E-52.47N.jpg

Date	Craft, Name, Event or Country	Event, Builder and Boat Specifications	City, River, Lake, Sea, Ocean
1944[1122]	December 18 *PT-77*	While US Army bombers were going after the Japanese bombardment flotilla off Mindoro, the PT-77 was caught by "friendly fire." Every member of the 17-man crew received wounds. **Type:** Higgins PT-71-class,78' **Speed:** 41 knots **Displacement:** 43 tons **Length:** 78' **Beam:** 19' **Draft:** 5' 3" **Engines:** 3 x Packard 1,200-hp, petrol **Armament:**4 x 21" torpedo tubes, 1 x 40-mm, 2 x .50-cal MG **Crew:** 17	Off Mindoro Is, Philippine Is.
1944[1123]	December 18/19 *PT 223, 220*	Off Western Mindoro the two PTs attacked a flotilla of Japanese destroyers and the *223* boat sank the DD *Kiyoshimo*. **Type:** Higgins PT-200class, 78'	Off Mindoro Is. Philippine Is.
1944[1124]	December 23 *MTB-722, 712* (RNonN)	In Bomlofjord, Norway, the two MTBs attacked a German surface group and sank the minesweeper *M 489.* **Builder***: (722)* Thornycroft, *(712)* Wallasea Yacht Station, Fairmile "D", 115	Norwegian Sea
1944[1125]	December 26 *MTB-717, 627* (RNoN)	In the early evening, the two MTBs attacked a small convoy of two steamships, escorted by the *V-5102* and *V-5114*, near Froysjoen, Norway. The small tanker *Biwi* is sunk in the action. **Builder:** *(717)* A.M. Dickie & Sons, *(627)* Harris & Sons, Ltd. Fairmile "D", 115'	Norwegian Sea
1944[1126]	December 29 MTB-782 "D"	Hit a mine off River Schelde and sank. Builder: Wallasea Yacht Station, Fairmile "D" 115'	Off Dutch Coast North Sea

---

[1122] Keating, *The Mosquito Fleet*, 237

[1123] Keating, *The Mosquito Fleet*, 237- 23.8.

[1124] Rohwer, *Chronology of the War At Sea*, 379 and Lambert, *The Fairmile 'D' Motor Torpedo Boat*, 11, 13.

[1125] Rohwer, *Chronology of the War At Sea*, 379 and Lambert, The *Fairmile 'D' Motor Torpedo Boat*, 11, 12.

[1126] Lambert, *The Fairmile 'D' Motor Torpedo Boat*, 15, and *Warship Supplement 65*, 22..

1944[1127]	December 28-29 *MAS-541*	(OSS Ops Clinton) Successfully landed two agents near Cap-du-Pinet (Cap-Camarat).	S.W. coast France
1944[1128]	December 29-30 *MAS-541*	(SIS Ops Furtive) The mission was to deliver a package of vital equipment and retrieve agent's reports of German installations and movements. Their first attempt failed due to agents not at the pinpoint location, Cap-du-Pinet (Cap-Camarat). The second attempt failed and it was not until January 28 that the mission succeeded. Additionally, two agents also embarked for return to the UK.	S.W. coast France

Figure 398. The Japanese destroyer *Kiyoshimo* was sunk partially by bombs during an attack by AAF Army bombers in the approaches to Mindoro. The *PT-223* finished off the ship 145 miles south of Manila. PT boats rescued five survivors.[1129]

[1127] Richards, *Secret Flotillas*, 688.
[1128] Richards, *Secret Flotillas*, 688.
[1129] http://en.wikipedia.org/wiki/File:IJN_Kiyoshimo_1944

Figure 399. The Italian MAS-549.[1130]

[1130] http://wapedia.mobi/en/MAS_(boat)

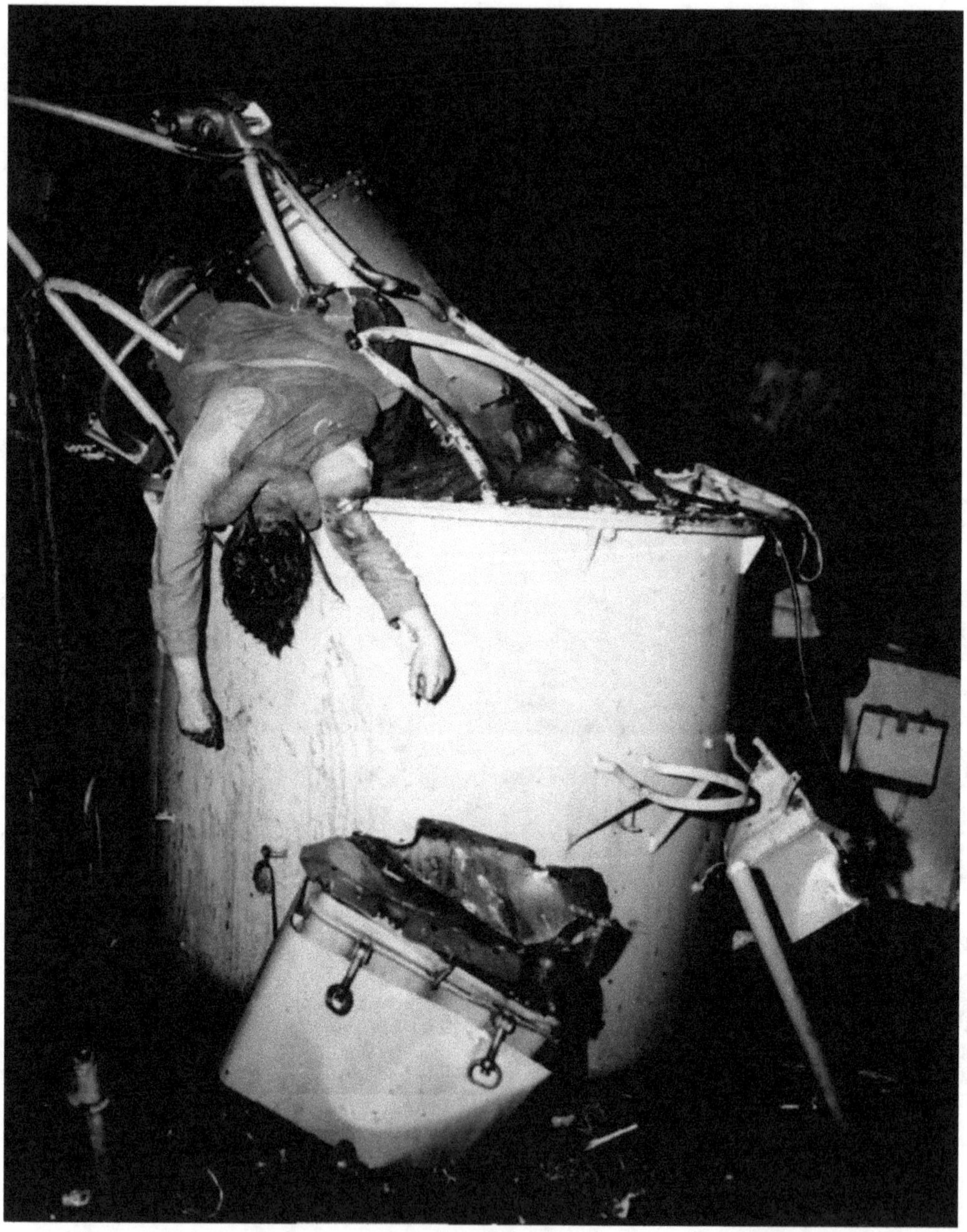

Figure 400. I mean no disrespect to the man or his family in the use of this photograph. This is a brutal image of what the war in small boats and the navies in general really meant. You need to lay aside the excitement of the rough and tumble stories and remember-this war was about people. Thousands of young men the world over gave their all for their countries. (U.S. Navy Historical Department)

THE SHIP KILLERS

THE DEFINITIVE ILLUSTRATED HISTORY OF THE TORPEDO BOAT

VOLUME IX: TIMELINE 1945

The Japanese were husbanding thousands of these primitive Shinyo suicide boats on Kyushu in preparation for the cataclysmic battle to defend the homeland that was only averted by the use of atomic bombs on Hiroshima and Nagasaki.

BY JOE HINDS

NIMBLE BOOKS LLC

CONTENTS (INCLUDING FIGURES)

DEDICATION

This volume is dedicated to all of the people who make up the web site, Coastal Forces. There must be thirty different people who have responded positively to me when I have posted questions about MTBs. It has always amazed me that this group is so readily forthcoming in assisting me to find answers to my questions and many times a note to cheer me on. I am forever pleased to be a part of this group of fine people. To you all, I dedicate this volume in your honor.

TIMELINE 1945

Date	Craft, Name, Event or Country	Event, Builder and Boat Specifications	City, River, Lake, Sea, Ocean
1945[1105]	January New-model S-boats	The second release of the new S-boats came into the fray known as the *700* class. These carried an additional two-stern torpedo tubes and 6 x 30-mm Luftwaffe cannons. These boats could match the new "dog" boats and destroyers for firepower. Unfortunately, they came too late. **Type:** Danziger Wagonfabrik- *S-701* **Speed:** 42 knots **Displacement:** 107 tons standard / 124 fully loaded **Range:** 740 nm **Length:** 114' 6" **Beam:** 16' 7" **Draft:** 9' 6" **Engines:** 3 x 2,500-hp Daimler MB 511, diesel **Armament:** 4 torpedo tubes, 6 x 30-mm Luftwaffe rapid-fire cannons	Germany
1945[1106]	January 4 *PT-78, PT-81*	The PT boats acted as escorts for the US invasion fleet headed to the landings at San Jose Harbor, Mangari Bay, Mindoro, Leyte Island. Japanese kamikazes had been wreaking havoc on the support ships since December 27th. On this day, the machine gunners of the *78* and *81* set fire to one of four IJN bombers; but it wasn't enough. This aircraft slammed into the ammunition ship *Lewis Dyche*, one-quarter mile away from the two boats. The merchant ship disappeared into vapor of fire and smoke, taking 71 sailors to the bottom. The two PT boats were lifted clear of the water, suffering great structural damage, killing two PT crewmembers and wounding ten others by concussion and falling debris.**Type:** *PT-71* **Class:**-Higgins, 78' **Speed:** 40 knots **Displacement:** 48 tons **Crew:** 17 **Length:** 78' 6" **Beam:** 20' 1" **Draft:** 5' 3" **Engines:** 3 x 1,500-hp V-12 Packard, 100-octane avgas **Armament:** 4 x 21" torpedo tubes, 4 x twin .50-cal. MG, 1 x 20-mm rapid-fire cannon	Leyte Island Philippines

[1105] Frank, *German S-Boats in Action in the Second World War*, 157.
[1106] Keating, *The Mosquito Fleet*, 243-244.

U.S.S. PT-78. Higgins type 78ft. m.t.b., 26th January 1943.

Figure 401. The *78* Higgins boat shows the scar of heavy fighting since she went into service on September 5, 1942 and was burned in the Philippines on November 24, 1945. A terrible end for a damned fine warrior.

Figure 402. *MTB-724* "Dog" boat. (John Lambert, Lambert Plans)

Date	Craft, Name, Event or Country	Event, Builder and Boat Specifications	City, River, Lake, Sea, Ocean
1945[1107]	January 6 *MTB-722*"D"	Off Stavfjord, the *722* scored two torpedo hits and sank the German iron ore carrier *Dora Dritzen* **Builder:***722:* Harris, Appledore; *709:* Dickie, Bangor & Tarbert, Fairmile "D", 115'	Norwegian Sea
1945[1108]	January 8/9 *MTB-712, 716* "D"	Coming south from Stavfjord, off Bommelofjord, the *712* hit the SS *Viola* with one torpedo; she sank with 1,000 tons of valuable chrome and zinc ore. **Builder:** *712:* Wallasea Bay Yacht Co., Fairmile "D", 115'; *716* Woodnut, St. Helens Low, Fairmile "D" 115'	Norwegian Sea
1945[1109]	January 8/9 *MTB-711, 623* "D"	Off Stavfjord, the *711* sank the merchant ship SS *Nikolaifleet* with two-torpedoes. The *709* did not sink this ship as some sources say. **Builder:** *711* Brooke Maritime, Lowestoft & Oulton Broad; *623:* Wallasea Bay Yacht Station, Fairmile 'D', 115'	Norwegian Sea
1945[1110]	January 10 *S-58, 60*	The two S-boats ran aground and were scuttled after suffering withering gunfire from British MGB's off the West coast of Unije. **Type:** *S-30*	Adriatic Sea
1945[1112]	January 13 5th SFltl.	Six boats set out to lay mines off Cromer and two, *S-98* and *67,* had to break off and return to base due to engine trouble. **Type:** *S-38*	North Sea English Channel
1945[1113]	January 14 6th SFltl.	During night operations off Cromer and the Thames estuary, *S-180* was lost to a mine close to Den Helder. The commanding officer and 11 crewmembers lost their lives. **Type:** *S-100*	North Sea

1107 Reynolds, *Dog Boats At War*, 113.
1108 Reynolds, *Dog Boats At War*, 113.
1109 Reynolds, *Dog Boats At War*, 113.
1110 Rohwer, *Chronology of the War At Sea*, 388.
1112 Rohwer, *Chronology of the War At Sea*, 386.
1113 Frank, German *S-Boats in Action*, 124.

Figure 403. Sailors around a 20-mm cannon.

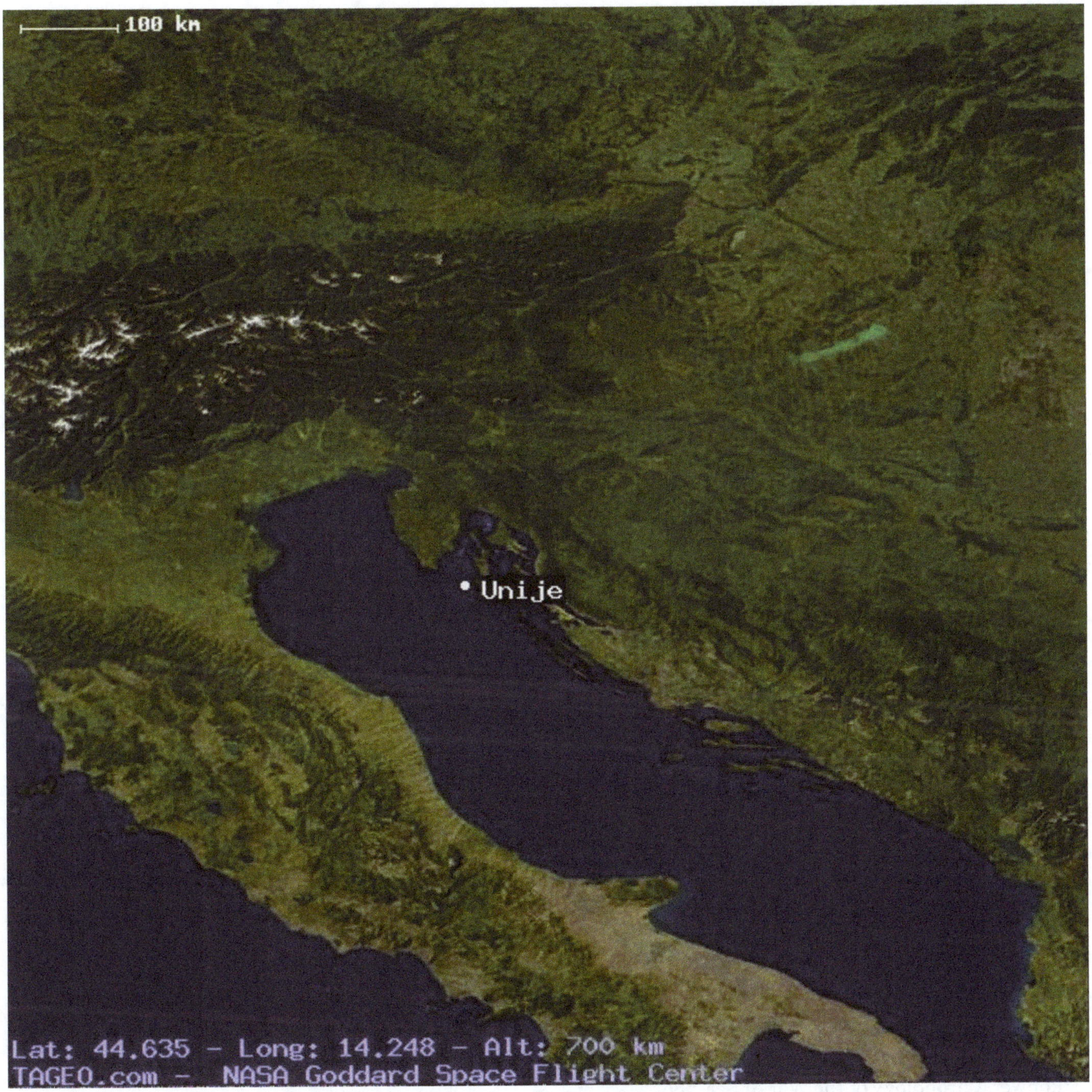

Figure 404. Unige on the coast of Yugoslavia. (Wikipedia)

Date	Craft, Name, Event or Country	Event, Builder and Boat Specifications	City, River, Lake, Sea, Ocean
1945[1114]	January 14 *PT-73*	Set on fire to prevent capture in Baliquias Bay, Mindoro **Type:** *PT-71* class Higgins, 78'	Philippines Islands
1945[1115]	January 14/15 2nd, 4th, 5th, 6th, 8th, 9th SFltl.	The six flotillas made a concerted effort to block all British shipping from moving. The 5th and 2nd flotilla's eight boats advanced s far as the mine-swept channels off the Humber. The British destroyer, HMS *Farndale*, fired on the 5th SFltl. The British destroyer HMS *Seymour* and British coastal force boats fired on the 9th SFltl. operating in the Scheldt estuary. The 6thSFltl. attacked an Allied convoy off the Scheldt. The *S-222* and the new *S-705* made claims for hits with FAT torpedoes. Before they could finish off their prey, the HMS *Curzon* (destroyer escort) and HMS *Cotswold* (destroyer) drove the flotilla away. The 4th SFltl. got in a gunfight with the HMS *Guillemont* (corvette). The 8th SFltl, with eight boats, attacked a convoy of LSTs near the Tongue Sand Fort with eight FaT and LuT torpedoes. The Lagenunabhänger Torpedo, or LuT, could be launched while the boat was on any bearing, and then change course towards its target. A LuT torpedo struck the *LST-415* (she did not sink) and the fire from the fort drove the flotilla away.	North Sea English Channel
1945[1116]	January 15 3rd, 8th SFltl.	*S-33* ran aground on Unije Island and had to be abandoned. **Type**: *S-10*	Adriatic Sea
1945[1117]	January 16 8th SFltl.	S-boats attacked and torpedoed *LST-415*, which was beached off Thurrock, Thames estuary, England. **Type:** *S-100, S-30, S-38*	Thames estuary North Sea
1945[1118]	January 17 *S-58, 60*	The two S-boats were run aground and scuttled after being shot up by British MGB's on the west coast of Unije. **Type :***S-38*	Adriatic Sea

[1114] Polmar, *PT-Boats At War,* 117.
[1115]Rohwer, Chronology of the War At Sea, 386.
[1116] Hümmelchen, *Warship 31,* 173.
[1117] *Dictionary of American Fighting Ships,* James L. Mooney, Ed., (Naval Historical Center Department of the Navy, 1981), Vol. VII, 622.
[1118] Rohwer, *Chronology of the War At Sea,* 388.

Figure 405. S-boat crews had an advantage in the forward 20-mm gun mount. It was power controlled for 360 degree rotation and could be raised and lowered by three feet in times of need. The armor plated "gun tub" saved many a sailor's life. (Collection of Hans Frank)

Figure 406. The HMS Farndale was one of many destroyers that fought the S-boats to a standstill on January 14th and 15th.

Figure 407. The photo shown shows the full size of a well loaded LST. LST were nor the small troop landing craft common mistaken for. The LST 415 would have looked quite similar after being attacked.[1121]

[1121] http://www.wlb-stuttgart.de/seekrieg/4307-bilder/lst318%20gr.jpg

Date	Craft, Name, Event or Country	Event, Builder and Boat Specifications	City, River, Lake, Sea, Ocean
1945[1122]	January 22 Admiral Dönitz	The admiral ordered Kriegsmarine ships of all types and sizes, as well as civilian German ships, in the Baltic region to begin immediate evacuation operations. The operations took four months and resulted in over 2.3 million people rescued from the Soviet armies.	Baltic Sea
1945[1123]	January 23 9th SFltl.	The *S-168* and *175* of the 9th flotilla sank the Allied freighter *Halo* north of Dunkirk. **Type***:* *S-100*	North Sea
1945[1124]	January 23 8th SFltl.	Five boats of the flotilla attack the Destroyer Escort HMS *Seymour*, corvette HMS *Guillemont* and several MGBs on The Thames estuary near North Foreland. The *S-199* suffered severe damage when *S-701* scuttled near Tongue Sand Fort rammed it. Later, the *MTB-495* collided with the already damaged *S-701* and the S-boat was non-operational for the rest of the war. **Type:** *S-100*	Thames estuary North Sea
1945[1125]	January 23/30 5th, 4th, 9th, 6th,, 2nd SFltl.	Eighteen S-boats lay extensive mine fields off East Dudgeon, off Orfordness and Humber. **Type:** *S-100, S-30, S-38*	North Sea
1945[1126]	January 25 Kriegsmarine	The German Navy undertook the single largest evacuation of refugees and soldiers in naval history. The Soviets had pushed the German forces to the edge of their borders and German S-boats, patrol boats, minesweepers and any ship of any size afloat were pressed into service for the job. More than 2.5 million people were evacuated between January 25th and March 10th.	Baltic Sea
1945[1127]	January 28 *MTB-690* "D"	Lost in collision with submerged wreck off Lowestoft, North Sea. **Builder:** British Power Boat Co. Falmouth, Fairmile 'D',115'	North Sea
1945[1128]	January 28 *PT-338*	The *338* boat in a firefight was run aground on Semirara Island, Mindoro.**Type:** *PT-103* class, Elco 80'	Philippine Islands
1945[1129]	January 31 *MTB-715* (RNoN)	The *MTB-715* sank the German minesweeper *M-382* off Ravnefjord, north of Molde. **Builder:** Woodnut, Fairmile 'D', 115'	Kristiansand Norwegian Sea

[1122] Frank, *German S-Boats in Action*, 134.
[1123] Rohwer, *Chronology of the War At Sea*, 388.
[1124] Rohwer, *Chronology of the War At Sea*, 388.
[1125] Rohwer, *Chronology of the War At Sea*, 388.
[1126] Rohwer, *Chronology of the War At Sea*, 389.
[1127] Lambert, John. The *Fairmile 'D' Motor Torpedo Boat, Anatomy of the Ship* (Conway Maritime Press, 1985), 14.
[1128] HyperWar: US Navy Torpedo Boats (PT-Boats), 1940-1945, 10.
[1129] Rohwer, *Chronology of the War At Sea*, 391.

Figure 408. An S-boat underway on a river.

Figure 409. Japanese ship wrecks littered the South Pacific islands as a testament of the ferocity of the war. (U.S. Navy Historical Department)

Figure 410. The Kriegsmarine, like every navy, put its best foot forward for the public.

Figure 411. Frank Andruss has more photos of the "Real" young men who served on Americas PT boats than anybody. This photograph is one of the most powerful in his collection. The young man on the front, far left, has his eyes on the sky. He would spot enemy planes that could chew up a PT boat. The man at the helm is fully focused ahead and still relaxed, the sign of a man long accustomed to being ever vigilant. The central figure is the kid, front right enjoying a quick smoke and ever so watchful. He is relaxed but wound tight and ready to spring into action when called upon. Right behind him is another crewman on the binoculars also looking out for aircraft. This is the one shot that sums up a PT crew on operations like no other. Frank has a treasure here. I thank him for sharing. (Frank Andruss, Sr., The Mosquito Fleet Exhibit)

Date	Craft, Name, Event or Country	Event, Builder and Boat Specifications	City, River, Lake, Sea, Ocean
1945[1130]	February 1 *PT-77, 79* *1st Attack of IJN Shinyo suicide boats*	The navy's invasion of Luzon, from Nasugbu, SW of Manila Bay, was attacked by an estimated 30 Shinyo Japanese explosive motor boats. The US Navy sub-chaser *PC-1129* was sunk and the destroyer *Claxton* and destroyer escort *Lough* fought a nerve-wracking battle to quell the swarm of enemy motor boats. In the confusion of the rapid pace of the engagement, the destroyer escort *Lough* and destroyer *Conyngham* sank the two PT boats, causing four killed and many wounded. The 30 survivors swam to enemy shores and hooked up with Pilipino guerrillas until their rescue by *PT-227* and *230* on February 3rd. **Type:** Higgins, 78' Japanese "*Shinyo*" explosive motor boat **Speed:** 25/30 knots **Displacement:** 1 ¼ tons standard, 2 tons fully loaded **Length:** 16' 6" **Beam:** 18' **Crew:** 1 **Engines:** 2 x 80-hp auto, two screws **Armament:** 551 pounds TNT, or 2 x 360-lb depth charges, 2 x 4.7" rockets	off Nasugbu Luzon, PI Philippine Sea
1945[1131]	February 5 *PT-308, 313*	Motor torpedo boats *PT-308* and *PT-313,* on a routine patrol from Leghorn, Italy, sighted two south-bound KT ships and one escort and engaged them a half mile northwest of the swept-channel off Portovenere, claiming at least one KT ship sunk.. They claimed hit on one KT ship and sank the other. **Type:** *PT-71* class, Higgins 78'	Mediterranean
1945[1132]	February 6 *S-36, 61*	The two S-boats were heavily engaged in a firefight when they collided off Pola. The boats were towed into port and abandoned. *S-36* **Type:** *S-14; S-61* **Type:***S-30*	Mediterranean
1945[1133]	February 7 *PT-373, 356*	The two PTs were the first American boats to enter Manila Bay since the defeat of 1942. **Type:** Elco *PT-103* class, 80'	Manila Bay

---

[1130] Keating, *The Mosquito Fleet*, 244.
[1131] HyperWar: The Official chronology of the U.S. Navy in World War II, Chapter VII: 1945 http://ibiblio.org/hyperwar/USN/USN-Chron/USN-Chron-1945.html.
[1132] Rohwer, *Chronology of the War At Sea*, 389.
[1133] HyperWar: The Official Chronology of the US Navy in World War II, http://ibiblio.org/hyperwar/USN/USN-Chron/USN-Chron-1945.html.

Figure 412. The "Shinyo" suicide boats were based on the very successful Italian models perfect by the Italians over the years. The Italian boats were not suicide boats. The design allowed for the pilot to aim at the target, lock in the steering, and use an ejection seat explosive to roll over the back of the boat and be picked up by a small craft. The Japanese meant for the man and the boat to vaporize together.

Figure 413. This photo from earlier in the war typifies the massive destruction suffered by the IJN all over the Pacific. Thousands of islands were surrounded by hundreds of wrecks lost during Word War II. To this day the evidence of great battles is still visible in the islands of the South Pacific. (U.S. Navy Historical Department)

Date	Craft, Name, Event or Country	Event, Builder and Boat Specifications	City, River, Lake, Sea, Ocean
1945[1135]	February 7 *PT-303,304* *MTB 422*	While patrolling off Savona, Italy, the three boats attacked an enemy convoy and sank at least two vessels. The patrols operated from Golfe Juan, France *PT-303, 304* **Type:** Higgins,78'; *MTB-422* Builder:Higgins,78'-6"	Mediterranean
1945[1136]	February 8 *PT-308*	The *308* boat along with two British MTBs attacked three northbound flak lighters off Point Moneglia. **Type:** Higgins, *PT-200* class, 78'	Ligurian Sea Mediterranean
1945[1137]	February 13 *MTB-394*	Off Great Yarmouth, the MTB sank a German two-man submarine, "Seehund" (Seal). **Builder:** Vosper, Wivenhoe, 73'	English Channel
1945	February 13/14 MTB-*710, 660*	Using radar, the flotilla spotted three F-Lighters in the Quarnero Channel, East cost of Istrian Peninsula. They sank the first F-Lighter with gunfire and torpedoed the third by radar vectoring. **Builder:** 710, Osborne, Littlehampton, Fairmile "D", 115'	North Adriatic Sea
1945[1138]	February 14 Many MTBs	All boats were lost to fire and explosions in Ostend harbor. **Builder:** British Power Boat Co., 71'-9" MTB-*438, 444, 459, 460, 461, 462, 465, 466* **Builders:** In order of listing. Lady Bee, Southwick; Risdon Beazley, Northam; Thompson & Balfour, Bo'ness; Austins, East Ham, 115'; *MTB-255,* White, 72'-6", *776, 789, 791, 798,255*	Ostend Harbor North Sea
1945[1139]	February 16 *PT-303*	The *303* boat and two British MTBs were operating out of Leghorn, Italy; they engaged two enemy barges off Vernazza with inconclusive results. **Type:** Higgins PT-200-class, 78'	Ligurian Sea
1945[1140]	February 17 *PT-605*	The *605* sank after striking a sunken object off Ostend, Belgium **Type:** Elco *PT-103* class, 80'	North Sea
1945[1141]	February 17/18 2nd, 5th, 9th SFltl.	The 2nd and 5th SFltl. laid mines off Humber; their mines sank the armed trawler HMS *Aquarius* and damaged two other ships. The 9th laid a mine barrage off the South East coast of England.	English Channel

1135 http://www.WarSailors.com
1136 http://www.WarSailors.com.
1137 Rohwer, *Chronology of the War At Sea,* 397 and Uboat.Net Types- Midget submarines http://uboat.net/types/Seehund.htm.
1138 Internet correspondence Roger Pearson, August 30, 2007.
1139 http://www.WarSailors.com.
1140 http://www.WarSailors.com.
1141 Rohwer, *Chronology of the War At Sea,* 394.

Figure 414. A rare shot of a Higgins PT before it suffered the ravages of war. (Collection of Mike Montana)

PT 302 SKIMS THE WATER IN 1944, WITH ENS. ROSS AT THE HELM. THE 2 GERMAN MARKS ON THE WIND SCREEN SHOW THIS BOAT HAD 2 KILLS, BOTH OF THEM BEING GERMAN AMMO SHIPS. ONE AMMO SHIP WAS PICKED UP ON RADAR JUST BEFORE CHRISTMAS SOUTH OF GENOA. PT 302 CREPT TO WITHIN 500 YARDS OF THE SHIP FIRING 2 TORPEDOES. THE SHIP WENT UP LIKE A CHRISTMAS TREE. PT 302 WOULD TAKE ON 7 GERMAN SURVIVORS.

Figure 415. The *302* was one of the original Mediterranean Flotilla boats. They were constantly interacting with each and all proved themselves quite well. The operations in the Med helped these crews learn valuable lessons about the enemy. No.1 was to truly fear F-Lighters and No. 2 was to have great respect for S-Boats. (Collection of Frank Andruss, Sr., The Mosquito Fleet Exhibit)

Figure 416. The mascot, Umbriago, was with the 209 boat until the crew was ordered home in 1945. No one seems to know what happed to the mascot.[1142] "Umbriago, whom Durante calls "that lifelong chum," is partly a creation of Jimmy's far-fetching imagination and partly a figment of Italian folklore. The folklore Umbriago is a friendly, lusty, happy little man who is always the life of the party. But in Jimmy's comedy, Umbriago may assume many shapes—clarinetist, bank president, farmer—according to whatever mischief Durante is up to."[1143]

1142 Courtesy James J. Stanton. http://www.pistolpackinmama.net/pb/wp_37d910a0/wp_37d910a0.html
1143 http://www.urbandictionary.com/define.php?term=umbriago

Figure 417. The *Seehund* (Seal) became a successful design for Germany's midget submarine attempt. It was not an ideal weapon but if it had been developed earlier it may have made a difference to the war effort. As it was, the small sub force accounted for sinking 120,000 tons of Allied shipping. Three deigns were made with the last one being the most successful. Still, the surface speed of 8 knots and 10.5 knots submerged were not enough for active warfare. The boats were mostly used for running supplies to stranded German troops throughout the Baltic Sea.[1144] The picture above shows a Seehund factory.

[1144] http://www.hnsa.org/ships/seehund.htm

Figure 418. A Seehund of German origin, now bearing French colors (S-622) in the *Musée national de la Marine* in Brest, France.[1145] Submersible torpedo boats share a lineage derived both from submarines and torpedo boats.

1145 http://en.wikipedia.org/wiki/File:Submarine_S622.jpg

ARTHUR FRONGELLO (LEFT) HOLDS ONTO RADIOMAN (WAGNER) AFTER BECOMING SEA SICK DURING THE TRIP. THE CREW WAS CRACKING JOKES ABOUT ARTHUR'S DELIEMA. ARTHUR SAID HE HAD NO CHOICE BUT TO LAUGH WITH THE GUYS, EVEN THOUGH HE FELT ROTTEN.

Figure 419. This is a terrific photo that captures the spirit and bravery of the men who ran the small boats. It was the American sense of, humor, stead fast dedication to duty, and outright ferocity when called upon that helped win that war. These were real men, or real kids. They did a real crappy job very well and we owe them so many thanks that can never be repaid. So anyway, thanks Seaman Frangello and Wagner. (Collection of Frank Andruss, Sr., The Mosquito Fleet Exhibit)

Date	Craft, Name, Event or Country	Event, Builder and Boat Specifications	City, River, Lake, Sea, Ocean
1945[1147]	March *S-302*	A complete operating radar set was removed from the wreck of a British bomber at Rotterdam in 1943. A copy of this device, called the *Berlin-S*, was hand-made for testing and placed on *S-302* in the Baltic Sea. **Type:** *S-100*	Baltic Sea
1945[1148]	March 2 *PT-373*	General MacArthur returns to the Philippines the same way he left it, on a PT boat. **Type:** Elco *PT-103* class, 80'	Philippine Sea
1945[1149]	March 2 4th SFltl. *S-220*	British destroyer's and RAF fighter attacked the flotilla while causing considerable damage and dispersing the flotilla. The *S-220* was sunk by RAF gunfire off Humber. **Type:** *S-100*	North Sea`
1945[1150]	March 6/7 *MTB-703, 705; 656* 'D'	Off the Arsa Channel the *656* sank a merchant ship with torpedoes. The *703* and *705* attacked the convoy with guns and suffered damage from the return fire of escorts. **Builder:** *703,* Tough Brothers, Teddington; *705*; Wallasea Bay Yacht Station; *656:* Risdon Beazley, Northam, Fairmile 'D', 115'	North Adriatic Sea
1945[1151]	March 11/12 British MTBs	A MTB force sank the German minesweeper depot ship *Paris* off Haugesund, Norway.	Norwegian Sea

[1147] Frank, *German S-Boats in Action,* 138.
[1148] Keating, *The Mosquito Fleet,* 246.
[1149] Frank, *German S-Boats in Action,* 128.
[1150] Reynolds, *Dog Boats At War*, 232.
[1151] Frank, *German S-Boats in Action,* 128.

Figure 420. This photo of the General MacArthur and his entourage coming ashore is pretty funny. This is a purely staged photo shot. MacArthur came back to the Philippines the same way he left, on a PT boat. (National Archives)

Figure 421. S-boats attacking a convoy. Sboats used tried and true maneuver measures to obtain maximum results from each patrol. Discipline was strict and helped the S-boat forces achieve tremendous results. (Wikipedia)

Figure 422. *S-100* side view. (Illustration by author)

Figure 423. The Japanese had begun to believe their own propaganda. This poster of Japanese bombers over New York was a great moral booster in Japan and a joke in America. The planes displayed are the very small Japanese bombers that were incapable of such an attack. (Library of Congress)

Date	Craft, Name, Event or Country	Event, Builder and Boat Specifications	City, River, Lake, Sea, Ocean
1945[1153]	March 16 MTB-675 'D'	The MTB sank a new "wonder weapon", the "*Seehund*" (Seal), off the coast of North Ireland. The Seal was one of six types of one- and two-man submarines made by Germany in the last days of the war. The boat had a crew of two weighed 17 tons and carried two very effective German torpedoes the G7e.Seehunds went on 142 sorties against the Allies, lost 35 craft (70 men), and sank 8 Allied vessels. **Builder:** Alex Robertson (Yacht Builders) & Sons Ltd. Sandbank, Scotland, Fairmile 'D',115'	Atlantic Ocean
1945[1154]	March 17 6th SFltl.	The flotilla boats sank two freighters on the Scheldt Route off Cromer England.	North Sea
1945[1155]	March 17/18 2nd SFltl.	The flotilla carried out additional mining operations off Smiths Knoll.	North Sea
1945[1156]	March 18/19 8th, 9th SFltl.	The flotilla attacked convoy FS 1759 off Lowestoft sank the freighters SS *Crichtoun* and SS *Rogate*. These werethe last Allied merchant ships torpedoed by S-boats. **Type:** *S-30, S-38, S-100*	North Sea
1945[1157]	March 18/19 2nd,4th SFltl.	The two flotillas laid additional mines along the Thames-Scheldt convoy route. As a result, the LST-80, freighters *Samselbu, Empire Blessing* and an armed trawler were sunk. The US Liberty ship *Hadley F. Brown* was seriously damaged by a mine.	North Sea
1945[1158]	March 21 *S-203*	The *S-203* was sunk by Allied aircraft off Texel, Netherlands after striking a mine. **Type:** *S-100*	North Sea

[1153] Frank, *German S-Boats in Action,* 397, and U boat Net-Types-Midget submarines http://uboat.et/types/seehund.htm.
[1154] Frank, *German S-Boats in Action,* 129.
[1155] Rohwer, *Chronology of the War At Sea*, 401.
[1156] Rohwer, *Chronology of the War At Sea*, 397.
[1157] Rohwer, *Chronology of the War At Sea*, 401.
[1158] Rohwer, *Chronology of the War At Sea*, 402.

Figure 424. German S-Boats in the old North carried a distinctive color pattern. This appeared to aid in the boats blending into the haze of the cold waters. (Illustration by author)

Figure 425. The American war effort's vast need for skilled technicians was matched by an ample supply of technically literate citizen-soldiers and sailors. (Library of Congress)

Date	Craft, Name, Event or Country	Event, Builder and Boat Specifications	City, River, Lake, Sea, Ocean
1945[1163]	March 22 *MTB-655* "D"	The *655* struck a mine while returning to Zara, off Arsa. She broke in two and sank. The *MTB-643* boat picked up 23 survivors. 7 men remained missing. **Builder:** Osborn, Littlehampton, Fairmile 'D', 115'	Adriatic Sea
1945[1164]	March 23 MTB-705 'D'	She struck a mine on the passage from Zara to Maknare Channel. Her stern was blown off; two men went missing and one was injured. Builder: Risdon Beazley, Northam, Fairmile 'D', 115'	Adriatic Sea
1945[1167]	April 7 *MTB-501*	The boat was sunk in a firefight with_S-boats. **Builder:** Camper & Nicholson 117'	North Sea
1945[1168]	April 7/8 *MTB 5001* 'D'	In the North Sea, the *MTB-5001* 'D' attacked and rammed the S-*177* boat, which in turn sank the *5001,* off Humber estuary. **Builder**: Woodnut, 115'	North Sea
1945[1169]	April 7/8 *MTB-494*	Lost, rammed, and sunk by an S-boat. **Builder:** British Power Boat Co. 71'9"	North Sea
1945[1170]	April 7/8 4th, 6th ,9th SFltl.	During mine-laying operations along the Thames-Scheldt convoy route, British MTBs made an attack, and the 4th flotilla lost the *S-202* and *703* to collision. The 9th withdrew when rough weather makes operation impossible. **Type:***S-100*	North Sea
1945[1171]	April 10/11 *MTB-710 "D"*	Sunk by a mine near Zara, NE Adriatic. **Builder:** Osborne, Fairmile 'D'. 115'	Adriatic Sea
1945[1172]	April 10 S-boats	In the confusion of the massive evacuation effort by the German armed forces from the Russian War Zone, the S-boats sank the German freighter *Newerk* by mistake.	Baltic Sea
1945[1173]	April 10/16 4th, 6th, 9th SFltl.	While laying mines along the Nore-Flushing-Channel area, S-boats engaged the British destroyer HMS *Elkins* and the destroyer escort	North Sea

[1163] Reynolds, *Dog Boats At War*, 233.
[1164] Reynolds, *Dog Boats At War*, 233.
[1167]Royal Navy Vessels Lost At Sea 1939-45-by Type www.naval-history.net/WW2BritishLosses3Coastal.htm
[1168] Lambert, The Fairmile 'D' Motor Torpedo Boat, 15
[1169] Royal Navy Vessels Lost At Sea 1939-45-by Type www.naval-history.net/WW2BritishLosses3Coastal.htm.
[1170] Rohwer, *Chronology of the War At Sea*, 407.
[1171] Lambert, *The Fairmile 'D' Motor Torpedo Boats*, 14.
[1172] Rohwer, *Chronology of the War At Sea*, 409.
[1173] Rohwer, *Chronology of the War At Sea*, 409.

		HMS *Hambledon* along with several MTBs. The mines laid by the flotillas damaged Elkins beyond repair off Ostend, sank the tanker *Gold Shell* and damaged the freighters *Conakrian, Benjamin H. Bristow,* and *Horace Binney.*	Strait of Dover
1945[1174]	April 12 POTUS died	President Franklin D. Roosevelt died. Vice President Harry S. Truman succeeded him.	America
1945[1175]	April 13 *MTB-670, 697* "D"	The two British MTBs sank the German fleet torpedo ship (TA 45, ex-Italian *Spica)* in Morlaca Channel, Croatia. **Builder:** *670,* Wallasea; 697, Woodnut Fairmile 'D', 115'	Morlacca Channel
1945[1176]	April 15 *TKA-131, 141*	The TKAs *131 and 141* attacked and severely damaged the German destroyer, *Z 34,* while escorting a convoy of ships full of German refugees from the Rixhöft-Nevstdate-Karthaus line in Danzig Bay. **Type:** *G-5,* 62.5'	Baltic Sea
1945[1177]	April 16 British MTB	A lone MTB sank a German Motor Minesweeper, *R 15,* in the North Adriatic Sea.	Adriatic Sea
1945[1178]	April 17 *MTB-697* 'D'	Hit a mine off Krk Island, NE Adriatic ,and sank. **Builder:** Woodnut, Fairmile 'D', 115'	Adriatic Sea
1945[1179]	April 17 S-boats	German S-boats engaged Allied destroyers off San Remo. An Italian explosive motor boat sank the French destroyer *Trombe.*	Western Mediterranean
1945[1180]	April 18/20 TKA (Unknown)	Soviet TKAs sank the refugee-laden freighter *Emily Sauber* off the coast of Neufahrwasser (Gdansk).	Gulf of Gdansk Baltic Sea
1945[1181]	April 19 *Ms-72*	The Italian motor torpedo boat and the Italian destroyer, *Grecale,* sank the almost completed aircraft carrier *Aquila* in the Genoa Harbor.	Adriatic Sea
1945[1182]	April 23-25 *MAS-549* et al.	MAS-*549, 561, 558, 557, 505, 554* and *551*, along with dozens of others under German command, were scuttled along the Italian Rivera.	Ligurian Sea Mediterráneas

[1174] Rohwer, *Chronology of the War At Sea*, 409.
[1175] Rohwer, *Chronology of the War At Sea*, 409.
[1176] Rohwer, *Chronology of the War At Sea*, 398.
[1177] Rohwer, *Chronology of the War At Sea*, 407.
[1178] Lambert, *The Fairmile 'D' Motor Torpedo Boat*, 14.
[1179] Rohwer, *Chronology of the War At Sea*, 410.
[1180] Rohwer, *Chronology of the War At Sea*, 410.
[1181] Rohwer, *Chronology of the War At Sea*, 411.
[1182] Rohwer, *Chronology of the War At Sea*, 411.

1945[1183]	April 26 TKA (Unknown)	The TKA force sank the German lighter MFP *F-248A* near Hela.	Arctic Sea
1945[1184]	April 27 Japanese *"Shinyo"*	The destroyer USS *Hutchins* was seriously damaged by a Japanese *"Shinyo"* suicide motor boat in Buckner Bay, Okinawa, and not repaired at war's end.	Okinawa Ryukyu Island
1945[1185]	May 1 Last Flag Parade	At Geltinger Bay the crews of the 3rd SFltl Training Flotilla and 10th SFltl., along with the crews of the S-boat depot ships *Hermann Von Wissmann, Tsingtau, Tanga, Carl Peters* and *Buea,* attended their last flag ceremony. At ceremony's end the operations the Schnellboote of the Kriegsmarine were finished with World War II.	Baltic Sea

Schnellbootbegleitschiff Tsingtau in 1934

Figure 426. The German depot ships carried food, fuel and ammo for the S-boat crews. All of these ships were latge enough and well armed enough to put up a stiff fight.[1187]

1183 Richard Worth, Internet correspondence with Roger Jordan, *Worlds Merchant Fleets, 1939,* 466 Internet correspondence to Author July 24, 2007.

1184 Casualties, Navy and Coast Guard Ships, WWII :Destroyer (DD) http://www.history.navy.mil/faqs/faq82-1.htm.Accessed by Author January 21, 2007.

1185 Frank, *German S-Boats in Action,* 135.

1187 "German Naval History" © Michael Emmerich 1997-2005 http://www.german-navy.de/kriegsmarine/ships/fleettender/tsingtau/index.html

Figure 427. By 1945, the primary mission for surface combatants in the Pacific was anti-aircraft fire.

Figure 428. **SPARS** was the United States Coast Guard Women's Reserve created 23 November 1942 with the signing of Public Law 773 by President Franklin Delano Roosevelt. The name is a contraction of the Coast Guard motto Semper Paratus and its English translation Always Ready. The name also refers to a spar in nautical usage.

Figure 429. The most infamous of the "Shinyo Class" was a direct copy of the very successful Italian Motor Explosive Boat. However, the Italians never in the designed this as a suicide craft. The Italians wanted trained and battle hardened crews to stay alive and fight many days. The Italian version had a small explosive charge under the seat so the pilot could eject safely prior to impact. The Japanese, fighting, as they saw it, for national survival, did not require the pilots back. [1188]

[1188] http://en.wikipedia.org/wiki/Tokk%C5%8Dtai

Date	Craft, Name, Event or Country	Event, Builder and Boat Specifications	City, River, Lake, Sea, Ocean
1945[1189]	May 3 *S-201*	The S-boat was sunk at it base by RAF bombers at Kiel Bight, Kiel Fjord, city of Kiel, Germany. **Type:** *S-100*	Kiel Bay S. Baltic Sea
1945[1190]	May 3 S-Boat surrender	The S-boats *S-30, 36, 61, 151, 152, 155,* and *156* moved from Pola (Pula), Croatia to Ancona, Italy for the surrender to the Allies.	Adriatic Sea
1945[1191]	May 4 *S-103*	Sunk by RAF Typhoon fighter aircraft in the Southern Baltic Sea; she was likely attempting to make it to Kiel, Germany, for the surrender. **Type:** *S-38*	S. Baltic Sea
1945[1192]	May 4 *MTB-720, 798*	Both boats were flying the White Ensign while returning from a Cherbourg patrol to, Portland, England. A group of RAF Beaufighters made a murderous rocket attack and shot up both boats badly. The *798* was sunk by gunfire from a British Hunt class DD and the *720* was towed to port and repaired. **Builder:** *798:* Hall, Glampton *720* King, Burnham on Crouch, Fairmile 'D' 115'	North Sea

Figure 430. The RAF Typhoon fighter was no glamorous airplane but it was plenty mean. The Typhoons destroyed more then their share of German and Italian ships . (Wikipedia)

[1189] Rohwer, *Chronology of the War At Sea*,414; Wikipedia http://en.wikipedia.org/wiki/Kiel Bight.
[1190] Rohwer, *Chronology of the War At Sea*,414; Wikipedia http://en.wikipedia.org/wiki/Pola/Ancona.
[1191] Rohwer, *Chronology of the War At Sea*, 414; Wikipedia http://en.wikipedia.org/wiki/Pola/Ancona.
[1192] Reynolds, *Dog Boats At War*, 100; *Warships* Supplement 65, 19.

Figure 431. This version of a suicide boat was made by the Japanese Army. While the Japanese navy version (Shinyo) was not a great product, the Army version was an ever bigger failure. Of the 3,000 built, there is no record of any of the Army models being used. The Japanese were husbanding their resources for the defense of the homeland. These models had the annoying habit of catching fire and blowing up in training sessions. (U.S. Navy Historical Department)

Figure 432. An S-boat flying a white flag. (Wikipedia)

Figure 433. The S-boat commands were among the most orderly and efficient arm of the German military to come to terms with the Allies. The boats were turned over in clean and well maintained condition. By 1946 many of these crewmen, officers and S-boats would be working for American and British intelligence in the Baltic and Arctic Seas. (Roderick Timms)

Date	Craft, Name, Event or Country	Event, Builder and Boat Specifications	City, River, Lake, Sea, Ocean
1945[1193]	May 5 S-Boat Operations	All S-boat operations officially ended in the English Channel on this date. A few boats made for the German Bight and the rest made for Den Helder or Rotterdam. The S-boat score for 1945 was six ships of 12,972 tons by torpedo, 25 ships of 75,999 tons by mines, and seven ships of 26,408 tons seriously damaged.	North Sea
1945[1194]	May 6 *S-226*	Scuttled after hopelessly damaged by air attacks East of Fehmarn Island, near Kiel Bight. **Type:** *S-100*	South Baltic Sea
1945[1195]	May 7 *S-301, 191*	The boats sank after colliding in Fehmarnsund, the arm of the sea of the Baltic Sea between Kieler Bay and Mecklenburger Bay, which the island Fehmarn separates from the mainland. **Type:** *S-100*	Baltic Sea
1945[1196]	May 7	Unconditional surrender of Germany and all its armed forces.	Germany
1945[1197]	May 10 S-Boat	A mine in Geltinger Bay sank the S-boat depot ship, *Carl Peters*.	Baltic Sea
1945[1198]	May 14 *PT-135, PT-343*	On night patrol, the two PT boats discovered a channel island at Piso Point across from Davo Island, Davo Gulf. By following the channel, they encountered six carefully concealed Shinyo suicide motor boats. The 2 PTs commenced to shoot up the boats, sheds, and ammunition dump and fuel farm with fire from their 40-mm cannon and twin .50 caliber machine guns. **Type:** Elco, *PT-103*-class, 80'	Davo Gulf, Mindanao Island
1945[1199]	May 19 MTB-*715, 709* "D" (RNoN)	An explosion in the engine room of the *715* while at Fosnavag, Norway, moored alongside the *709*, burned out of control and both boats were lost. *709* **Builder:** Dickie,Bangor & Tarbert Fairmile "D", 115'; *715,* Woodnut, Fairmile "D", 115'	North Sea

[1193] Frank, *German S-Boats in Action*, 130.
[1194] Rohwer, *Chronology of the War At Sea*, 414.
[1195] Rohwer, *Chronology of the War At Sea*, 416.
[1196] Rohwer, *Chronology of the War At Sea*, 417.
[1197] Rohwer, *Chronology of the War At Sea*, 417.
[1198] Hoyt, *MacArthur's Navy*, 195.
[1199] The Norwegian MTB Flotilla in Shetland, 7.

Figure 434. The R-Boot minehunter has been terribly underrated in the previous years of naval history works. The R-Boote work in packs and each one came with a deadly amount of fire-power. Many carried: 1 x 37-mm Luftwaffe cannon, 2 x twin barreled 20-mm cannon, 2 x ingle barrel 20-mm cannon, 8- x 12 7.92 MG, an enormous load of star shells, and six DC's. They were also the first military craft to use the "Voith Schneider Propellers" which enabled the boat to make seemingly 90% turns without cutting speed. Germany built 424 of these boats and 140 survived the war.[1202]

Figure 435. Pinup girls were the most popular theme for nose art of aircraft, tanks, and small boats. The crew of the *171* chose the January 1943 girl (right) for their nose art (left). They added the phrase "Good girls won't, but Dinah Might." The original pinup calendar remained with the *171*, and is now at Battleship Cove Museum in Fall River, MA. (Mike Montana)

[1202] "German Naval History" is (C) by Michael Emmerich 1997-2005, http://www.german-navy.de/kriegsmarine/ships/minehunter/rboat/index.html and http://en.wikipedia.org/wiki/R_boat

Figure 436. This picture and the two on the following page show Rear Admiral Breuning arriving Felixstowe at the head ofFPBs of the 9th Flotilla just after the surrender of Germany. He was not directly involved in FPB-operations but had a great knowledge about minefare. His job was to give the British all information about German mine barriers or areas were the Germans (also FPBs) had laid mines. (U.S. Navy Historical Department)

His record: Rear Admiral Erich Alfred Breuning 09.24 – 05.43 Chief of the 3rd Escort Division (as Captain) 06.43 – 09.44 Commander Escort Command West (as Rear Admiral) 10.44 – 05.45 Chief of staff of in the Armed Force National Socialist Leadership staff in the German Armed Forces Staff.[1203]

[1203] Note by Hans Frank.

Figure 437. Breuning aboard *S-216*.

Figure 438. Breuning arriving at Felixstowe.

Figure 439. MacArthur aboard PT-490 in summer 1945. En route to Negros, Philippines.[1204]

Figure 440. The crew displaying the personal flag of General of the Army Douglas MacArthur. While MacArthur was no paragon of civil-military relations, the "personal flag" is not a "red flag" all "flag officers in the U.S. military have personal flags.[1205]

1204 http://www.navsource.org/archives/12/05490.htm

1205 http://en.wikipedia.org/wiki/Flags_of_the_United_States_armed_forces#Personal_flags

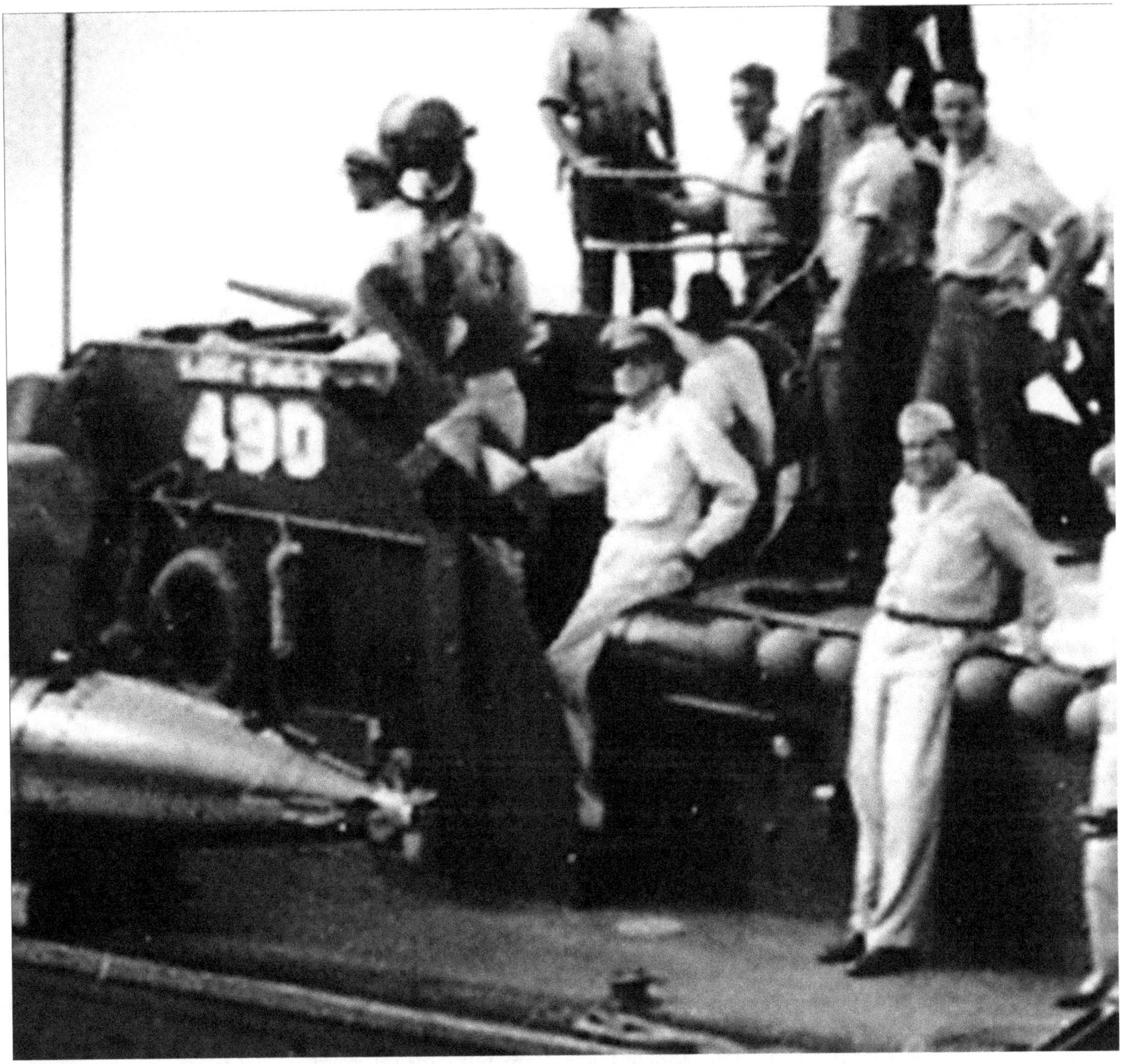

Figure 441. MacArthur aboard *PT-490* in summer 1945. (close up).

Date	Craft, Name, Event or Country	Event, Builder and Boat Specifications	City, River, Lake, Sea, Ocean
1945[1206]	August 26 MTB-261 (ex-PT-12)	The *261* foundered and sank off Alexandria, Egypt. The hulk became a navigation hazard and was destroyed. **Type:** PT 10-class, 70′	Alexandria Mediterranean
1945[1207]	September Surrender	The final formal Japanese surrender took place aboard the battleship USS *Missouri.*	Japan

Figure 442. The final surrender was not the final surrender. Japanese units and individuals would continue to fight on in isolated pockets all over the Pacific. The last American plane shot down by the Japanese was a B-25 in November. (U.S. Navy Historical Department)

[1206] Polmar, *PT Boats At War*, 114
[1207] Rohwer, *Chronology of the War At Sea*, 430

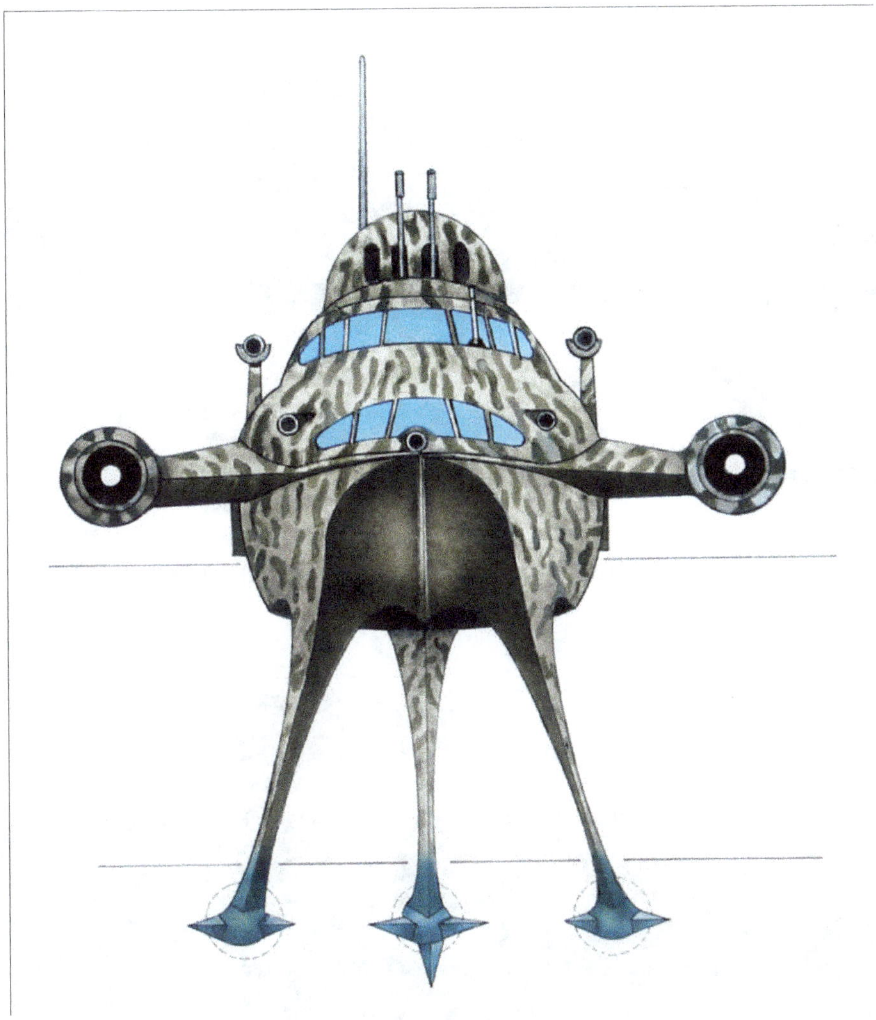

Figure 443. Turbojet Tragflügelboot Projekt (front view). This late-war German design was never built, but shows an interesting glimpse of a future that might have been. "Being based on the design of a small fast attack boat (Typboot 4), this design of a K-Verband project was created with the focus on implementing a method for faster target approaches and disengagement. Therefore the previous design was modified to a hydrofoil, together with a 400-hp power increase. The result was a projected speed increase of another 8 knots. Since the Kriegsmarine did not have any operational experience for hydrofoils, this would have been the first hydrofoils in Kriegsmarine service, but none of these boats was ever built.[1209] (Illustration by author)

[1209] http://www.german-navy.de/kriegsmarine/ships/fastattack/tb5b/index

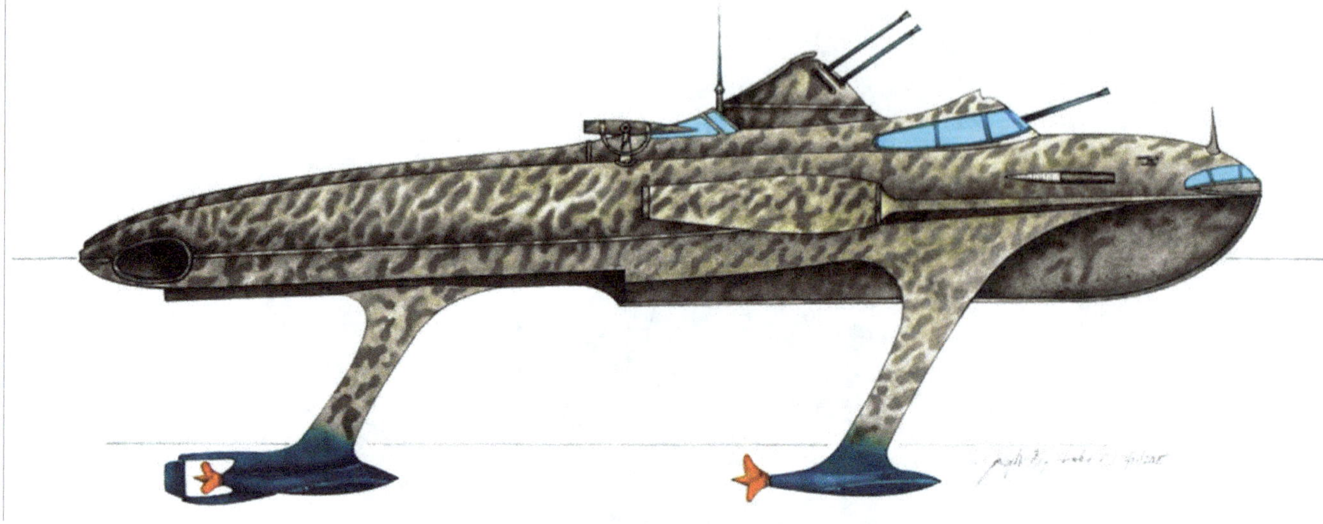

Figure 444. Side view of the Turbojet Tragflügelboot Projekt. (Illustration by author)

Figure 445. Overhead view of the Turbojet Tragflügelboot Projekt. These designs prefigured postwar hydrofoils, but hydrofoil technology never really caught on with the world's navies. Aircraft and helicopters had superior speed. Similarly, although torpedoes remained the most effective ship-killers, especially when detonated underneath the target's keel, they were gradually replaced on fast attack craft by surface-to-surface missiles, which had superior range and speed. (Illustration by author)

THE SHIP KILLERS

A DEFINITIVE ILLUSTRATED HISTORY OF THE TORPEDO BOAT

VOLUME X: BONUS MATERIAL AND APPENDIXES

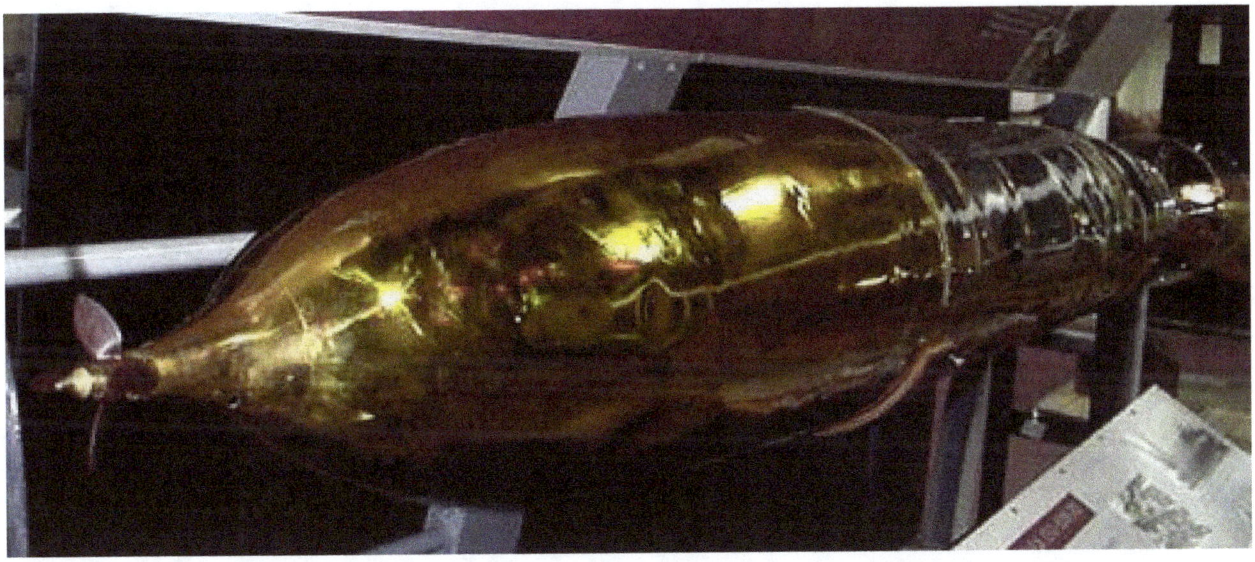

The Howell Torpedo that was mounted on the U.S. Navy's first torpedo boat, the Herreshoff Brothers' Stiletto. *(Naval Undersea Museum)*

BY JOE HINDS

NIMBLE BOOKS LLC

CONTENTS

FIGURES

CSS DAVID-CLASS TORPEDO BOAT: 1863

The Confederate States Ship (CSS) *David* class of torpedo boat was born of desperation. The people of Charleston, South Carolina, had been pounded by incessant shelling for months by the U.S. Navy. Charleston Harbor was full of U.S. Navy ships sent to block the port from all shipping. The main cause of the immediate fear of the civilian citizens of Charleston was the withering fire from one ship, the USS *(United States Ship) New Ironsides*. She was one of the largest and most powerful ships in the world. The other two ships like her belonged to Great Britain and France.

The USS *New Ironsides* was a behemoth. Covered with tons of thick iron over a heavy wood frame, she was invincible. The armor on her sides was 4" and at the upper deck level 10" thick. The ship was 232' long, with a beam of 15'-6", and a complement of over four hundred men. She carried sixteen of the largest cannons in the U.S. Navy arsenal. Two of the guns were massive 150 pound Parrot rifled cannon and fourteen were 11" Dahlgren smoothbore cannons. These were among the largest cannon carried by any navy in the world. Her massive shells had been pounding the buildings of Charleston to rubble since her arrival.

A reward by the Confederate authorities of $100,000.00 was offered up as an incentive, to any individual or group, who could drive off or sink the *New Ironsides*. A group of Charleston businessmen, with an eye to making a tidy profit, formed a company called the Southern Torpedo Company to build a torpedo boat that they called *David.*

The actual creation of the *David's* design has become wrapped in Southern myth and glorification. The original design came from a source in the North from a person as yet unknown. The *David* spar-torpedo boat was made from a small model that was based on previously established designs that had been known for fifty years. These designs were, in part, based on the efforts of Robert Fulton and his successful experiments in France in 1810.

Heading the enterprise was an unlikely man named Dr. St. Julien Ravenal plus Theodore Stoney, a South Carolina boat builder, and a few friends.[1210] Dr. Ravenal had a plantation on the Cooper River, just north of Charleston. He was able to supply the facilities, most materials and slaves to build the boat.

[1210] *Infernal Machines: The Story of Confederate Submarine and Mine Warfare* by Milton F. Perry (Baton Rouge LA, 1965).

The men of the new company hired David Ebaugh to be in charge of construction. The Stoney Landing Plantation supplied the construction setting in the form of a large shade tree, near the riverbank of the Cooper River. The hull of the *David* was built around an eighteen-foot-long railroad boiler that was a bit over four feet, six inches in diameter.[1211] The plantation slaves did the intricate work of carving of the boats oak staves and fitting them to the boiler. The job of assembly was even made more difficult by the tiny size of the craft and its confined spaces that had to hold a steam engine, gearing for the propeller, and fore and aft ballast tanks. The work was so well done that the original *David,* after years of hard use, survived the war and wound up on display in the naval yard at West Point. It was the slaves' craftsmanship that ultimately led to the success, and longevity, of the *David.*

THE NIGHT ATTACK

The *David's* fame was made on the night of October 5, 1863. Four men—Lt. W. M. Glassel, officer in charge; James H. Tomb, engineer; James Sullivan, fireman, and J. Walker Cannon, pilot—made their way through the U.S. Navy ships on blockade duty and arrived at about nine p.m. near the *New Ironsides.*[1212]

The small size of the cockpit forced Lt. Glassel to sit on the rear of the cockpit, with his feet on the small ships wheel just below the level of the deck. He rested a loaded shotgun across his lap. Engineer Tomb had to lie down next to the engine because he could work no other way. The internal space around the engine, at its maximum width, was less than three feet in diameter. The pilot and fireman knelt near the front of the cockpit at the door to the boiler. There was a small built-in table holding their tiny coal supply behind them. As with all torpedo boats ever built, crew comfort was never a factor. Four grown men worked in a hot, steamy space; 7' x 3' x 3'. Into this was crammed the engine, dual gearing mechanism, coal supply, the ship's wheel and the miscellaneous collection of small tools needed in any engine room. The men could barely fit the food for the mission. They were lucky to have cold bacon, a few biscuits and a canteen or two of water. With luck they would return from their mission before they got too hungry.

---

[4] Solomon, Roberts S., M.D., *The C.S.S. David; The Story of the First Successful Torpedo Boat,* COMPRINtons, Charleston, S.C., 1970.

[1212] *Ibid.*

Engineer Tomb laid alongside of the engine with a lantern at hand. He had to kneel or lay prone to operate the water-feed for the boiler, add oil to the twin crank shafts and control the speed changes as directed by Lt. Glassel. Like all ship's engineers, he worked in a dark, dirty, steam-filled space while breathing live steam the entire time. The twin piston rods that drove the propeller slid back and forth with great force. Any wrong move on the engineer's part could result in a smashed hand, finger or bad burns from the machinery.

In the *David's* night attack, her torpedo struck the *New Ironsides* at mid-ships, but too shallow to cause her to sink. The powerful underwater explosion damaged the *New Ironsides* badly. Large guns, nearest the point of detonation, were knocked off their trucks. Deck beams were fractured or split. Armor plate was knocked lose in many places; the gigantic gun port shutters, weighing several tons, were jammed shut; and the crew members were momentarily stunned.

The *David* itself was almost sunk by the tower of water from the explosion that rose up beside the *New Ironsides.* Her open smoke stack acted as a tube to feed gallons of water into her tiny fire box. She was swamped and the engine died. Now the boat was at the mercy of the current and out of control. Two of the crew, Glassel and Sullivan, were captured and two, Tomb and Cannon escaped.[1213]

Tomb was able to relight the boiler fires, get up steam and get back to the Charleston pier. They had made naval history. The *David* would forever be on record as the first spar-torpedo boat to disable a capital ship of the line.

What follows is a litany of damages to the structural integrity of the *New Ironsides.* The damages range from broken stanchions, split beams, cracked ribs, hanging knees broken lose, deck joints opened, guns jumped off their trucks, and more. Rear-Admiral Dahlgren, Commanding Officer of the *New Ironsides* in his report of November 24, 1863 wrote:, "In my opinion this ship ought to be docked as soon as she can possibly be spared from this harbor." The *New Ironsides* was no longer a threat to Charleston. It simply could no longer fight. Doctor Ravenal won fame throughout the South and was paid $29,000.00 by Theodore Stoney for his efforts. Genius comes with a price in any era.

For the first time the U.S. Navy, which previously had full command of the initiative, was placed on the defensive. Expensive precautions were now required to guard against

[1213] *Ibid.*

the spar-torpedo boats. A complete system of anti-torpedo boat warfare was now needed, although the role of torpedo boat had not yet been clearly defined.

Conversely, the Rebels had a new weapon but no clear strategy to maximize its use. In the whole of the Confederate Navy, there were less than a dozen people who knew of the existence of the new weapons system. Crew training would forever be inadequate and new construction would be hampered by the plethora of inventors and their "new and improved" designs that that were based on fancy and engineering fact.

There were a number of *David*-class torpedo boats built during the Civil War, about eighteen in all, although the exact number is not known.[1214] There were also numerous variations of fantastic and even silly designs, most of which were never put in the water. The incentive for this massive effort that consumed so many Southerners was the standing rewards offered by their government. The development continued into the hopeless last days of the war. Methods to make the boats ride lower in the water were experimented on that included tons of iron ballast, water ballast tanks, and dive planes mounted forward of the conning tower to force the bow down. A fairly sophisticated yet simple block and tackle pulley system was used to raise and lower the spar torpedo. This removed the danger of the boat running around with a deadly torpedo in the water.

There were no formal plans drawn for the *David*. Among all the surviving documents, particularly those from the men who had her built, there is great discrepancy in measurements and most everything else. What are quoted below are the most historically acceptable dimensions as mentioned *The Confederate Navy, the Ships, Men and Organization, 1861-65*.

The original *David* was lost sometime after 1865; her exact construction dimensions can no longer be found. There is a sketch of the *David* made by a Confederate officer in the Museum of the Confederacy, Richmond, Virginia. This is the only truly accurate depiction of the *David* known to historians.

[1214] Dictionary of American Fighting Ships, Confederate Ships, http://www.hazegray.org/danfs/csn/d.txt.

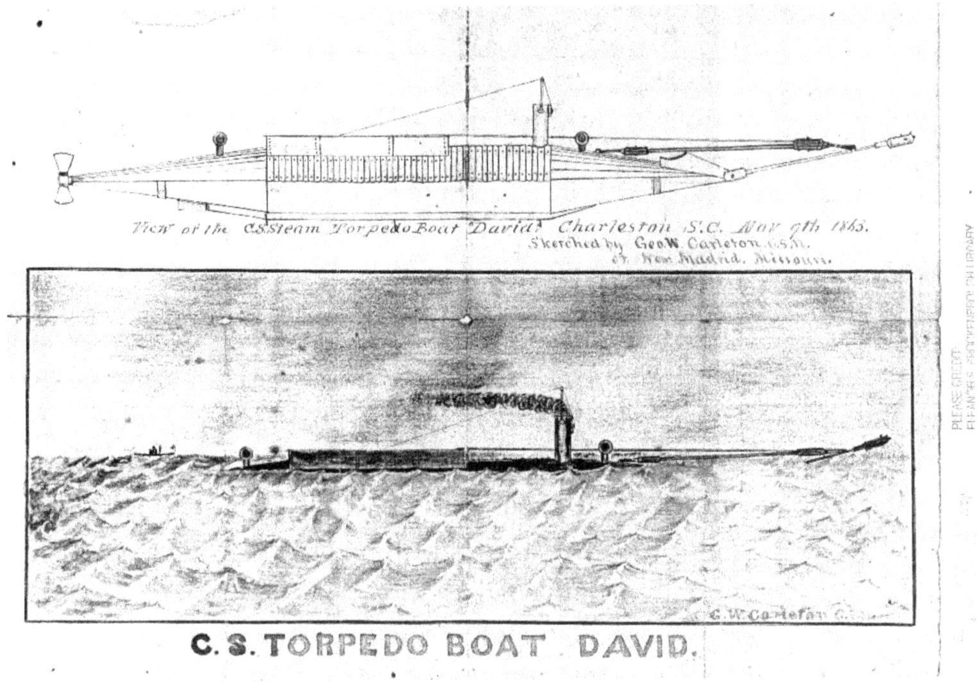

Figure 456. Profile of *CSS David.* The hurriedly built *David* proved successful when it put the mighty *New Ironsides* out of action for the balance of the Civil War. (Museum of the Confederacy, Richmond, Virginia)

SPECIFICATIONS

Hull length: 48' 6" .

Beam: 5' 0".

Draft: 4' 0".

Armor: 1" iron plate over deck, engine and boiler.

Armament: 1 15' spar with 70-pound contact torpedo.

Engines: Small, double-cylinder industrial type team engine. Single propeller, 24" diameter.

Fuel: Coal.

Endurance: Five hours.

Crew weapons: Two pistols and two shotguns.

Crew: 4.

Paint: "Blockade Runners Gray."

USN TORPEDO BOAT NO.5, *SPUYTEN DUYVIL*

The U.S. Navy needed to fight the Confederate Navy with a torpedo boat of its own. Unlike the Confederate boats, these would be large enough to sustain themselves for long periods and carry enough armor to stand up to any ship in a fight. A force of torpedo boats could push the Rebel fleet into a defensive posture.

DESIGN

After the success of Lt. Cushing's successful mission against the *Albemarle,* the Navy immediately went to work to create their own torpedo boat fleet. Once again, Engineers Wood and Lay became heavily involved in the enterprise. Their design implemented the first underwater launching system for their spar torpedo. Up until this time, an attack craft sailed out with one spar torpedo and it when completed the mission, or aborted and exploded the torpedo on purpose, it returned to base. The *Spuyten Duyvil* and her sister boats could carry up to twenty torpedoes and during an attack destroy several ships. The boats had fearsome potential. The South had no vessels that came close to matching the sophisticated engineering, destructive power, endurance, or strength of the *Spuyten Duyvil* and her many sisters.

The *Spuyten Duyvil* was the first spar-torpedo boat to carry an underwater door for launching its spar. Behind this door was a twenty-foot iron tube that could hold the iron spar and torpedo. When the moment of attack came, the bow door opened and the torpedo engine's chain-drive ran the spar out to its full length at a 30-degree downward angle. The torpedo would be automatically disengaged from the spar at this point, and then would begin to float towards the underbelly of the enemy ship. As it rose, a thin line attached to the tip of the spar, with the other end attached to a pin on the torpedo, tightened, pulling out the pin. When the pin released, a steel ball in a small tube, fell a short distance, breaking a small glass vial of acid. When the acid reached its maximum temperature , the torpedo exploded. The explosion would happen under the ship's powder magazine, or other vital parts, insuring the total destruction of the enemy ship.

This design was an offshoot of the John Ericsson design for his original *Monitor* class ship of the 1830's. Of all the lessons learned in the naval war one of the most important was that the torpedo, to inflict the most damage, needed to hit the enemy ship deep beneath her hull. Many of the U.S. Navy ships lost to stationary torpedoes (stationary mines) were not sunk by direct contact with the mine. The mines were several feet underwater and it was the hydro-aquatic force of the water actuated by the explosion that was far more deadly than the explosive itself.

With almost unlimited manufacturing resources at hand, the U.S. Navy could make torpedo boats from tugboats under construction. The hulls were sturdy and, as an added benefit, had their engines already installed. This brilliant idea of using existing small craft saved the Union Navy months of design, testing and construction time. They six tugboats armed with spar torpedoes were the *Alpha, Belle, Delta, Gamma, Hoyt,* and *Martin.*[8] In 1871, two more tugs were converted into experimental spar-torpedo boats; these were the *Fortune* and *Tirana.*

The tug *Stromboil* was converted into a state-of-the-art torpedo boat in 1864 by Pook Company and renamed *Spuyten Duyvil* ("speedy devil"). She had plating to well below her waterline, as did the pilothouse. This heavy iron armor would withstand the heaviest guns of the Rebel fleet. The entire crew, except for the two officers in the command tower, would be below decks, several feet below the water line. Her most unusual features were an internal torpedo tube for a retractable spar torpedo and a set of ballast tanks that could be filled or emptied in a few minutes.

This tremendous weight imposed impossible expectations on its one horsepower engine and single screw. In the best of conditions, she could only make five knots and proved to be slow and highly unwieldy. She did make an appearance for a combat engagement at Trent's Reach on the James River Virginia, on January 24, 1865. Here the crew of the vessel ran head first into the debilitating ignorance of their senior naval officers.

Thanks to an inept Fleet commander, the *Spuyten Duyvil* did nothing but steam up and down the James River on fruitless expeditions to nowhere. She was finally able to help destroy the Confederate obstructions in the James River after the war ended. At least this finally proved that the spar mechanism would have worked as intended. She was used as an experimental vessel until 1880 and then sold.

CONCLUSION

The American Civil War was over, and naval and army liaison officers from several countries rushed home to report their findings about this new invention called the torpedo boat. These observations provided the nucleus of a new industry. The development of the torpedo boat reshaped the strategy for the world's navies. While the battles at sea would still be between big ships with big guns, big ships at harbor or in shallow waters were now at risk.

Stealthy small combatant boats, properly armed with a well-trained crew, could sink any ship in harbor. Big ships were just better targets. So far, the torpedo boats had only attacked ships at anchor. The concept of small boats pursuing big ships at sea or laying in ambush on the open water had not yet arrived.

During the Civil War, the Confederate Government ordered engines, boilers and boats from England. Thus, the English had boat, spar torpedo and engine plans from 1862 onwards. There are records of six torpedo boats being built and delivered, in sections, for assembly in America. Engines were sent separately and installed on Confederate torpedo boat *David* and *Squib* class boats. The bulk of the craft stayed in Scotland and England at war's end. It is likely that the Scottish and English boat builders who gained experience from this construction effort provided the basis for the designs of Thorneycroft and Yarrow boat yards.

An English official explained the design of these vessels in a letter to {Navy Secretary} Mallory in 1864. [1215]

> *The six torpedo boats, ordered in your dispatch of July 18, No. 2, have been put in the hands of the builder. These will be built on the lines furnished by Mr. Graves, and there will be no alterations except to strengthen the stem and to change slightly the rake of the stern to suit the material (iron) of which they must be built. I have had a model made upon Mr. Graves' designs, and have had the displacement calculated for the three water lines shown on the sketch, which will be as follows: First water line on 1-foot = 1.5 tons; 2-feet line = 5.37 tons; 3-feet line = 10.6 tons. The weight of the boat, if of steel, including boiler, water in boiler, engine's, screws, and shafts complete, crew, say, 5 men, and coal for 24 hours, will be 9.18 tons, so that they will just carry the necessary weights on the given draft of water; but, owing to the exceeding*

[1215] *Official Records of the Civil War,* Series II, vol. II, pp. 724-725.

sharpness of the lines and shortness of floor, the boiler, will of necessity be rather small for the engines recommended by Chief Engineer Williamson in his letter of July 19. The remaining three boats will be kept nearly to Mr. Graves' design, only having increased length of floors so as to accommodate larger, boilers; draft and. other qualities remaining the same. The two twin-screw steamers, for the port of Wilmington, alluded to in your dispatch of July 30, are also in the builders' hands. I found some difficulty in getting the order taken, the principal builders being much pressed with work, but have contracted on the 14th instant with a leading Scotch firm for both the boats, each to be of the following general description:

Length, 170 feet; breadth, 25 feet; depth, 12 feet 6 inches; tonnage, 515 tons; engines two pair of 120-horsepower collective. They will be very strongly built, with heavy decks and beams, so as to carry a gun at each end, of from 5 to 6 tons weight, and the internal and deck arrangements will be such as to suit them for vessels, and yet not to excite suspicion. They were designed for a speed of 12 knots on a draft of 7 feet 6 inches aft, but can be loaded to 8 feet 6 inches with safety as seagoing boats. The price of each boat, complete for sea, with the exception of coal and stores, will be £17,500. One to be delivered on January 1, 1865, and the other six weeks later. I will send further particulars and drawings when plans in detail are finally adopted."

From these early Confederate, U.S. and English designs came the engineering requirements of torpedo boats. The technological development would rest on three equally important factors: the development of the automatic torpedo, high-speed engines and high-speed hull designs.

SPECIFICATIONS

Length: 84'- 2".	**Beam:** 20'- 8".
Draft: 7'- 5"	**Engines:** Three: one engine to supply power for propulsion, one engine to operate the four ballast tank pumps and one engine to operate the spar launching system.
Armament: One reusable iron spar operated by a steam engine to push the spar out to launch the torpedo and then mechanically retract. Twelve Wood-Lay Torpedoes, 70 to 85 pounds each, carried on board.	
Endurance: 8 days of fuel, water and food.	**Armor:** 5" sides and conning towers, deck was 1" iron throughout.
Crew: 23.	**Speed:** 6.5 knots.

THE HERRESHOFF BROTHERS

In the middle of the Civil War, 1863 to be exact, two brothers in Bristol Rhode Island started a ship yard that would make their name, Herreshoff, one of the most respected names of boat building in the world. By 1868, the Herreshoff Manufacturing Company, Bristol, Rhode Island, had built their first steam-powered boat, the *Annie Morse*.

John Brown Herreshoff was completely blind at age 15. He managed his own sail boat building company until his brother, Nathaniel, joined him in 1878. His ability to work with this monumental obstacle led him to receive commission for boats that were renowned for their seaworthiness, speed and beauty.

Nathanael Greene Herreshoff, John's brother, was known as the "Wizard of Rhode Island." Nathan had worked long at building boats, and through his uncanny photographic memory, he could dictate to his blind brother the steps necessary to correct the design. His brother, John, through the feel of his hands alone, would correct the design by running his hands over a half-hull or full-hull model. They made a remarkable team and they left their mark on American and world maritime history.

In 1878, the Herreshoff brothers launched themselves into designing and perfecting not only small boat engines but fast hulls and light-weight engines with their boilers. In the town, the brothers were known as, "Blind John" (John Brown Herreshoff) and "Captain Nat" (Nathanael Greene Herreshoff). The clean designs and efficient engines of their ships captured the attention of the U.S. Navy. They were asked to place bids for the new torpedo boats the Navy wanted.

The Herreshoff steam generator boilers were a significant breakthrough in small steam-powered boat design. Nathanael Herreshoff's unique design enabled them to fire up a boiler to a full head of steam in minutes. In addition, his design, compared to normal boilers of the time, was extremely light. This was revolutionary news in the world of small boats. Not only were these water craft lighter, they were faster. During this time, the brothers created *Hull No. 44*, which became for the British Navy, *H.M. 2nd Class Torpedo Boat No. 63*. It was an all steel boat (1/8" thick) that performed quite well in its trials

The *Lightning*, a double-ender craft, was ordered by the US Bureau of Naval Ordnance, to be built at a cost of $5,000.00. This was the first boat purchased by the U.S. Navy since the *Spuyten Duyvil* boats of 1863. The entire boat was so well built and so light that it could be stopped within her own length, while moving at full speed. She was a great test bed, but just too small to be a torpedo boat.

In 1879 the brothers built *Hull Nos. 53, 55* and *60* for the Peruvian government. *Hull No. 53* was the Peruvian vessel, *Republica*. She was involved heavily in the "Nitrate Wars" between Peru, Bolivia, and Chile of 1879–1881. The 55 boat became the became the *Allay*. The boat was captured by the Chilean forces in December of 1879. The boat was then renamed the Guacolda. She participated in the attack at Callo, April 19, 1880.

Hull No. 64 was produced in 1880 for the Russian Government and more than likely used as a test bed for that Navy since there is no record of her being commissioned into the Russian Navy. However, there is ample evidence that the boat was a major influence on the Russian-designed line of torpedo boats used successfully against their old enemy, Turkey and the Ottoman Empire.

By 1885 the brothers were prosperous from building foreign torpedo boats. They built the high-speed motor yacht *Stiletto* and included a new engine design that put them far ahead of other boat builders. This gave her a remarkable speed of a sustained 20 knots and top speed of 26.5 knots, which was unheard of in this time. The hull was light-weight and wooden with five watertight bulkheads, and large compartments for engine room and crew. She had a forward conning tower that would be a hallmark of the external design of all torpedo boats for years to come.

The *Stiletto* was purchased by the U.S. Navy on March 3, 1887, then ordered to the Naval Torpedo Station at Newport Rhode Island and later designated by the Navy as, Wooden Torpedo Boat No. 1. The *Stiletto* was the U.S. Navy's first torpedo boat that launched a self-propelled torpedo. She fired a Howell Mark I automobile torpedo in 1892 from a deck mount. The Navy later experimented with using fuel oil to burn in her engines. *Stiletto* was the first serious effort by the Navy to enter the global torpedo boat race.

Figure 457. The Howell Torpedo. (Naval Undersea Museum)[1216]

The Herreshoff company built six torpedo boats for the U.S. Navy from 1890 to 1897. They were *Cushing, TB-1; Porter, TB-6; Dupont, TB-7; Morris, TB-14; Talbot, TB-15; and Gwin, TB-16*. All of the boats served in the Spanish-American war. The brothers experienced great difficulty in dealing with the Federal government and naval officials. They faced a constant battle to receive funds allocated and complained of hundreds of hours negotiating with clerks who had no idea of naval construction. It became so difficult that the brothers were actually losing money on government orders. After the *Gwin* was completed, they simply refused to build any more boats for the Federal government, or any government, and turned their talents to building racing and pleasure yachts.

[1216] http://www.history.navy.mil/museums/keyport/html/howell_torpedo.htm.

THE TORPEDO BOAT *WINSLOW*, *TB-5*

When the Spanish-American war broke out, it was the first chance for America to test out its brand new torpedo boat fleet. As in the American Civil War, senior officers had control of a weapon they did not understand or believe in. To see small boats getting center stage attention was an affront to their prestige. These same men did not understand the actual tactical advantage that the small boats brought to the fight.

The *Winslow* was tasked with patrolling the northern coast of Cuba and nearly met her end on May 11th when struck by shells from Spanish gunboats and from both Spanish forts in Cardenas Bay[1217]. The *Winslow's* armament consisted of three 1-pound guns, which were not much use against stone forts and heavily armed gunboats. The *Winslow* was raked by gunfire, killing three men immediately and mortally wounding two other. Out of a crew of twenty. only eleven were left to man the guns and control the ship.[1218] There can be no question of the bravery of these men to stand and fight against overwhelming odds. They did not run into a clever trap set by the Spanish. Rather, the cause was the lack of a clear mission for the torpedo boat. It was the Spaniards' luck that their heavy guns were concealed on a point of land jutting out from the Cardenas docks. The three Spanish gun boats had placed floating markers on the bay to serve as guides for the heavy artillery. In a piece of especially bad luck, *Winslow* steamed right up to the markers shortly before the Spanish gunners opened fire.[1219]

Winslow's crew lived in squalor and filthy conditions as did all the men of the torpedo boat fleet. One contemporary observer remarked: "If an old-fashioned wind sail is then rigged in the hatch or tower, as best serves, a good supply of air will be insured in these two compartments in practically any weather, a condition that has often been denied our

---

[1217] [Clerk of Joint Committee on Printing, The Abridgement of Message from the President of the United States to the Two Houses of Congress, Washington: Government Printing Office, 1899. 4 vols. Cited at http://www.spanamwar.com/winslow2.htm].

[1218] Jackson, R.H. Lt. U.S.N.," Torpedo Craft: Types and Employment",(*Proceedings*, Vol. X XVI, No.1, March,1900).

[1219] Dictionary of American Fighting ships: http://www.history.navy.mil/ danfs/w10/winslow-i.htm.

boats. A scant supply of air at 115 degrees, and that mixed with salt water, will lay low the toughest mariner, and will result in a short time in a disabled crew and boat."[1220]

The thin steel skins of the boats were without any insulation. The result was constant "sweating" of the boat. The same observer remarked, "More permanent discomfort has existed on board the boats from a failure to give a thin canvas lining to all living spaces then from any other cause." He goes on and talks about the men of the *Winslow* sleeping under rubber sheets and their clothes lockers full of small vile insects and water. The result of cramped, wet, conditions led to many men being debilitated by rheumatism. Naval records point to a high rate of medical discharges being granted to men as young as eighteen for this crippling disease

The storage of food and clean water was never given much attention by torpedo boat designers in any navy. Fresh foods were unheard of and fresh bread was a waste as it would mold almost overnight. The foods stored in boxes on deck caught the full brunt of the Caribbean sun and the spray from the ocean and quickly became a feeding haven for insects and rats. The fresh water tanks were not treated any better than any other tank on the small boats. They rusted and collected green pond scum within a few days.

When crews turned to canned foods, their options were no better. Giant meat packing houses had huge government contracts for the troops of the war. There were little or no government inspections of their products. Sick and diseased animals were processed along with insects, rats, mice and floor sweepings. The operators saw no reason to be picky about what they put into their cans. Hundreds of American servicemen became deathly ill and spent the war in tent field hospitals all over southern Florida.

Men serving time on torpedo boat duty always lived in less than ideal conditions. The following is an extract from the parliamentary report of the British maneuvers of 1890: "The radius of action of a torpedo boat is limited less by her coal and feed water supply than by the physical endurance of her crew, especially her commander."[1221] This would remain a truism for all the years of torpedo boats in all naval history.

The majority of the world navies had been busy building different classes of torpedo boats and developing tactics that integrated the small boats into fleet actions and strategic

---

[1220] Jackson, R.H., Lt. U.S.N." Torpedo Craft: Types and Employment," *Proceedings,* Vol . XXVI, No. 1 ,March, 1900).

[1221] *Brassey's Annual,* 1891.

planning. Up to the Spanish-American war, the U.S. Navy Department had never attempted to do the same with its torpedo boats.

In March 1900 U.S. Navy office Lieutenant R.H. Jackson wrote that year's prize essay for *Proceeding of the United States Naval Institute.* The lieutenant's article was forty-five pages of scathing reporting on the inside story of what happened to the American torpedo boat fleet. Jackson says, "... during the Spanish war, [the torpedo boats] performed almost every function except the one which properly belonged to them—that of attacking an enemy's fleet." Jackson goes on to point out the many deficiencies of our torpedo boats with which other navies had already dealt. Not only did we possess the smallest torpedo boat fleet in the world, it was also the worst built and among the worst managed.[1222]

Another U.S. Navy officer, Lieutenant Edward L. Beach (himself later a legendary figure, and father of the author of *Run Silent, Run Deep*), wrote a great essay that ran for 42 pages in the *Proceedings,* March 1903. His title is interesting and gives a picture of how we looked to the rest of the world. "OUR TORPEDO-BOAT FLOTILLA. The Training Needed To Ensure Its Efficiency. Motto: We question the utility of torpedo boats because we don't know how to utilize them-W.W. Kimball."

Beach says on page 120, "...in the Spanish war they were never in a single instance put to legitimate torpedo-boat work. The Spanish war proved neither their destructiveness nor their war time utility, but demonstrated their inherent weakness. During the war the tendency was not to test them but to preserve them ... has not existed in this country, either in its legislators or in its naval officers, a well defined, well agreed upon torpedo-boat policy ...."[1223]

SPECIFICATIONS FOR *USS WINSLOW, TB-5*

Length: 161' 6-3/4".	**Armament:** 3 1-pounder guns, 3 18" torpedoes.
Beam: 16'-3/8".	**Crew**: 20.
Draft: 5'-0".	**Speed:** 24.82 knots.

[1222] Jackson, R.H., Lt. U.S.N. "Torpedo Craft: Types and Employment", *Proceedings,* Vol. XXVI. No. 1 March, 1900.

[1223] Beach, Edward L. Lt. "Our Torpedo-Boat Flotilla. The Training needed to Ensure Its Efficiency" (*Proceedings*, Vol. 1 , XXIX, No. 1, March, 1903.

THE RAVEN'S WRATH

In the dark, you could hear the low rumble of a monstrous being in the far distance. It was in fact the sound of massive diesel engines. With three engines per boat, at 2,500 horsepower each, the nine boats coming out of Cherbourg brought their own storm of the sound of 22,500 horsepower that could make men's blood run cold.[1224]

The deep resonance of the engines, and the acoustics of the water in the English Channel, made them sound like they were coming from everywhere at once. It was impossible to get a directional fix on the boats by human hearing alone.

The sound gradually increased as the white boats glided into the English Channel. These were the killer S-boats of the 9th and 5th Schnellboot Flotilla from the harbor of Cherbourg, France. It was 2300 hours on the night of April 27, 1944 and those involved would never forget this night. These nine magnificent motor torpedo boats were about to change a day in modern day world history.

The S-boats maintained a 20 knot speed until they were well beyond the British MGBs (Motor Gun Boats) and MTBs (Motor Torpedo Boats) lying in wait for them off the coast. They had slipped by these English boats almost every night for months.[1225]

A young naval officer, Oberleutnant zur See Gunter Rabe, was known to the men of his Flotilla as the *Raven*. His *S-130* had been assigned to the 9th S-boot Flotilla in mid-November 1943[1226]. High on the bridge on of each side of the bridge was a coat of arms with a silhouette of a Raven to match her code name "Rabe" (Raven). The man called the *Raven* was known for his skill and his boat had the reputation of being a "lucky" boat.

From the bridge of the *S-130*, Rabe passed the order to punch the engines up to their cruising speed of 36 knots (or 41.34 mph). This was not the boats' fastest speed. They would not use that speed until it was absolutely necessary to make a getaway from British destroyers or the large Fairmile "D" (115') or "dog boats" as they were called.

[1224] Fock, Harald *Fast Fighting Boats 1870-1945* (Naval Institute Press, 1978) Table 12 German Schnellboote 1926-1945 (compiled from (3) and other sources,

[1225] Exercise Tiger, a disaster of WW2-what did happen? http://www.mikeekemble.com/ww2/tiger1.html. Accessed by author March 13, 2006

[1226] Laarman, Maurice. *S-130, The Old Solider* PrinzEugen.com Schnellboot Archive http://prinzeugen.com/S130.htm. Accessed by author March 22, 2006.

The nine *Schnellboote* (Speedboats) were superior in speed, firepower and sea-keeping abilities to the Allied MTBs. These S-boats were unlike any other motor torpedo boats in the world. At twice the size of any Allied motor torpedo boat, and with far heavier weapons, larger more powerful engines, and faster, they were an awesome opponent to face any night.

The "Raven" was on course to attack two British convoys off the coast of southern England. Earlier that day a Luftwaffe reconnaissance flight had reported in on the convoy's position. This information was carefully recorded at headquarters and men of the Intelligence branch added snippets of information gathered during the day from the radio traffic of English ships. By the combination of air reconnaissance, radio traffic, known patrol areas and known convoy routes, they could set traps for the convoys running the East coast of England. The force was under the command *Fuhrer der Schnellboote (FdS) Korvettenkapitan*, 36 year old Rudolf Peterson who had the uncanny ability to be right most of the time.[1227]

The man assigned to be the cook on the *S 130* was down below putting on a large pot of cocoa for the crew. The S-boat was the only motor torpedo boat that could run at 35 knot cruising speed and still make coffee or cocoa! It was her deep hull design that allowed for the stability and smooth ride the boat gave the crew. The S-boats could run at near cruising speed in Force 5 seas while most Allied boats were staying in harbor at anything over Force 3 sea conditions

The S-boat was 115 feet in length with a crew of twenty-one to thirty men and officers. The number of personnel was determined by the weapons carried on a particular boat. The boats on this night packed two large torpedoes, 53.3-cm, and a bow mounted 20-mm rapid-fire cannon, the 2-cm C/38 (1,000 rounds per gun; 2 x 2-cm LM44 (twin-barrel 1,500 rpg) mounted mid-ships; one 4-cm Flak 28 or a 3.7-cm gun (500 rpg) mounted aft.

Once they were about to engage an enemy, the commander could call for eight machine guns, 8 x MG15 (9,000 rpg) t9 be mounted on the bridge and about the deck. He also had the MG 151 (1,500 rpg) and one 8.6-cm RAG M42 rocket launcher with 30 R Sprgr 800-meter-range and 60 RLg rounds (for AA defense). On the aft end of the boat were two chemical smoke boxes that could throw out more smoke than a destroyer. In stern racks

---

[1227] Frank, Hans. *German S-boats in Action in the Second World War* (Naval Institute Press,2007), 159.

were six depth charges which were good for discouraging any ship following too close. The crew also had a store of 30 hand-grenades for boat to boat fighting. On top of all of this was the enormous number of rounds of machine gun bullets and cannon rounds for the 20-mm and 40-mm ammo that they carried. The S-boats could stay in a firefight far longer than the Allied boats.

Another well known *S-Boote* officer, *Korvettenkapitan* Bernd Klug, commanded the 5th *Flottille* of six S-boats, *the S-136, 138, 140, 142, 143,* and *100*. The 9th *Flottille* was commanded by the able *Korvettenkapitan* Gotz Frfr. von Mirbach who brought his three boats to the fight, the *S-130, 145 and 150*. Both flotillas were under the *Schnellboot* group command of a true naval genius, *Kapitan zur See* Petersen, more commonly known as *FdS* Petersen. He operated from his headquarters in Cherbourg where he had proven masterful in plotting enemy convoy movements from bits and pieces of radio transmissions, reconnaissance reports, and creative insights. From this data, Peterson and his team could plot the course that convoys might take and position his boats at points that offered the best opportunity for sinking convoy ships.

Peterson was sending his force back into territory that had proven fruitful in sinking Allied shipping in the area near Lyme Bay. On two other occasions S-boats had attacked two convoys at rest in Lyme Bay, February 26/27 1942, and got away after sinking four ships and causing great damage to the other ships at anchor[1228]. Again on April 13/14, 1942 the Norwegian destroyer HNoM**S** *Eskdale,* was sunk by S-boats in Lyme Bay[1229].

Now, as the boats drew closer to Lyme Bay, they broke into two groups of 4 pairs each (*Rottens*) for their attack, with the ninth boat standing by and ready to move in as back up.

Each S-boat carried three radiomen on all patrols, which was unheard of in any other navy. One radioman assigned to the bridge operated the VHF type radio Lo 1 UK 35, which was ship-to-ship communications. Another radio operator manned the RDF, Radio Direction Finder. This was the man who could pinpoint the enemy boat's position by triangulation from radio transmitters on shore or track enemy ship movements at sea. On this patrol, he would be focused on enemy ship communications. A third radioman, below

---

[1228] Combined Operations Combinedops http:// www.combinedops.com/Op_Tiger.htm. Accessed by author March 13, 2006.

[1229] Combined Operations Combinedops Op_Tiger.htm http:// www.combinedops.com/Op_Tiger.htm. Accessed by author March 13, 2006.

the bridge, manned the FuG V transmitter and receiver. The operators had become skilled at jamming the radio signals of enemy ships with their powerful radios. It was the larger ship-to-shore radio that was used to communicate with Headquarters. The Germans did not have effective radar, at least not in large quantities, but they made up for the loss with this system of radio signals and air recon.[1230]

The vast majority of *Kriegsmarine* sailors on this mission had been at war since 1939 and some even before that. Many had been part of the German crews who trained the Spanish sailors during the Spanish Revolution of 1934. The Spanish had been placing large orders for S-boats from the Germans since 1932 and part of the contract called for active combat training. The Germans were equally pleased as this gave them a chance to test the boats, weapons, and men under live-fire conditions. The Spanish were pleased to upgrade their coastal forces with state-of-the-art hardware and weapons systems.

The S-boats of 1944 had new and more powerful engines, Daimler MB 501 (2,000 hp) and MB501A (2,500 hp). Some S-boats packed the even newer 3,000-horsepower engines. The speed of the S-boat had dramatically improved, but Allied intelligence had not: at least not American naval intelligence. The British knew that the only boats in the area that were a match for the S-boat were the Fairmile '"dog" boats... The American PT boats lacked sea-keeping abilities in rough Channel weather, and their crews had little combat experience. The American planners seemed to believe there are no major combatants of the *Kriegsmarine* around and no naval threat in the Channel. They had completely ignored the good information provided by their British friends who had been fighting against the *Kriegsmarine* S-boots since 1939. For all of their skills this was fatal short sidedness. They mistakenly put the *Schnellboot* into the same category as the PT boat. This mistake would cost them dearly.

The American Army had been holding exercises all over England in preparation for D-Day. The Slapton Sands beach in Lyme Bay proved invaluable in training troops for the invasion of Normandy. Lyme Bay had an uncanny resemblance to Utah Beach.

On this night, this particular convoy had been participating in a multiple-ship training exercise. The convoy, T-4, consisted of the Plymouth section: LST (Landing Ship, Tank) Group 32 with LSTs *515, 496, 511, 531,* and *58*. The Brixham section included LSTs *499, 289*

[1230] Radio & Signals PrinzEugen.com Schnellboot Archive http://prinzeugen.com/SchnellSignals.htm. Accessed by author February 27, 2002.

and 507. Other troops had been landed on the beaches of Slapton Sands and set up firing positions before the last convoy ever set sail. The British Fighter Direction Tender *FDT 217* was dispatched out of Portland harbor top provide radar and communications cover for the convoy. She mistakenly received a radio transmission, "Make Port All Haste", which she did. The only source for defensive radar was gone because of a gross mistake[1231]. The practice "beach bombardments" had to follow the fake "invasion." The weapons on shore included small artillery pieces and heavy machines guns along with 20-mm anti-aircraft guns. Live ammunition was issued for the second phase of the exercise. "*Operation Tiger*" was about to begin. No one realized that the Luftwaffe had spotted the convoy.

The soldier on the LSTs played cards, groused about another "exercise" when they could have been in a comfortable pub. They were components of the 1st Engineer Special Brigade, troops of the 4th Infantry Division and a mix of personnel from VII Corps. These men were composed of Combat Engineers destined for Utah Beach on D-Day. The individual units included the 462nd and 478th Amphibious Truck companies; Headquarters, 1st Engineer Support Brigade; and the 531st Engineer Shore Regiment. In addition to the Engineers were the 3206th Quartermaster Company, a Chemical Decontamination Company, and Engineer Map Detachment and of course two fully manned Graves registration companies[1232]. In all, it was over 3,500 men with everything needed to support the seizure of their section of Utah Beach.

Some men wrote letters while most grabbed sleep on this boring and slow transit to the exercise grounds. Most felt that once this silliness was done, they could get back to their barracks, shower, don clean uniforms, and go on the hunt for young (or not) lonely British women. Others would chose to go sightseeing in the lovely summer English countryside. For now, they were on their way to join elements of the 4th and 29th Infantry, 82nd Airborne and 188th Field Artillery Group camped on the beach at Slapton Sands. If nothing else, they would have some great poker games and maybe time for a football game or baseball game.

[1231] Exercise Tiger, a disaster of WW2-what did happen? http://www.mikeemble.com/ww@/tiger1.html.

[1232] Tiger, The E-Boat Attack http://qmmuseum.lee.mil/d-day/tiger.htm. Army Quartermaster Museum-Fort Lee, Virginia; Extracted from: *The Administrative And Logistical History Of ETO Part VI Neptune Training, Mounting, The Artifical Ports, Historical Division, United Sates Army Forces, European Theater*, Lt. Clifford L. Jones, March 1946.

The standing US Navy orders had every gun on all LSTs manned during all exercises. This exercise was as much for them as for the soldiers. The British Admiralty had admitted privately that the German *Schnellboote* had operational qualities very superior to anything used by the Royal or American Navy. This information was not well received by the Americans. The British knew the actual danger these boats represented. They had already experienced a great loss of life and ships in learning this lesson. The Americans never took this warning seriously. For all the accumulated skills and knowledgeable men on the D-Day planning effort, they overlooked the *S-boat* threat.

Men become complacent when military exercises go on too long. Discipline becomes slack at this point. It is not just the enlisted men who fall under the spell of compliancy; it also affects equally inexperienced officers. This exercise was the last of the lot so the higher-ups decided to inject a feeling of importance and urgency the other previous exercises lacked. All infantry troops were issued full allotments of rifle, machine gun, and mortar ammunition. All vehicles; tanks, jeeps, trucks, DUKWs (amphibious trucks[1233]) and LCVPs (Landing Craft, Vehicle, and Personnel) would carry full tanks of gas. They would also carry their full allotment of 5-gallon Jerry cans of gasoline strapped in place. This exercise was to be as real as possible; it would involve 300 ships and 30,000 men.

The unsuspecting convoy sailed on at less than six knots, as if on a Sunday parade. There had been a typographical error in the written orders and the LST radios were set to a different frequency then the HMS *Azalea* and again different from British Naval Headquarters. A picket ship reported S-boot sightings in the Channel just before midnight; the message was received by the *Azalea* but not the LSTs nor British Naval Headquarters[1234]. The relentless training had the effect of lulling men to a lax awareness. The sailors of the LST had the watch and the soldier's mood rubbed off on them. While the ships guns were manned 24/7, there was no vigorous watching this night. All LSTs had been well armed with 20-mm, 37-mm, twin and single .50 caliber machine guns. They were not like the helpless merchant ships.

[1233] DUKW is a General Motors naming designation. "**D**" indicates a vehicle designed in 1942, the "**U**" meant "utility (amphibious)", the "**K**" indicated all-wheel drive and the "**W**" indicated two powered rear axles.

[1234] MacDonald, Charles B. "Slapton Sands: The Cover-up That Never Was" (Extracted from Army 38, No. 6 (June 1988):64-67; Naval Historical Center http:// history.navy.mil/faqs/faq20-2.htm

The T-4 convoy was to have three protective destroyers and an assortment of other smaller combat vessels. Through mishaps and miscommunication, there was only one vessel to lead and protect the convoy. It was the aged corvette HMS *Azalea* with Lt. Cmdr. Geddes blissfully leading the way this night. Instead of riding on the flanks of the convoy where he belonged, the Commander chose to ride a full mile in front of the first ship, *LST-515*. LST-515 was commanded by Cmdr. Bernard Skahill, who was also the convoy's Flag Commander, and he had never experienced combat. The other seven ships of the convoy were spaced at 700 yards each. So the total length of the convoy was over 660 yards or three and three-quarter miles. The last ship in line was the LST-507 with a group of 36' LCVPs and many trucks. Other LSTs were full of tanks, trucks, Jeeps, DUKWs, small mobile-artillery, and medical units.

Unknown to the Raven, one of the LST's was equipped with surface radar. The LST picked up two targets but wrongly assumed these to be the British screening force of MTB's and MGB's[1235]. Now the Raven and other S-boat commanders closed to assault the unsuspecting convoy. One of the LST's was equipped with surface radar and picked up two targets, but wrongly assumed these to be a British screening force of MTBs and MGBs[1236]. The S-boats fanned out into pairs while some distance from the convoys. The eight LST's on line were beginning to make a complicated turning maneuver to line themselves up to their landing sites. This was a critical moment in the exercise and all efforts were focused towards the shore. Nobody was looking out behind them into the blackness of the Channel.

[1235] Tiger, The E-Boat Attack http://qmmuseum.lee.mil/d-day/tiger.htm. Army Quartermaster Museum-Fort Lee, Virginia; Extracted from: THE ADMINISTRATIVE AND LOGISTICAL HISTORY OF ETO PARTVI; NEPTUNE TRAINING,MOUNTING, THE ARTIFICAL PORTS, Historical Division, United Sates Army Forces, European Theater, Lt. Clifford L. Jones, March 1946.

[1236] Tiger, The E-Boat Attack http://qmmuseum.lee.mil/d-day/tiger.htm. Army Quartermaster Museum-Fort Lee, Virginia; Extracted from: *The Administrative And Logistical History Of Eto Partvi; Neptune Training,Mounting, The Artifical Ports*, Historical Division, United Sates Army Forces, European Theater, Lt. Clifford L. Jones, March 1946, The E-Boat attack supra.

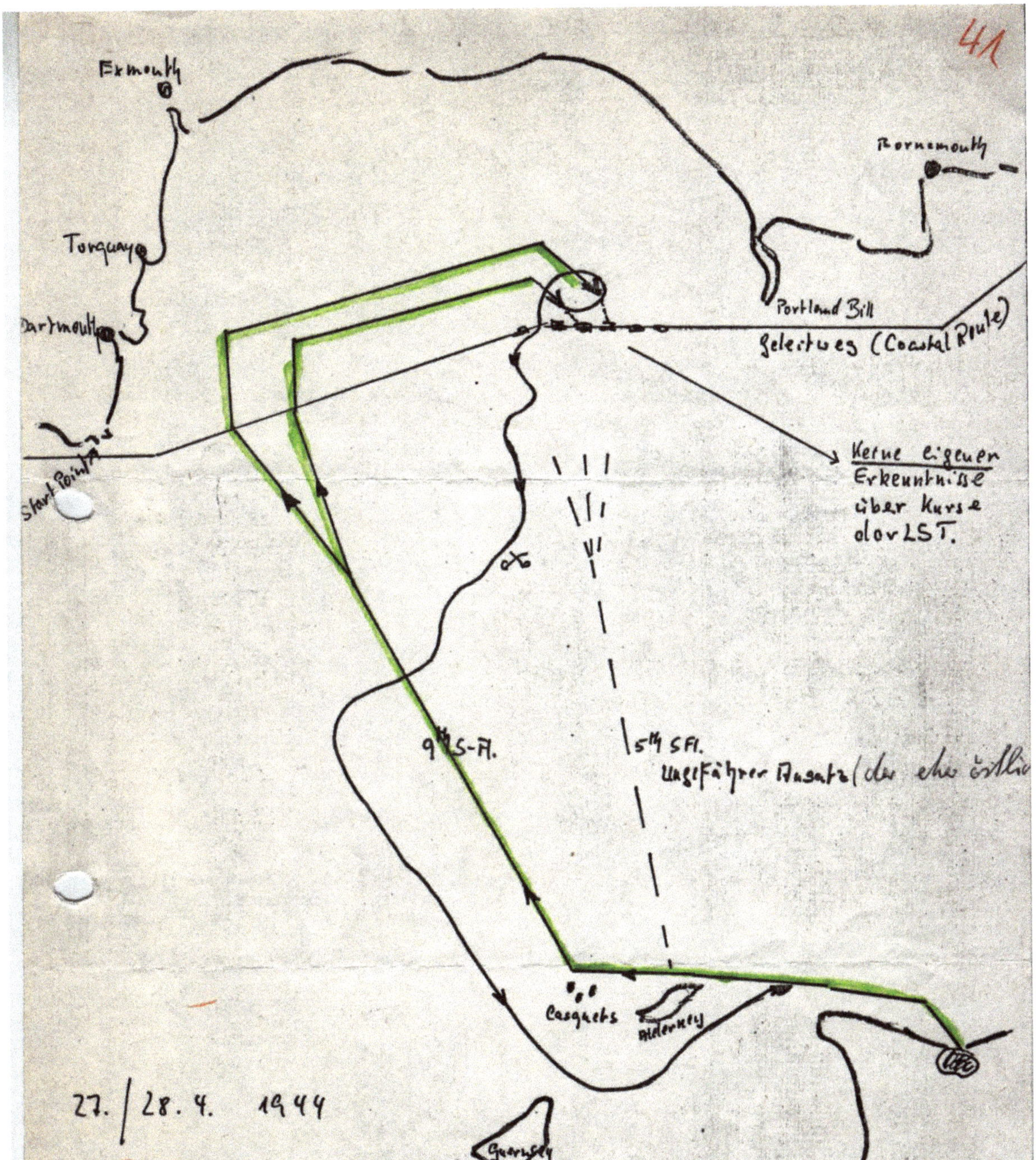

Figure 458. Never before published: the correct track the S-boats took to attack Operation Tiger. (Hans Frank)

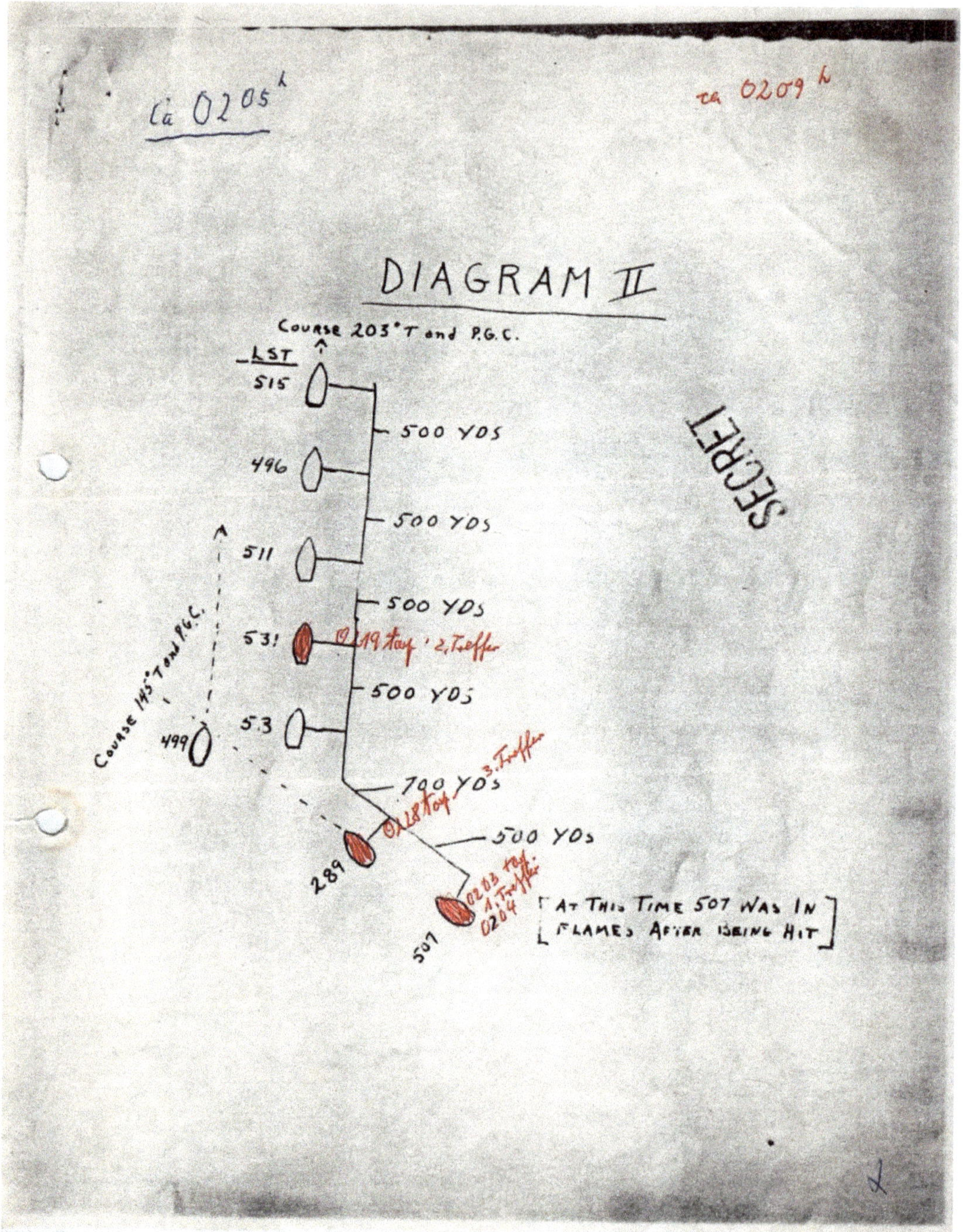

Figure 459. Never before published: correct positions of S-boats during their successful attack on the LSTs in Exercise Tiger. (Hans Frank).

As the S-boats moved into position, the CO (Commanding Officer) on each boat selected targets through powerful binoculars. The binocular bearing finders were mounted in the center of the bridge for the skipper to use. As he followed a target, bearing and range passed mechanically onto an analog torpedo computer. The torpedo computers could also automatically make fine adjustments to the helm as required. Once the heading, range and course was coordinated with the computers firing solution, the two torpedo men, one port and one starboard, set the depth and speed of the torpedoes. The torpedo man would fire on command. With trained and experienced men, this whole operation took a minute or less.

At this moment the first two S-boats of the 5th *Flotille, Rotte-3* (formation 3) *S-136* and *138* sighted what they thought were two destroyers. They moved to attack speed, and they each fired two 53.3-cm torpedoes. The first target was hit and an explosion seen. Seconds later the second target received two torpedoes and exploded in a fireball. *Rotte* 2's S-*140* and *142*, fired double salvos at their targets at almost the same moment. No explosions were heard or seen so it was correctly reasoned that the vessels were shallow-draft landing craft. The radioman, on both boats, passed the message on to the other boats so they could correct their torpedo depth settings. *Rotte 1*, S-100 and 143, saw what they thought was a "tanker" completely ablaze to the north. They sighted another ship in the light of the flames and both S-boats fired salvos. Within seconds, a satisfying explosion was heard and seen. The three boats of the 9th *Flotille* moved to the same area where S-*130* and *150* engaged another target while *S-145* went chasing after what looked like small escort vessels. These were actually the LCVPs lowered into the water from *LST-507*. The other S-boats had fired their torpedoes and now headed for the designated assembly area. It was time to head home. When the sun came up, enemy fighter aircraft came with it, the one natural enemy of torpedo boats.

Behind them lay hell on the water. The first LST to be hit had been the hindmost ship, *LST-507* with its load of fuel laden tanks and trucks. *LST-531* sank like a rock in six minutes. It went down with 424 out of 492 soldiers and sailors. The 3206th Quartermaster Service Company was wiped out. Of its force of 251 men and officers, 201 men and officers were killed or wounded. *LST-289* took a torpedo straight into the aft section and immediately started taking on water. This was the only ship to start immediately lowering their small landing craft. The *507* boat had time to get some men off but many died trapped below decks with their trucks. Several young, green officers hampered the life-saving efforts in their confusion and panic. Trucks were now burning wildly from all of the gasoline held by their gas tanks and extra gas cans. The gunners on all the LSTs were widely firing their heavy weapons trying to stop the unseen enemy. Instead, the 20-mm, 37-mm, and twin -.50 caliber machine guns raked the decks of the other *LSTs* with withering fire. There is no telling how many men lost their lives to "friendly fire" this night. On *LST-496*, an excited solider sitting in his half-track on deck cut loose with volumes of twin-.50 caliber machine gun fire against the *LST-511*. He not only shot up the LST but created many more wounded and dead American soldiers in his uncontrolled rampage of fear[1237].

A young, Navy medical officer, Lt Eugene E. Eckstam, MC, USNR (Ret.) tells the story best:[1238]

> *"General Quarters rudely aroused us about 0130. I remember hearing gunfire and saying that they had better watch where they were shooting or someone would get hurt. At 0203 Y was stupidly trying to go topside to see what was going on and suddenly "BOOM"! There was a horrendous noise accompanied by the sound of crunching metal and dust everywhere. The lights went out and I was thrust violently in the air to land on the steel deck on my knees, which became very sore immediately thereafter. Now I knew how getting torpedoed felt. But I was lucky.*
>
> *The torpedo hit amidships starboard in the auxiliary engine room, knocking out all electric and water power. We sat and burned. A few casualties came into the wardroom for care and, since there was ample help, I checked below decks aft to be sure no one required medical attention there. All men in accessible areas had gone topside.*
>
> *The tank deck was another matter. As I opened the hatch, I found myself looking into a raging inferno it pushed me back. It was impossible to enter. The screams of those many army troops in there still haunt me. Navy*

[1237] Greene, Ralph C. and Oliver E. Allen "What Happened Off Devon" American Heritage Magazine 36, Issue 2 February/March 1985 Volume 36, Issue 2 http://www.americanheritage.com/articles/magazine/ah/1985/2/1985_2_26.shtml.

[1238] Eckstam, Eugene, LT, MC, USNT, (ret.) Recollections by Lt Eugene E. Eckstam, MC, USNT,(Ret.), a medical officer on USS LST-507; Adapted from: "The Tragedy of Exercise Tiger," *Navy Medicine* 85, No. 3 (May-June):5-7; Oral History-Exercise Tiger, 28 April 1944, Oral Histories of World War II: LT Eckstam, Exercise Tiger http://www.history.mil.navy/faqs/faq87-3g.htm. Accessed by author March 13, 2006 \.

> *regulations call for dogging the hatches to preserve the integrity of the ship, and that's what I did".*

LSTs 507 and 531 sank that night. The 289 managed to make it back to port with its LCVPs pushing from both sides of the bow as a makeshift steering mechanism. It would finish the war in an English shipyard undergoing repairs.

A memorial park stands on Slapton Sands and tells the full gruesome story. It needs to be mentioned that the U.S. Army official history reports a disdainful appreciation of the quality of all officers, Navy and Army, that night. Colonel Eugene M. Caffey, 1st Engineer Special Brigade commanding officer, commented, "Officers and NCOs cannot expect their men to remain cool when they themselves seem to go completely crazy."[1239]

While this was a bloody and violent night, the real damage came on the morning of June 6, 1944. At the opening of the invasion of Europe, the missing engineers of Lyme Bay and their tanks, trucks, tanks and weapons destined for Utah Beach were absent. Utah Beach was a scene of bloody carnage. This is where the S-boats truly scored a victory. It was not just about lost LSTs and equipment. It was about the almost 1,000 trained and combat ready men who were to play a key role in the invasion plans.

Among the dead were ten officers classified as "Bigots.". These were officers who were known to have had copies of plans for the main invasion then planned for June 4, 1944. It would take days for the intelligence people to shift through the hundreds of the dead still trapped in the two LST's. Other men would be searching for bodies washed out to sea. Eventually it was determined that all ten officers could be accounted for and there was no danger of the invasion plans falling into enemy hands[1240].

What also came to light was the terribly inadequate training for the soldiers on the LSTs. On the 507 there had never been an abandon-ship drill.[1241] On all the LSTs,, none had given the Army troops full instructions for wearing their life belts. The Navy personnel had kapok life jackets which supported the whole body in the water. The Army issue life belts could only work when they were placed high under the arm pits. As it was, most soldiers threw on the life belts over their regular belts. When they jumped into the water the low center of gravity of the inflated belts forced the weakened soldier's head

---

1239 Tiger, The E-Boat attack, *supra*.

1240 Combinedops Op_Tiger.htm http://www.combinedops/Op_Tiger.htm. Accessed by author March 13, 2006.

1241 Greene & Allen, *supra*.

into the water and held him there. Hundreds died in this fashion. Others stripped off all of their clothing before jumping over board. Hypothermia claimed these men in just a few minutes in the 42°F water. The skipper of the 507 gave orders to abandon ship at 230 hours only to find most of the allotted life rafts were frozen solid with rust and in effect, welded to the deck.

Meanwhile, the S-boats, the Raven's *S-130* among them, made their way back to their Cherbourg unmolested. The cook put on another pot of cocoa for the return trip. Many of the crew gathered in the crew's quarters to drink cocoa, relax and talk about the night's excitement. None of the *Schnellboote* had been touched by all of the firing of the LSTs or of the several salvos fired by the DD, HMS *Onslow.*[1242]

Of the more than 3,000 men who participated that night, not one reported seeing an S boat. They all spoke of hearing the motors, they saw the color of tracers and gun fire from low on the water to mark the boats position, but nobody saw an S boat. This gave rise to another myth of World War II, "The German boats are painted black." Nothing could be further from the truth. The S-boats were painted in their standard *"Schnellbootewise"*, speed-boat white.

Nine S-boats came very close to stopping 6,000 ships and 1.5 million men before they ever set foot on the shores of France. The numbers of ships required by the planners were on a tight margin with no room for spares or losses. The exact number of LSTs was a critical to the success of the invasion. If losses had been worse by just two more LSTs, Eisenhower may have had no choice but to set back the invasion. SHAEF knew there were no more large landing craft anywhere in the European theatre. Replacements of these ships would have to come from the Pacific theatre and that would take months. Luckily, there were three LSTs in the Mediterranean Sea that could be rushed to England to replace enough of lost LSTs. It was a narrow race.

The invasion came off as planned and it was quickly evident that the lack of Army engineers on Utah Beach put the success of the entire invasion to be in doubt. The men who died at Lyme Bay were sorely missed on D-Day.

[1242] *Ibid.*

APPENDIX A. ABBREVIATIONS AND TERMS USED IN BOAT SPECIFICATIONS

Note that many of the abbreviations used in volumes I through XI have been spelled out in those volumes. This list is provided both for information and for interest.

CONSTRUCTION AND MANNING

Builder: shipyard, site of build and constructor (only for sailing warships) example: John Smith & Co. - Portsmouth (James Williams)

Build Dates: 1/2/3, e.g. 02.1972/05.1974/11.1974

- Beginning (usually keel laid)
- Launch[ing
- Commission[ing] (into first owner)
- Yard number (whenever available)

When precise dates are unknown:

- 1972-1975: ship was built in these years (often means "ordered"-"commissioned"),
- 1971;1976: ship was built between these years,
- 1972: ship was built in this year (probably launched or commissioned).

Size (possible data)

- T - (local) tonnage (sailships)
- bm - Builder's Old Measurement (sometimes approximate calculation; some pre-1850 vessels)
- t - displacement (standard/full load, surfaced/submerged for submarines)
- Grt - Gross Register Tonnage (merchant ships)
- Dwt - Deadweight (merchant ships)

Propulsion (only for self-propelled vessels):

- Horsepower (hp)
- Indicated horsepower (ihp)
- Nominal horsepower (nhp)
- Speed in knots (kt), surfaced/submerged for submarines

ARMAMENT:

- Number of different guns with pound rating (pdr) or caliber (mm)
- Number of torpedo tubes (with caliber in mm)
- Oher weapons
- Deck aviation

SHIP NAMES

I try to give most accurate names of ships in national language. Below I list some rules concerning non-Latin alphabets or scripts:

- Chinese (pre-1949): Traditional Chinese (when available), Pinyin and sometimes Wade-Giles.
- Chinese (PRC): Simplified Chinese and Pinyin.
- Chinese (Taiwan): Traditional Chinese and official transcription.
- Greek: Greek alphabet for warships (e.g. Πάνθηρ).
- Japanese: Kanji and Romaji.
- Korean (North Korea): Han'gŭl and McCune-Reischauer romanization.
- Korean (South Korea): Hangeul and Revised Romanization.
- Russian, Ukrainian and Bulgarian: Cyrillic alphabet (e.g. Ардаган)
- Serbian (Serbo-Croatian): Latin script.
- Turkish: modern Turkish alphabet, sometimes image of name in elifbâ (Ottoman Arabic) added.
- Other alphabets and scripts (Arabic, Hebrew, Georgian, Bengali, Thai etc.): official (usually English-based) name and name in original script (whenever available).
- Ccivilian (merchant) vessels: Latin name (as in e.g. LRS), in exception of Cyryllic alphabets (Russian, Bulgarian, Ukrainian) and Japanese.

Abbreviation "N.S." means "Nuestra Señora" or "Nossa Senhora", a common part of Spanish/Portuguese names for sailing ships.

Transcription of names written in some non-Latin scripts or alphabet should be visible with moving the cursor over them; e.g.: Σμύρνη, Свияга, 龍湍.

Appendix B. Identification Codes by Country from *Jane's Fighting Ships* 1901

Country	Abbreviation
America	U.S.A. (États Unis)
Austria	O.N. (Oesterrich-Ungarn)
Chile	C.I.
China	C.E. (Chine)
Denmark	D.K. (Danemarck)
England	E.D. (Angleterre)
France	R.F. (Republique Française)
Germany	D.D. (Deutschland)
Greece	G.E.
Holland	N.S. (Netherlands)
Italy	I.A.
Japan	J.N.
Norway	N.E.
Portugal	P.I.
Russia	R.E. (Russia before 1917)
Spain	E.A. (Espuma)
Turkey	T.E. (Turquie)

APPENDIX C. WORLD NAVY PREFIXES BY YEAR

Year	Country/Naval Code	Year	Country/Naval Code
1901-1939	U.S.A, USS, USN	1939-1945	USS, USN
1901-1939	ED (Angleterre) GB, HM British Royal Navy	1939-1945	HMS British Royal Navy
1901-1939	HMAS Royal Australian Navy	1939-1945	HMAS Royal Australian Navy
1905-1939	HMCS Canadian Navy	1939-1945	HMCS/NCSM Canadian Navy
1901-1926	CI, Chil. Chilean Navy	1927- Present	CN Chilean Navy
1901-1939	CE, ICN, CNS Chinese Navy Ship	1939-1945	ROCS Republic of China Ship
1901-1939	DK, HDMS/KDM Royal Danish Navy	1939-1945	Royal Danish Navy
1906-1939	RF: Republique Française French Navy	1939-1945	FS French Navy
1901-1939	FI Finnish Navy	1939-1945	FNS Finnish Navy
1907-1929	SMS Austro-Hungarian Navy	1929	Ceased to exist
1901-1939	RHS, HHMS Royal Hellenic Navy	1939-1945	HS Hellenic Navy (Greece)
1905-1939	RN (*Regia Nave*) Italian Royal Navy	1939-1946	RN(Regia Nave) RS (*Regio Sommergibile*) Royal Submarine Italian Royal Navy
1901	J, JN, IJN, HIJMAS Japan Japanese Navy Imperial Japanese Navy His Imperial Majesty's Ship	1939-1945	IJN Imperial Japanese Navy
1901-1939	HMIS Indian Navy		INS Indian Navy (Pre-Republic)

1802-1939	NS Netherlands (Holland)		HNMS Her Netherlands Majesty's Ship
1901-1939	NE, HNorMS (KNM) Nor Norge (Norvege)Navy	1939-1945	HNLMS, Hr. Ms., Zr. Ms. Netherlands
1848-1852	SMS, DD *Reichsflotte*-Imperial Fleet	1866-1871	*Norddenutsche Bundesmarine* Northern German Federal Fleet
1872-1918	*Kaiserliche Maritime* SMS, DD Imperial Navy	1919-1921	*Vorlaufige Reichsmarine* Temporary Imperial Navy
1935-1945	DKM, KMS *Kriegsmarine* War Navy		
1905-1939	BRP Philippine Navy	1939-1945	Philippine Offshore Patrol
1800-Present	PN Paraguay Navy		
1905-1917	RN, R.E. Russian Imperial Navy		Ceased to exist in 1917
1917-1939	USSR Soviet Navy	1939-1945	USSR Soviet Navy
1901-1939	E. A. (Espana) Spain SpN Spanish Navy	1939-1945	SP Spanish Navy
1905-1939	HSwMS Swedish Navy	1939-1945	HSwMS Swedish Navy
1901-1939	TE Turquie TCG Turkey	1939-1945	TK Turkey
1901-19??	ORP *Okret Rseczy Posdilitej* Polish Republic Ship	1939-1945	ORP *Okret Rseczy Posdilitej* Polish Republic Ship
1901-1929	ON *Oesterrich-Ungarn* (Austria)		

APPENDIX D. FOREIGN WARSHIP PREFIXES

Prefix	Full version
HHMS	His Hellenic Majesty's Ship
HIJMAS	His Imperial Japanese Majesty's Ship
HMAS	His Majesty's Australian Ship
HMCS	His Majesty's Canadian Ship
HMIS	His Majesty's Indian Ship
HMNZS	His Majesty's New Zealand Ship
HMS	His (Britannic) Majesty's Ship
HMSAS	His Majesty's South African Ship
HNMS	Her Netherland Majesty's Ship
HNorMS/KNM	His Norwegian Majesty's Ship
L-Lse	Lend-Lease (U.S. built for Great Britain or France)
ORP *(Okret Rseczy Posdilitej)*	Polish Republic Ship
RF	Republique Française
RN *(Regia Nave)*	Royal Ship (Italy)
RS *(Regio Sommergibile)*	Royal Submarine (Italy)
RN (RAN, RCN, etc.)	Royal Navy & Dominion navies
RSN	Royal Spanish Navy
SPN	Spanish Navy

APPENDIX E. GERMAN NAVAL TERMS

Prefix	Term	Role	Ship number range
F	*Flottenbegleiter*	Escort vessel	
M	*Minensuchboot*	Large mine hunter, ocean-going, used as escort vessels	
R	*Raumboot* or *Minenraumboot*	Mine hunter	*R-1* to *R-545*
RA	*Raumboot Ausland*	Mine hunter, captured	*RA-1* to *RA-232*
SAT27	*Ostee Schwere Artillerie Trager*	Heavy artillery barge operating in the Baltic Sea	
S	*Schnellboot*	Fast attack boat (MTB)	
Uj or *UJ*	*Ubootjager*	Submarine hunter	*UJ-A* to *UJ-K* *UJ 101- UJ 6094*
V	*Vorspostenboot*	Patrol boat	*V 101* to *V-7036*
Z	*Zerstorer*	destroyer	
ZG	*Zerstorer Griechenland*	Greek destroyer	

APPENDIX F. LIST OF BRITISH MTBs BY MANUFACTURER

1-19[1243]	British Power Boat	259-268	Elco
20-23	Vosper	269-274	US Navy
24-28	Thorneycroft	275-306	Vosper Built USA
29-30	Camper & Nicholson (Vosper type)	307-326	Elco
31-40	Vosper	327-331	Thorneycroft
41-48	White	332-343	Canadian Power Boat[1244]
49-56	Thorneycroft	344-346	Thorneycroft
57-66	Vosper	347-378	Vosper et al including USA
67-68	Thorneycroft	379	Vosper experimental
69-98[1245]	Vosper and others	380-411	Vosper et al including USA
99	Not used	412-418	BPB
100	British Power Boat	419-423	Higgins
101	White (hydrofoil)	424-429	White
102-103	Vosper Experimental	430-500	British Power Boat
104-107	Thorneycroft Experimental	501	C&N Experimental
108[1246]	Chantiers Navales de Meulan	502-509	British Power Boat
109	McGruer, incomplete	510	Vosper Experimental
110-200	Not used	511-518	Camper & Nicholson
201-212	White	519-522	British Power Boat
213-217	Thorneycroft	523-539	Vosper, not all completed
218-245	Vosper and others	540-600	Not used
246-257	White	601-800	Fairmile "D"
258	BPB Experimental (ex-PT 9)	5001-5029	Fairmile "D"

[1243] *MTB-19* was the same as *MTB-1;* there was an *MTB-13*, renamed *MTB-7* after the first *MTB-7* became the second *MTB-1*.

[1244] Plus a related boat, *CMTB-1*.

[1245] *MTB-75* was destroyed while building and was replaced by another hull given the same name.

[1246] The first *MTB-108* was the confusing *MTB40K*. It was renamed, then Vosper started another *MTB-108* (hydrofoil?) which was destroyed incomplete.

APPENDIX G. AMERICAN MANUFACTURERS OF PT BOATS IN WWII

Name	**Built**
Annapolis Yacht Yard	128
Electric Launch (Elco)	379
Fisher Boats Works	2
Harbor Boat Building	12
Herreshoff Manufacturing	28
Higgins Industries	213
Huckins Yacht	19
Robert Jacob	22
Miami Shipbuilding	2
Philadelphia Naval Ship Yard	2
Scott-Paine	1
Wilmington Boat Works	11
Total	**808**

APPENDIX H. US PT BOATS BY NUMBERS AND MANUFACTURERS

Designators	Length	Manufacturer
PT-1 &2	58’	Fogal Boat Yard
PT-3 & 4	58’	Fisher Boatworks, Detroit,. Michigan
PT-5, 6(1), 6(2)	81’	Higgins Industry, New Orleans, LA
PT-7 & 8	81’	Philadelphia Navy Yard, Philadelphia, PA
PT-10 to *19*	70’	Elco, Bayonne, Fl
PT-20 to *68*	77’	Elco, Bayonne, NJ
PT-69	72’	Huckins Yacht Co., Jacksonville, Fl
PT-70	76’	Higgins Industry, New Orleans, LA
PT-71 to *94*	78’	Higgins Industry, New Orleans, LA
PT-95 to *102*	78’	Huckins Yacht Co., Jacksonville, FL
PT-106 to *196*	80’	Elco, Bayonne, NJ
PT-197 to *254*	80’	Elco, Bayonne, NJ
PT-255 to *264*	78’	Huckins Yacht Co., Jacksonville, FL
PT-265 to *313’*	78’	Higgins Industries, New Orleans, LA
PT-314 to *361*	80’	Elco, Bayonne, NJ
PT362 to *367*	80’	Harbor Boat Building, Terminal Island, CA
PT-368 to *371*	70’	Canadian Power Boat Co., Montreal, CD
PT-373 to *383*	80’	Elco Boat Works, Bayonne, NJ
PT-384 to *399*	70’	Robert Jacobs Shipyard, City Island, N.Y.
PT-400 to *429*	70’	(Vosper) Annapolis, MD
PT-430 to *449*	70’	(Vosper) Herreshoff Manufacturing, Bristol, RI
PT-450 to *485*	78’	Higgins Industry, New Orleans, LA
PT-486 to *562*	80’	Elco, Bayonne, NJ
PT- 564	70’	HELLCAT, Higgins Industry, New Orleans, LA
PT-565 to *624*	80’	Elco, Bayonne, NJ
PT-626 to *660*	78’	Higgins Industry, New Orleans, LA
PT-661 to *730*	70’	(Vosper’s) Annapolis, MD

Designators	Length	Manufacturer
PT-731 to *790*	80′	Elco, Bayonne, NJ
PT-791 to *808*	78′	Higgins Industry, New Orleans, LA
PT-809	98′	Electric Boat, Groton, CY (Aluminum Hull)
PT-810	89′	Bath Iron Works, Bath, ME
PT-811	94′	John Trumpy & Sons, Annapolis, MD
PT-812	105′	Philadelphia Navy Yard, Philadelphia, PA

PT BOATS WITH REASSIGNED NUMBERS

PT-3 to 9	Retained in Canada for Air-Sea rescue
PT-41	Used as gunboat on Lake Lanao, Mindanao, Philippines after all torpedoes and gasoline were expended.
PT-49 to *58*	Boats became *BPT 1-10.*
PT-59 to *68*	Were ordered as *BPT 11-20,* and became *PT-59* to *68*
PT-85 to *94*	Became *RPT-1-10*, not all transferred to USSR.
PT-197, 198	Became *RPT-11 & 12*. Assigned to UK and not USSR.
PT-216	Assigned to Turkey.
PT-201, 204, 207, 208, 213, 217	To Yugoslavia.
PT-385	Sold with *PT-387* to Egypt.
PT 546 to *563*	Ordered as *RPT-13* to *30.*
PT-813 to *822*	Built in Denmark under MDAP for Royal Danish Navy. Not U.S. design.

US PT Boat List: Numerical List of Boats, http://www.hazegray.org/navhist/pt/list.htm

APPENDIX I. VESSELS LOST TO GERMAN HUMAN TORPEDOES

Name	Type	Date	Location
Cato	Minesweeper	5 Jul 19444	Normandy
Magic	Minesweeper	6 Jul 1944	Normandy
Pylades	Minesweeper	8 Jul 1944	Normandy
Dragon	Cruiser	8 Jul 1944	Normandy (operating under Polish command)
Isis	destroyer	20 Jul 1944	Normandy
Garsay	trawler	3 Aug 1944	Normandy
Colsay	trawler	2 Nov 1944	Ostend Belgium

APPENDIX J. VESSELS DAMAGED OR SUNK BY JAPANESE SUICIDE EXPLOSIVE MOTOR BOATS

Ship	Type	Date	Event	Location
USS *Hutchins (DD-476)*	destroyer	27 Apr 1945	Damaged	Buckner Bay, Okinawa, Ryukyu Islands.

APPENDIX K. ALLIED MERCHANT SHIPS LOST TO S-BOATS 1940-1945

Date	Name	Type	Tons	Sunk by
28 May 1940	*Abukir*	Steamship	694	Torpedo
19 Jun 1940	*Roseburn*	Steamship	3,103	Torpedo and gunfire
24 Jun 1940	*Albuera*	Steamship	3,477	Torpedo
24 Jun 1940	*Kingfisher*	Motor vessel	276	Torpedo
4 Jul 1940	*Elmcrest*	Steamship	4,343	Torpedo
11 Jul 1940	*Mallard*	Steamship	2,162	Torpedo
26 Jul 1940	*Lulonga*	Steamship	821	Torpedo
26 Jul 1940	*London Trader*			
26 Jul 1940	*Broadhurst*	Steamship	1,013	Torpedo
8 Aug 1940	*Holme Force*	Steamship	1,216	Torpedo
8 Aug 1940	*Fife*	Coast motor vessel	367	Torpedo
4 Sept 1940	*New Lambton*	Steamship	2,709	Torpedo
4 Sept 1940	*Joseph Swan*	Steamship	1,571	Torpedo
4 Sept 1940	*Fulham V*	Steamship	1,562	Torpedo and gunfire
24 Sept 1940	*Continental Coaster*	Steamship	555	Torpedo
17 Oct 1940	*Hauxley*	Steamship	1,595	Torpedo
15 Dec 1940	*N. C. Monberg*	Steamship	2,301	Torpedo
6 Feb 1941	*Angularity*	Motor vessel	501	
19 Feb 1941	*Algarve*	Steamship	1,355	Torpedo
26 Feb 1941	*Minorca*	Steamship	1,123	Torpedo
7 Mar 1941	*Dotterel*	Steamship	1,385	Torpedo
7 Mar 1941	*Kenton*	Steamship	1,047	Torpedo
7 Mar 1941	*Corduff*	Steamship	2,345	Torpedo
7 Mar 1941	*Boulderpool*	Steamship	4,805	Torpedo
7 Mar 1941	*Rye*	Steamship	1,048	Torpedo
8 Mar 1941	*Togston*	Steamship	1,547	Torpedo
8 Mar 1941	*Norman Queen*	Steamship	957	Torpedo

Date	Name	Type	Tons	Sunk by
18 Mar 1941	*Daphne II*	Steamship	1,970	Torpedo
17 Apr 1941	*Effra*	Steamship	1,446	Torpedo
28 Apr 1941	*Ambrose Fleming*	Steamship	1,555	Torpedo
11 Aug 1941	*Sir Russell*	Steamship	1,548	S-boat torpedo
7 Sept 1941	*Duncarron*	Steamship	478	Torpedo
17 Sept 1941	*Teddington*	Steamship	4,762	Torpedo
12 Oct 1941	*Chevington*	Steamship	1,537	Torpedo
19 Nov 1941	*Aruba*	Steamship	1,159	Torpedo
19 Nov 1941 (sank on 20th)	*Waldinge*	Steamship	2,462	Torpedo
24 Nov 1941	*Virgilia*	Tanker	5,723	Torpedo
29 Nov 1941	*Asperity*	Tanker	699	Torpedo
29 Nov 1941	*Cormarsh*	Steamship	848	Torpedo
30 Nov 1941	*Empire Newcomen*	Steamship	2,840	Torpedo
9 Jul 1942	*Pomella*	Motor vessel / tanker	6,766	Torpedo
13 Aug 1942	*Glenorchy*	Motor vessel	I 8,892	Torpedo
13 Aug 1942	*Wairangi*	Motor vessel	12,436	Torpedo
7 Oct 1942	*Sheaf Water*	Steamship	2,730	Torpedo
7 Oct 1942	*Ilse*	Steamship	2,874	Torpedo
7 Oct 1942	*Jessie Maersk*	Steamship	1,972	Torpedo
11 Mar 1942	*Horseferry*	Steamship	951	Torpedo and gunfire
19 No 1942	*Yewforest*	Steamship	815	Torpedo
19 Nov 1942	*Birgitte*	Steamship	1,595	Torpedo
3 Dec 1942	*Gatinais*	Motor vessel	383	Torpedo
12 Dec 1942	*Avonwood*	Steamship	1,056	Torpedo
12 Dec 1942	*Glen Tilt*	Steamship	871	Torpedo
12 Dec 1942	*Lindisfarne*	Steamship	999	Torpedo
12 Dec 1942	*Knitsley*	Steamship	2,272	Torpedo
27 Feb 1943	*Modavia*	Motor vessel	4,858	Torpedo

Date	Name	Type	Tons	Sunk by
14 Apr 1943	*Stanlake*	Steamship	1,742	Torpedo
2 Nov 1943	*Dona Isabel*	Steamship	1,179	Torpedo
2 Nov 1943	*Storaa*	Steamship	1,967	Torpedo
2 Nov 1943	*Foam Queen*	Steamship	811	Torpedo
6 Jan 1944	*Polperro*	Motor vessel	403	Torpedo
6 Jan 1944	*Underwood*	Motor vessel	1,990	Torpedo
31 Jan 1944	*Emerald*	Steamship	806	Torpedo
31 Jan 1944	*Caleb Sprague*	Steamship	1,813	Torpedo
24 Feb 1944	*Philipp M*	Steamship	2,085	Torpedo
10 Jun 1944	*Dungrange*	Steamship	621	Torpedo
10 Jun 1944	*Brackenfield*	Steamship	657	Torpedo
10 Jun 1944	*Ashanti*	Motor vessel	534	Torpedo
30 Jul 1944	*Sanware*	Steamship	7,219	Torpedo
2 Nov 1944	*Rio Bravo*	Tanker	1,141	Torpedo
22 Jan 1945	*Halo*	Steamship	2,365	Torpedo
22 Jan 1945	*Goodwood*	Steamship	2,780	Torpedo
22 Jan 1945	*Blacktoft*	Steamship	1,109	Torpedo
19 Mar 1945	*Crichtoun*	Steamship	1,097	Torpedo
19 Mar 1945	*Rogate*	Steamship	2,871	Torpedo

APPENDIX L. US NAVY OFFICIAL LISTING OF ALLIED SHIPS SUNK BY S-BOATS 1944

USS *LST-289* severely damaged in Lyme Bay by S-boats and put of action for the duration of WWII.[1247]

USS *Osprey* (*AM-56 Auk* class minesweeper) struck a presumed S-boat mine as she headed for the Normandy invasion beaches, 6 June, 1944. First US ship lost on D-Day. [1248]

USS *Corry* II *(DD-463)* was in position to offer fire support for the invasion beaches when she struck a presumed German S- Boat mine and sank 6 June 1944.[1249]

USS *Meredith (DD-726)* on 6 June 1944 while giving fire support to Utah Beach the *Meredith* hit what could have been an S-boat mine. She was towed to the Bay of Seine to undergo repairs but German aircraft attacked and the ship was lost.[1250]

USS *Susan B. Anthony* (AP-7, transport) sunk 7 June 1944 by striking two presumed S-boat mines off Omaha Beach. Unloaded 3,000 Army and Navy personnel with no loss of life.[1251]

USS *Tide* (*AM-125 Auk* class Minesweeper) sank after striking a presumed S-boat mine while clearing lanes for fire-support ships 7 June 1944, invasion route.[1252]

PT-505 suffered mine damage on 7 June 1944 and was towed to England for extensive repairs.[1253]

USS *LST-499* sunk by presumed S-boat mine off Normandy, France, 8 June 1944.[1254]

USS *Rich (DE-695)* sank after running over three presumed S-boat mines off Normandy 8 June 1944.[1255]

USS *Glennon* I *(DD-620)* while fulfilling vital fire support for the invasion troops the Glennon struck a presumed S-boat mine. She was finally sunk on 9 June 1944 by German shore batteries.[1256]

USS *LST-314* sunk by German motor torpedo boats off Normandy, France, 9 June 1944.[1257] [1258]

[1247] http://www.ibiblio.org/hyperwar/USN/ships/LST/LST-289.htm.
[1248] http://www.history.navy.mil/photos/sh-usn/usnsh-o/am56.htm.
[1249] http://www.history.navy.mil/danfs/c14/corry-ii.htm.
[1250] http://www.history.navy.mil/danfs/m9/meredith-iii.htm.
[1251] http://www.history.navy.mil/faqs/faq82-2.htm#anchor472938.
[1252] http://www.history.navy.mil/danfs/t5/tide.htm.
[1253] http://www.ibiblio.org/hyperwar/USN/CloseQuarters/PT-7.html.
[1254] http://www.history.navy.mil/danfs/l18/lst-499.htm.
[1255] http://www.history.navy.mil/photos/sh-usn/usnsh-r/de695.htm.
[1256] http://www.history.navy.mil/danfs/g5/glennon-i.htm.

USS *LST-376* sunk by German motor torpedo boats off Normandy, France, 9 June 1944.[1259]

USS *Partridge (ATO-138)* sunk after being torpedoed by S-boats off Normandy, France, 11 June 1944.[1260]

USS *LST-496* sunk by presumed S-boat mine off Normandy, France, 11 June 1944.[1261]

USS *LST-523* sunk by presumed S-BoatS-boat mine off Normandy, France, 19 June 1944.[1262]

USS *Nelson (DD-623)* sunk by S-boat torpedoes off Normandy.[1263]

USS *Harding (DD-625,* later *DMS-28)* received some damage on 17 August 1944 USS while giving fire support to invasion forces in the south of France.[1264]

YMS-*406* (District Motor Mine Sweeper) Contradictions in names and craft number.

HMS Kellett (J-05) damaged.

HMS *Tadoussac (J220)* damaged.

PT-555 sunk 8 September 1944 off Cape Couronne, Mediterranean Sea by German mine.[1265]

[1257] http://www.history.navy.mil/danfs/l15/lst-314.htm.
[1258] http://www.history.navy.mil/faqs/faq82-1.htm#anchor398241.
[1259] http://www.history.navy.mil/danfs/l16/lst-376.htm.
[1260] http://www.history.navy.mil/danfs/p2/partridge-i.htm.
[1261] http://www.history.navy.mil/danfs/l18/lst-496.htm.
[1262] http://www.history.navy.mil/danfs/l19/lst-523.htm.
[1263] http://www.history.navy.mil/danfs/n3/nelson.htm.
[1264] http://www.history.navy.mil/danfs/g5/glennon-i.htm.
[1265] Bulkley, *At Close Quarters*, p. 488.

APPENDIX M. BRITISH LOSSES 1940-1945—SUNK OR CAPTURED BY S-BOATS

Date	Ship	Location	Event
19 Oct 1939[1266]	HMS *Venetia* (D 53, destroyer	Knob Buoy, Thames estuary.	Sunk by S-boat laid mine.
30 Apr 1940[1267]	HMS *Dunoon* (J-52 *Hunt*-class minesweeper)	E-NE of Great Yarmouth	Sunk by S-boat laid mines.
29 May 1940[1268 1269]	HMS *Grafton* (H-89 destroyer),	Off Dunkirk in Operation Dynamo	Conflict in records whether sunk by torpedo by German sub *U-62* or by S-boat *S-38. Grafton* stopped to rescue troops from *Wakeful.*
29 May 1940[1270]	HMS *Wakeful* (i) (H-88 destroyer),	Off Dunkirk in Operation Dynamo	Sunk by *S-30* as she was rescuing survivors.
01 Jun 1940[1271]	HMS *Argyllshire* (trawler)	Off Dunkirk	Sunk by *S-34* .
01 Jun 1940[1272]	*Stella Dorado* (FY-131 (trawler)	Off Dunkirk	Sunk by *S-34.*
31 Aug 1940[1273]	HMS *Esk*_(H-15, ASW trawler)	Off Texel Island, Netherlands	Lost to mine.
17 Oct 1940[1274]	HMS *Dundalk* (J 60 *Hunt*-class Minesweeper)	Off Harwich	Sunk by S-boat mine.
24 Dec 1940[1275]	HMS *Pelton* (MS trawler)	Off Great Yarmouth	Sunk *by S-28.*

1266 http://www.naval-history.net/xGM-Chrono-10DD-09VW-Venetia.htm.
1267 http://www.uboat.net/allies/warships/ship/6521.html.
1268 http://www.uboat.net/allies/warships/ship/4395.html.
1269 http://www.naval-history.net/WW2BritishLossesbyName1.htm.
1270 http://www.uboat.net/allies/warships/ship/4249.html.
1271 http://uboat.net/allies/warships/ship/6301.html.
1272 http://uboat.net/allies/warships/ship/6480.html.
1273 http://uboat.net/allies/warships/ship/6480.html.
1274 http://www.uboat.net/allies/warships/ship/6522.html.
1275 http://uboat.net/allies/warships/ship/7102.html.

Date	Ship	Location	Event
10 Jun 41[1276]	HMS *Pintail* (sloop)	Off Humber	S-boat mine.
09 Jan 1942[1277]	HMS *Vimiera* (L-29 destroyer V&W class)	Thanes Estuary	Sunk in by S-boat mine
15 Mar 1942[1278]	HMS *Vortigren* (D-37-destroyer)	Off Cromer	Sunk by *S-104*
24 Mar 1942[1279]	HMS *Southwold* (L-10, destroyer	off Malta	Sunk by S-boat mines.
06 May 1942[1280]	HMS *Urge* (submarine)	Off Malta	Sunk by S-boat mines.
07 May 1942[1281]	HMS ML. 13 (Motor Launch)	Off Malta	Sunk by S-boat mines.
08 May 1942[1282]	HMS *Olympus* (submarine)	Off Malta	Sunk by S-boat mines.
10 May 1942[1283]	HMS *Moor* (trawler)	Off Malta	Sunk by S-boat mines.
10 May 1942[1284]	HMS *C 308* (Tug)	Off Malta	Sunk by S-boat mines.
10 May 1942[1285]	HMS *Eddy* (trawler)	Off Malta	Sunk by S-boat mines.
27 May 1942[1286]	HMS *Fitzroy* (i) (J 03-Hunt-Class Minesweeper)		Disputed whether it hit a hitting British mine or an S-boat mine.
30 May 1942[1287]	St. *Angelo* (Tug)	Off Malta	Sunk by S-boat mines.
04 Jun 1942[1288]	HMS *Cocker* (trawler))	Off Malta	Sunk by S-boat mines.
10 Jun 1942[1289]	HMS *Trusty Star* (Drifter)	Off Malta	Sunk by S-boat mines.

1276 http://uboat.net/allies/warships/ship/7102.html.
1277 http://uboat.net/allies/warships/ship/7102.html.
1278 http://uboat.net/allies/warships/ship/7102.html.
1279 Hans Frank correspondence, 01/29/2010, per War Diary 3rd *SFltl*
1280 *Ibid.*
1281 *Ibid.*
1282 *Ibid.*
1283 *Ibid.*
1284 *Ibid.*
1285 *Ibid.*
1286 http://www.uboat.net/allies/warships/ship/6521.html.
1287 Hans Frank correspondence, 01/29/2010 War Diary 3rd SFltl.
1288 *Ibid.*
1289 *Ibid.*

Date	Ship	Location	Event
15 Jun 1942[1290]	HMS *Hasty* (destroyer)	Off Malta	Sunk by torpedo from *S-55.*
16 Jun 1942[1291]	HMS *Justified* (Drifter)	Off Malta	Sunk by S-boat mines
16 Jun 1942[1292]	Polish destroyer *Kujawiak* (ex-HMS *Oakley*)	Off Malta	Sunk by S-boat mines
21 Jun 1942[1293]	South African minesweeper *Parktown*	Off Tobruk	Sunk by S-boat gunfire
21 Jun 1942[1294]	HMS *J.T.A.I.* (Tug)	Off Malta	Sunk by S-boat gunfire
21 Jun 1942[1295]	HMS *Alaisia* (Drifter)	Off Malta	Sunk by S-boat gunfire
21 Jun 1942[1296]	HMS *Highland Queen* (trawler)	Off Malta	Sunk by S-boat gunfire
21 Jun 1942[1297]	HMS *LCM-110*	Off Malta	Sunk by S-boat gunfire
21 Jun 1942[1298]	HMS *LCM-113*	Off Malta	Sunk by S-boat gunfire
21 Jun 1942[1299]	HMS *LCM-145*	Off Malta	Sunk by S-boat gunfire
21 Jun 1942[1300]	*HMS LCT--119*	Off Tobruk	Captured by 3rd SFltl.
21 Jun 1942[1301]	*HMS LCT--150*	Off Tobruk	Captured by 3rd SFltl.
21 Jun 1942[1302]	HMS *LCT- 146*	Off Tobruk	Captured by 3rd SFltl.
21 Jun 1942[1303]	HMS *LCA 193*	Off Tobruk	Captured by 3rd SFltl.
21 Jun 1942[1304]	HMS *LCP-65*	Off Tobruk	Captured by 3rd SFltl.

1290 *Ibid.*
1291 *Ibid.*
1292 *Ibid.*
1293 *Ibid.*
1294 *Ibid.*
1295 *Ibid.*
1296 *Ibid.*
1297 *Ibid.*
1298 *Ibid.*
1299 *Ibid.*
1300 *Ibid.*
1301 *Ibid.*
1302 *Ibid.*
1303 *Ibid.*

Date	Ship	Location	Event
21 Jun 1942[1305]	HMS *Parktown* (Whaler)	Off Tobruk	Sunk by S boat
09 Jul 1942,[1306]	HMS *Manor* (FY-333-trawler)	English Channel south of Lyme Bay in position 50º19'N, 03º00'W	Sunk by S-boat
13 Aug 1942[1307 1308]	*HMS Manchester* (light cruiser)	Near Cape Bon, Tunisia	**Not** sunk by German "E-Boats". Sunk by Italian MS-16 and MS-22
7 Oct 1942[1309]	HMS *Cariline Moller* (W-09, rescue tug)	Off Cromer	Sunk by S-boats
19 Nov 1942[1310]	HMS *Ullswater* (ASW whaler, *Lake* class)	English Channel	Sunk by *S-115*
1 Dec 1942[1311]	HMS *Jasper* (trawler)	English Channel	Sunk by *S-81*
7 Jan 1943[1312 1313]	HMS *Horatio* (T-153 trawler)	Off Bone, Tunisia	Sunk by *S-58*
27 Jan 1943[1314]	HMS *Lord Hailsham* (FY 109 trawler)	English Channel	Sunk by S-boat
5 Feb 1943[1315]	HMS *Stronsay* (trawler)	Off Philippevile, France	Sunk by S-boat mines
27 Feb 1943[1316]	*LCT- No. 381* (MKII Landing craft tank)	English Channel	Sunk by S-boat
12 Mar1943[1317 1318]	*HMS Lightning* (destroyer)	North of Bone, Tunisia	Sunk by *S-158* torpedo north of Bone

1304 Ibid.
1305 http://www.naval-history.net/WW2BritishLosses2Auxiliary.htm.
1306 http://uboat.net/allies/warships/ship/6417.html.
1307 http://www.uboat.net/allies/warships/ship/1234.html.
1308 http://www.naval-history.net/WW2BritishLossesbyName2.htm.
1309 http://uboat.net/allies/warships/ship/7342.html.
1310 http://uboat.net/allies/warships/ship/6573.html.
1311 http://www.naval-history.net/WW2BritishLossesbyName1.htm.
1312 http://www.naval-history.net/WW2BritishLossesbyName1.htm#h.
1313 Hans Frank correspondence, 01/29/2010, War Diary 3rd SFltl.
1314 http://www.naval-history.net/WW2BritishLossesbyName2.htm.
1315 Frank, supra.
1316 http://www.navalhistory.net/WW2BritishLosses4Amphib.htm#lcm.
1317 Frank, supra.
1318 http://www.uboat.net/allies/warships/ship/4466.html.

Date	Ship	Location	Event
14 Apr 1943[1319]	*HMS Eskdale (L-36 Hunt III* class destroyer)	Off the Lizard	On escort duty when was damaged from two torpedo hits by *S 90* and later sunk by *S 112*
15 Apr 1943[1320 1321]	HMS *Adonis*	Off Lowestoft	Sunk by S-boat
12 Mar 1943[1322]	HMS *Lightning* (*G 55*, L-class destroyer)	Off Bizerte, Tunisia	Sunk by *S-158*
15 Apr 1943[1323]	*HMS Red Gauntlet (*FY 900-trawler)	Off Felixstowe	Sunk by *S-86*
10 May 1943[1324]	*MTB 264*	Off Bizerta	Sunk by S Boat mines
12 May 1943[1325]	*HMS M.S. 89* (minesweeper)	Off Bizerta	Sunk by S Boat mines
14 May 1943[1326]	HMS *M.L 1154* (Motor Launch)	Off Bizerta	Sunk by S Boat mines
10 Sep 1943[1327]	HMS *Abdiel* (minelayer)	Off Taranto	Sunk by mines laid by *S 54* and *S 61*
25 Sep 1943[1328]	HMS *Franc Tireur* (FY-1560)	Off Harwich	Sunk by torpedo from *S-96*
23 Oct 1943 [1329]	HMS *Charybdis (Dido*-class light cruiser)	Off north coast of Brittany, France in position 48º59'N, 03º39'W	Sunk by two torpedoes from the German torpedo boats *T-23* and *T-27*
25 Oct 1943[1330]	HMS *William Stephen*	Off Cromer	Sunk by S-boats
01 Dec 1943[1331]	HMS *Avanturine* (FY 1886*)*	Off Beachy Head	Sunk by *S 142*

[1319] http://www.naval-history.net/xGM-Chrono-10DE-Eskdale.htm.
[1320] http://warsailors.com/forum/read.php?1,33984,38854#msg-38854.
[1321] http://www.naval-history.net/WW2BritishLossesbyName1.htm.
[1322] http://www.uboat.net/allies/warships/ship/4466.html.
[1323] http://www.naval-history.net/WW2BritishLossesbyName3.htm
[1324] Frank, supra.
[1325] Ibid.
[1326] Ibid.
[1327] Ibid.
[1328] http://uboat.net/allies/warships/ship/6933.html.
[1329] http://www.uboat.net/allies/warships/ship/4017.html.
[1330] http://www.naval-history.net/WW2BritishLossesbyName3.htm.

Date	Ship	Location	Event
05/06 Jan 1944[1332]	HMS *Polperro* (collier)	49º56'-36"N- 05º 28' 14" W	While in convoy WP 457 she was attacked and sunk by S Boats of the 5th SFltl
05/06 Jan 1944[1333]	HMS *Underwood*	49º56'-36"N- 05º 28' 14" W	While in convoy WP 457 she was attacked and sunk by S Boats of the 5th SFltl
05/06 Jan 1944[1334]	HMS *Solstad*	49º56'-36"N- 05º 28' 14" W	While in convoy WP 457 she was attacked and sunk by S Boats of the 5th SFltl
06 Jan 1944[1335]	HMS *Wallasea* (T 345-trawler, Isles Class)	Off Mounts Bay, Cornwall	Sunk by S-boats
31 Jan 1944[1336]	HMS *Pine* (T 101, trawler)	English Channel, southeast of Bagnor	Sunk by *S-142*
13 Feb 1944[1337]	HMS *Cap d'Antifer* (FY 350	North Sea	Sunk by S-boats
24 Apr 1944[1338 1339]	HMS *Roode Zee* (W 162, rescue tug)	Off Dungeness	Sunk by *S-100*

1331 http://www.naval-history.net/WW2BritishLossesbyName1.htm.
1332 http://www.polperro.org/ships.html.
1333 Rohmer, *Chronology of the War at Sea, 1939-45*, at 297. http://goo.gl/FzkWH.
1334 *Ibid.*
1335 http://uboat.net/allies/warships/ship/6677.html.
1336 http://uboat.net/allies/warships/ship/6580.html.
1337 http://www.naval-history.net/WW2BritishLossesbyname3.htm.
1338 http://www.naval-history.net/WW2BritishLossesbyname3.htm.
1339 http:uboat.net/allies/warship/ship/7464.html.

Date	Ship	Location	Event
06 Jun 1944[1340] [1341]	HNoMS *Svenner* (G 03) (ex-HMS *Shark*, destroyer)		The use of the term "E-Boats" has caused many problems in vessel identification. In this case the original report said,"3 E-Boats attacked the *Svenner*…" While in fact it was three German torpedo boats (destroyer sized ships) that sank the destroyer.
11 Jun 1944[1342] [1343]	*Sesame* (W 144, Rescue Tug)	Off Normandy	Sunk by S-boat
21 Jun 1944[1344]	HMS *Fury* (H 76)	Off Humber	Sunk by S-boat mines
24 Jun 1944[1345]	HMS *Swift* (G 46)	Off Sword Beach	Sunk by S boat mine
25 Jun 1944[1346]	*LCH* No. 185, Landing Craft Headquarters,	Off Normandy	Sunk by possible S-boat mine or pattern-running torpedo
02 July 1944[1347] [1348]	HMS *Empire Broadsword,* LSI (L) Landing Ship Infantry (Large)	Off Normandy Beach	Struck an S boat laid mine

[1340] http://www.naval-history.net/WW2BritishLossesbyname3.htm.
[1341] http:uboat.net/allies/warship/ship/7464.html.
[1342] http://www.naval-history.net/WW2BritishLossesbyname4Amphib.htm.
[1343] http:uboat.net/allies/warship/ship/7385.html.
[1344] http:www.naval-history.net/WW2BritishLossesbyName3.htm.
[1345] http://www.naval-history.net/WW2BritishLossesbyName3.htm.
[1346] http://www.naval-history.net/WW2BritishLosses4Amphib.htm.
[1347] http://www.naval-history.net/WW2BritishLosses4Amphib.htm#lcm.
[1348] Frank, Hans. *German S-boats in Action in the Second World War*, 170-171.

Date	Ship	Location	Event
Jun/July 1944[1349]	LCI *No. 99* Landing craft Infantry (Large)		May have been struck by Dackel, FAT I, pre-set running program torpedo, which ran straight for 16 miles then 16 miles in random re-set patterns or by LUT programmed torpedoes with acoustic self-steering that were drawn to sounds of boats propeller[1350]
21 Aug 1944[1351]	HMS *Orchis* (K 76)	Off Juno Beach	Severely damaged by S boat mine

1349 http://www.naval-history.net/WW2BritishLosses4Amphib.htm#lcm.
1350 http://www.naval-history.net/WW2BritishLosses4Amphib.htm#lcm.
1351 http://www.uboat.net/allies/warships/ship/798.html.

APPENDIX N. JAPANESE WARSHIPS AND VESSELS SUNK OR DAMAGED BY PT BOATS

Date	Japanese ship	Result	PT boat responsible
11 Jul 1942 (damaged)	*Mochizuki* (destroyer)	Damaged	*PT-37*
12 Oct 1942	*Tenuzuki (destroyer)*	Sunk	*PT-44*
11 Nov 1942	I-3 *Togami* (submarine)	Sunk	*PT-59*
12 Nov 1942	*III Terutsuki* (destroyer)	Sunk	*PT-40 , 110*
10 Jan 1943	*Hatsukaze* (destroyer)	Damaged	*PT-43, 112*
3 Mar 1943	*Oigawa Maru* (cargo ship)	Sunk	*PT-143, 150*
14 May 1943	RP-102 *Kanemoto*	Sunk	*PT-150, 152*
21 Sep 1944	Minelayer (name undetermined)	Sunk	*PT-342*
28 Nov 1944	Patrol Boat No. 105:	Sunk	*PT-127, 331*
28 Nov 1944	Sub-Chaser No. 53:	Sunk	*PT-127, 331*
28 Nov 1944	*Uzuki*	Sunk	
26 Dec 1944	*Kiyoshimo* (destroyer)	Sunk	*PT-223, 221*

Appendix O. List of S-boat Losses During World War II

By Hans Frank

Boat	Date	Location	Event
17	5 Sep 1939	North Sea	Heavily damaged by sea, broken up
32	21 Jun 1940	North Sea	Mine
23	12 Jul 1940	North Sea	Mine, scuttled
37	12 Oct 1940	North Sea	Mine
38	20 Nov 1940	Channel	Destroyers HMS *Garth, Campbell*
43	27 Jun 1941	Baltic East	Mine
106	27 Jun 1941	Baltic East	Mine
41	20 Nov 1941	Channel	Scuttled after rammed by *S-47*
53	20 Feb 1942	Channel	British escorts and collision with *S-39*
111	15 Mar 1942	North Sea	Captured by British MTBs, recaptured by Germans, scuttled
31	10 May 1942	Mediterranean Sea off Malta	Mine
34	17 May 1942	Mediterranean Sea off Malta	Coastal artillery
27	05 Sep 1942	Black Sea	Torpedoed by *S-72*
104	09 Jan 1943	North Sea	Mine
71	18 Feb 1943	North Sea	Destroyers HMS *Garth, Montrose*
35	28 Feb 1943	Mediterranean Sea off Tunisia	Mine
56	28 Feb 1943	Mediterranean Sea, Sicily harbor	Sunk by RAF bomber, refloated, towed to Toulon, damaged by Allied bomber 20 Nov 1943, scuttled
70	05 Mar 1943	Channel	Mine
75	05 Mar 1943	North Sea	RAF fighter-bomber
119	08 Mar 1943	Channel	Destroyer HMS *Mackay, MGB-20* and *21,* scuttled after collision with *S-114*
29	29 Mar 1943	Channel	*MGB-333, 321*s scuttled
59	06 Jul 1943	Mediterranean Sea, Sicily harbor	Allied bomber
102	08 Jul 1943	Black Sea	Mine

44	25 Jul 1943	Baltic West	Kiel harbor, Allied bomber
66	25 Jul 1943	Baltic West	Kiel harbor, Allied bomber
77	25 Jul 1943	North Sea	*MGB-40, 42,* scuttled
137	29 Jul 1943	Baltic West	Kiel harbor, Allied bomber
46	10 Sep 1943	Black Sea	Soviet fighter bomber
121	11 Aug 1943	Channel	RAF fighter bomber
96	25 Sep 1943	North Sea	Rammed *ML*-145, scuttled
88	25 Oct 1943	North Sea	*MGB-607*
63	25 Oct 1943	North Sea	Destroyer HMS *Mackay* and *MGB-603*
74	05 Nov 1943	North Sea	RAF fighter bomber, scuttled
55	11 Jan 1944	Adriatic Sea	RAF fighter bomber, scuttled
94	23 Feb 1944	Channel	Destroyer HMS *Garth*
128	23 Feb 1944	Channel	HMS destroyer Garth
93	26 Mar 1944	North Sea	Ijmuiden harbor
129	26 Mar 1944	North Sea	Ijmuiden harbor
147	26 Apr 1944	Channel	French destroyer *La Combattante*
141	13 May 1944	Channel	French destroyer *La Combattante*
87	20 May 1944	North Sea	RAF fighter bomber
139	07 Jun 1944	Channel	Mine
140	07 Jun 1944	Channel	Mine
136	11 Jun 1944	Channel	Escort HMS *Duff*
153	12 Jun 1944	Adriatic Sea	Destroyer HMS *Eggesford*
178	13 Jun 1944	Channel off Boulogne	RAF fighter bomber
179	13 Jun 1944	Channel off Boulogne	RAF fighter bomber
189	13 Jun 1944	Channel off Boulogne	RAF fighter bomber
84	14 Jun 1944	Channel, Le Havre harbor	RAF Lancaster bomber
100	14 Jun 1944	Channel, Le Havre harbor	RAF Lancaster bomber
138	14 Jun 1944	Channel, Le Havre harbor	RAF Lancaster bomber
142	14 Jun 1944	Channel, Le Havre harbor	RAF Lancaster bomber
143	14 Jun 1944	Channel, Le Havre harbor	RAF Lancaster bomber

144	14 Jun 1944	Channel, Le Havre harbor	RAF Lancaster bomber
146	14 Jun 1944	Channel, Le Havre harbor	RAF Lancaster bomber
150	14 Jun 1944	Channel, Le Havre harbor	RAF Lancaster bomber
169	14 Jun 1944	Channel, Le Havre harbor	RAF Lancaster bomber
171	14 Jun 1944	Channel, Le Havre harbor	RAF Lancaster bomber
172	14 Jun 1944	Channel, Le Havre harbor	RAF Lancaster bomber
173	14 Jun 1944	Channel, Le Havre harbor	RAF Lancaster bomber
187	14 Jun 1944	Channel, Le Havre harbor	RAF Lancaster bomber
188	14 Jun 1944	Channel, Le Havre harbor	RAF Lancaster bomber
190	23 Jun 1944	Channel	British escorts
182	27 Jul 1944	Channel	Scuttled after collision with *MTB-430*
39	02 Aug 1944	Channel, Le Havre Harbour	RAF bomber
114	02 Aug 1944	Channel, Le Havre Harbour	RAF bomber
57	19 Aug 1944	Adriatic Sea	British MGBs
26	19 Aug 1944	Black Sea, Sulina harbor	Soviet bomber
40	19 Aug 1944	Black Sea, Sulina harbor	Soviet bomber
72	19 Aug 1944	Black Sea, Sulina harbor	Soviet bomber
42	20 Aug 1944	Black Sea, Konstanza harbor	Soviet bomber
52	20 Aug 1944	Black Sea, Konstanza harbor	Soviet bomber
131	20 Aug 1944	Black Sea, Konstanza harbor	Soviet bomber
148	21 Aug 1944	Black Sea	Mine
28	25 Aug 1944	Black Sea, Konstanza harbor	Scuttled after Soviet bomber attack on 20 Aug 1944
149	25 Aug 1944	Black Sea, Konstanza harbor	Scuttled after Soviet bomber attack 20 Aug 1944
91	25 Aug 1944	North Sea	Destroyer HMS *Talybon,* frigate *Retalick*
45	29 Aug 1944	Black Sea off Varna	Scuttled, crew returned to Germany
47	29 Aug 1944	Black Sea off Varna	Scuttled, crew returned to Germany
49	29 Aug 1944	Black Sea off Varna	Scuttled, crew returned to Germany

51	29 Aug 1944	Black Sea off Varna	Scuttled, crew returned to Germany
80	01 Sep 1944	Baltic East	Mine
184	05 Sep 1944	Channel	Dover coastal batteries, scuttled
145	15 Sep 1944	Channel, Brest Harbor	Self-destroyed
183	19 Sep 1944	North Sea	Frigate HMS *Stayner* and *MTB-724, 728*
200	19 Sep 1944	North Sea	Frigate HMS *Stayner* and *MTB-724, 728*
702	19 Sep 1944	North Sea	Frigate HMS *Stayner* and *MTB-724, 728*
158	25 Oct 1944	Adriatic Sea	Allied bomber
54	31 Oct 1944	Mediterranean Sea	Scuttled after mine damage incurred 22 Apr 1944
203	10 Nov 1944	North Sea off Norway	Scuttled after collision with *R-220*
198	15 Dec 1944	North Sea, Ijmuiden harbor	USAAF bomber
185	23 Dec 1944	North Sea	HMS destroyers *Walpole, Torrington, Curzon,* Sloop *Kittiwake*
192	23 Dec 1944	North Sea	HMS destroyers *Walpole, Torrington, Curzon,* Sloop *Kittiwake*
33	10 Jan 1945	Adriatic Sea, Isle Unie	Grounded, self-destroyed
58	10 Jan 1945	Adriatic Sea, Isle Unie	Grounded, self-destroyed
60	10 Jan 1945	Adriatic Sea, Isle Unie	Grounded, self-destroyed
180	14 Jan 1945	North Sea	Mine
199	23 Jan 1945	North Sea	Scuttled after collision *with S-701*
181	21 Feb 1945	North Sea	RAF fighter bomber
193	22 Feb 1945	North Sea	British escorts
167	24 Feb 1945	North Sea	Scuttled after collision with own boats
220	01 Mar 1945	North Sea	Destroyer escort HMS *Seymour*
186	30 Mar 1945	North Sea, Wilhelmshaven harbor	USAAF bomber
194	30 Mar 1945	North Sea, Wilhelmshaven harbor	USAAF bomber
224	30 Mar 1945	North Sea, Wilhelmshaven harbor	USAAF bomber
176	07 Apr 1945	North Sea	*MTB-494, 493, 497;* scuttled
177	07 Apr 1945	North Sea	*MTB-494, 493, 497;* scuttled
223	08 Apr 1945	North Sea	Mine

202	08 Apr 1945	North Sea	British MGBs; scuttled after collision with *S-703*
703	08 Apr 1945	North Sea	British MGBs; scuttled after collision with *S-202*
157	01 May 1945	Adriatic Sea	Coastal artillery
201	03 May 1945	Baltic West, Kiel harbor	Allied bomber
103	04 May 1945	Baltic West	RAF fighter bomber
226	06 May 1945	Baltic West	Allied bomber, scuttled
19	07 May 1945	Baltic West off Isle of Fehmarn	Scuttled after collision with *S-301*
301	07 May 1945	Baltic West off Isle of Fehmarn	Scuttled after collision with *S-191*

REMARKS:

S-21 was never sunk, still in action throughout the war.

S-7 – S-18 (with the exception of *S-17* – broken up 1939) were removed in 1940 from active flotillas and served later on as submarine chasers, fast target tugs or coastal patrolers.

S-501 – 507 were originally Italian MAS (*Motobarca Armata Svano o Silurante*) boats in the Black Sea. On May 20, 1943 they were taken over by Germany as coastal patrollers, broken up October 1943.

S-510 – 513 were also originally Italian MAS, taken over by Germany on September 9, 1943. when Italy surrendered, but never used in combat action.

S-601 – 604 were originally Yugoslavian MTBs, taken over by Italy April 17, 1941 and later on by Germany as coastal patrolers after Italy's surrender September 9, 1943. All boats sunk by Allied Bomber or scuttled in September / October in the Saloniki Area (Aegean Sea).

S-621 – 630 were originally built for Italy at the dockyard Monfalcone. Not finished. After Italy's surrender Germany completed the boats but they never came in action.

S-666 was originally the British *MGB-666,* captured July 5, 1944 by coastal patrolers off Ijmuiden, and exploded at Ijmuiden on July 8, 1944.

In the North Sea / Channel the boats were often scuttled after combat action or collision due to the fact that these engagements took place close to the British coast. Therefore it was necessary to destroy the boats (otherwise they could be captured by the British) and to bring the crews back to man new boats.

APPENDIX P. SUMMARY OF PT BOAT LOSSES WWII

The US Navy lost 69 craft out of the total 531 boats put into service.

Count	Cause
5	Destroyed by enemy surface ship gunfire
1	Rammed by enemy ship
1	Rammed enemy ship
1	Enemy aircraft strafing
4	Enemy bombings
2	Kamikaze attacks
5	Enemy shore batteries
4	Enemy mines
1	Damaged by enemy fire, then destroyed
2	lost in transit, tanker torpedoed by enemy sub
26	**Total by enemy action**

18	Grounded in enemy waters and destroyed to prevent capture
3	Destroyed to prevent capture
3	Destroyed by US aircraft
2	Destroyed by Australian aircraft
2	Destroyed by US ships
1	Destroyed by enemy shore fire or wild shot from US warship
5	Grounded/destroyed outside enemy waters or in storms
6	Fire or explosions in port
3	Collisions
43	**By accident, friendly fire, or sea conditions**

APPENDIX Q. PT BOATS DAMAGED AND LOST BY COORDINATES

Boat number	Event	Date	Lat/long	Location
PT-21	Scuttled	1943		
PT-22	Abandoned	1943		Heavily damaged and abandoned off Adak, Aleutian Islands during a storm.
PT-27	Damaged	1943		Dora Harbor, Unimak Island
PT-28	Abandoned	1943		Dora Harbor, Unimak Island
PT-31	Sunk	1941	14º 15' N-120º 13' E	Subic Bay, Luzon, Philippines
PT-32	Destroyed	1942	10º 58' N-12º 12' E	Southwest Taguayan Island
PT-33	Destroyed	1941		Cape Santigao, Luzon, Philippines
PT-34	Destroyed	1941	10º 16' N-123º 52'E	Cavit Island, Philippines
PT-35	Scuttled	1942	10º 18' N-123º 54'E	Cebu Shipyard, Philippines
PT-37	Sunk	1943		Cape Esperance, Guadalcanal
PT-41	Scuttled	1941		Mindanao, Philippines
PT-43	Lost	1942	09º 15'S-159º 42'E	Guadalcanal
PT-44	Destroyed	1941	09º 10'S-159º 45'S	Guadalcanal
PT-63	Destroyed	1944	01º 45'S-150º 45'E	Savo Island
PT-67	Destroyed	1943	09º 02'S-149º 20'E	Tufi Island, New Guinea
PT-68	Sunk	1943	05º 56'S-147º 18'E	Vincke Point, Huon Peninsula, New Guinea
PT-73	Scuttled	1945	13º 50'N-120º 10'E	Baliquias Bay, Mindoro, Philippines
PT-77	Sunk	1945	13º 55'N-120º 36'E	Talin Point, Luzon, Philippines
PT-79	Sunk	1945	13º 55'N-120º 36'E	Talin Point, Luzon, Philippines
PT-107	Destroyed	1943	01º 45'S-150º 01'E	Hamburg Bay, E. Mirau Island, Solomons
PT-108	Destroyed	1943	01º 45'S-150º 01'E	Hamburg Bay, E. Mirau Island, Solomons
PT-109	Sunk	1943	08º Mar'S-156º 58'E	Blackett Strait, Solomons
PT-110	Sunk	1944	06º 17'S-150º 09'E	
PT-111	Sunk	1943		Guadalcanal
PT-112	Sunk	1943	09º 15'S-159º 42'E	
PT-113	Grounded	1943		

Boat number	Event	Date	Lat/long	Location
PT-114	Grounded	1942	09º 120'S-146º 29'E	Off Veale Reef, Tufi, New Guinea
PT-117	Destroyed	1943	08º 24'S-157º 19'E	Rendova Harbor, Lumbari Island
PT-118	Grounded	1943	07º 34'S-165º 35'E	Vella Lavella Island, Solomons
PT-119	Destroyed	1943	09º 02'S-149º 20'E	Tufi Island, New Guinea
PT-121	Sunk	1943	05º 17'S-151º 01'E	Bangula Bay
PT-123	Sunk	1943		Guadalcanal
PT-130	Damaged	1944		
PT-133	Destroyed	1944	03º 28'S-143º 34'E	Cape Pus, New Guinea
PT-135	Destroyed	1944	05º 21'S-152º 09'E	Crater Point. New Britain
PT-136	Destroyed	1943	05º 55'S-148º 01'E	Malai Island New Guinea
PT-145	Destroyed	1944	05º 34'S-146º 10'E	Mindiri Island, New Guinea
PT-147	Grounded	1943	05º 55'S-147º 20'E	Off Teliata Point, New Guinea
PT-153	Grounded	1943	08º 20'S-157º 15'E	Munda Point, New Georgia, New Guinea
PT-164	Sunk	1943	08º 25'S-157º 20'E	Rendova Harbor, Solomon
PT-165	Sunk	1942	23º 45'S-166º 30'E	Lost while deck cargo on board tanker *Stanvac Manila* 100 nm south of Nouméa, New Caledonia
PT-166	Sunk	1943	08º 15'S-156º 53'E	Sunk by B-25, Ferguson Passage, Solomon
PT-172	Grounded	1943	07º 34'S-165º 35'E	Off Vella Lavella Island Solomon
PT-173	Sunk	1942	23º 45'S-166º 30'E	Lost while deck cargo on board tanker *Stanvac Manila* 00 nm south of Nouméa, New Caledonia
PT-193	Scuttled	1944	00º 55'S-134º 52'E	Off Noemfoor Island, New Guinea
PT-194	Damaged	1944		
PT-200	Destroyed	1944		
PT-201	Destroyed		41º 23'N-71º 01'W	Newport R.I.
PT-202	Sunk	1944	43º 23'N-06º 43'E	Mine, Gulf of Frejus, France
PT-218	Sunk	1944	43º 23'N-06º 43'E	Mine, Gulf of Frejus, France
PT-239	Destroyed	1943	07º 42'S-156º 47'E	Lambu Cove, Vella Lavella, Solomon

Boat number	Event	Date	Lat/long	Location
PT-247	Destroyed	1944	06º 38'S-156º 01'E	SE Coast Bougainville
PT-251	Sunk	1944	06º 30'S-155º 10'E	Empress Augusta Bay, Solomon
PT-279	Sunk	1944	05º 30'S-154º 15'E	Collision *PT-282*, Off Bougainville
PT-28	Damaged	1944	05º 30'S-154º 15'E	Collision *PT-279,* Off Bougainville
PT-293	Sunk	1944	06º 27'S-155º 08'E	Sunk by USS destroyer *Guest* Off Bougainville
PT-300	Sunk	1944	12º 19'N-121º 05'E	Off Mindoro
PT-301	Destroyed	1944		Mois Eoendi Biak
PT-311	Sunk	1944	43º 41'N-09º 37'E	La Spezia Italy
PT-320	Sunk	1944	11º 11'N-125º 05'E	Off Leyte, Philippines
PT-321	Grounded	1944	11º 11'N-124º 19'E	San Isidro Bay, Leyte
PT-322	sunk	1943	06º 09'N-147º 36'E	Hardenberg Point, New Guinea
PT-323	Sunk	1943	10º 33'N-125º 14'E	Off Leyte, Philippines
PT-337	Sunk	1944	04º 09'S-144º 50'E	Hansa Bay, New Guinea
PT-338	Grounded	1945	12º 06'N-121º 23'S	Seminimara, Luzon
PT-339	Grounded	1944	04º 01'S-144º 41'E	Pur Pur, New Guinea
PT-346	Grounded	1944	04º 01'S-144º 41'E	Pur Pur, New Guinea
PT-347	Abandoned	1944	08º20'S-157º 15'E	Off Munda Point, New Guinea
PT-353	Sunk	1944	05º 17'S-151º 01E	Aussie A/C Bangula Bay, New Britain
PT-363	Grounded	1944	00º 55'N-127º 50'E	Off Cape Gorango, Halmahera Islands
PT-368	Grounded	1944	01º 59'N-127º 57'E	Halmahera Islands Netherlands East Indies
PT-371	Scuttled	1944		Tagalasa, Halmahera Islands
PT-490	Grounded	1944		
PT-493	Sunk	1944	10º 15'N-125º 23'E	Battle of Surigao Strait
PT-505	Damaged	1944		
PT-509	Sunk	1944	49º 11'N-02º 15'W	Off St. Malo, France. Rammed by M-boat
PT-511	Damaged,	1944		
PT-520	Damaged	1944		
PT-555	Sunk	1944	43º19'N-05º 30E	Off Cape Couronne, France

APPENDIX R. CAPTURED ALLIED TORPEDO BOATS

MTB 631. Beached off Floro (west coast, Norway), left by crew and on 18 March 1943 towed to Bergen, Norway. Entered German service 12 December 1943 as *SA-11* (*SA-Schnellboote-Ausland,* SBoote Foreign Service)[1352] [1353] [1354] [1355]

MTB 314. Captured 14 September 1942 off Tobruk by *R-10*. Entered German service as *RA-10* (*Raumboot Ausland,* mine clearance boat Foreign Country).Earmarked for Operation "Reisernte" ("rice harvest"), a disguised attack by Special Force"Brandenburg" against merchant traffic off West coast of Africa. The operation was called off and RA-10 was sunk by Allied aircraft off Tunis 30 April 1943. [1356] [1357] [1358]

MTB-345. Beached off Aspoy (close to Bergen), Norway 27 July 1943. Captured by *V 5301* 28 July 1943, towed to Askevold. Crew executed by Gestapo (secret state police). Entered German service 29 September 1943 as *SA 12*, detached to "Gruppe Nord" for security operations against MTBs and MGBs.[1359] [1360] [1361][1362]

MGB-666. Captured 5 July 1944 by 14th V Flotilla off Ijmuiden (Netherlands), towed to Ijmuiden 8 July 1944. Exploded by fuel-gas in S-boat bunker at Ijmuiden.[1363] [1364]

MGB-335. Captured off Brown Bank, English Channel by *S-80 and S 105*. The *335* boat carried a wealth of intelligence information including the newest radar set, newest radios, plotting charts of minefields, signal books and two telegraphists who were captured for extensive interrogation. This gold mine of information gave the German surface forces a complete edge on the course of the war for over a year. British sources still list this boat as having been sunk.[1365]

[1352] The Norwegian MTB Flotilla in Shetland http://shetlopedia.com/The_ Norwegian_ MTB_ Flotilla_in_Shetland.

[1353] Ships in Enemy Service-World Navy Ships Forums http://www.worldnavalships.com/forums/showthread.php?t=5604.

[1354] Eric Groner, *Die deutschen Kriegsmarine 1815-1945*, Vol.2, pg 174.

[1355] Hans Frank, Internet correspondence 02/12/2001.

[1356] Ships in Enemy Service. World Navy Ships Forums. http://www.worldnavalships.com/forums/showthread.php?t=5604.

[1357] Groner *supra* at 174.

[1358] Frank *supra.*

[1359] Shetland *supra.*

[1360] Forum *supra.*

[1361] Groner *supra.*

[1362] Frank *supra.*

[1363] Shetland *supra.*

[1364] Forum *supra.*

[1365] Hans Frank, *German S-boatsin World War II* at 54 and Internet correspondence 02/12/2001.

MGB-508. Used as blockade runner between UK and Sweden. Captured 02.11.43 off Lysekil (west coast of Sweden). Entered German service 10.11.43 as German *RA-11*. Detached to "Gruppe Nord" (Group North) for security operations against British MGBs/MTBs. After war returned to England. [1366]

[1366] Gruner *supra* and Frank, Internet correspondence 2/12/2001.

APPENDIX S. SUBS SUNK BY MOTOR TORPEDO BOATS DURING WW II

BY ENGLISH MTBS

On 19 Jan 1943 MTB-*260* sank Italian *Santorre Santarosa* at 37-10 N, Mar-15E.[1367]

On 11 Jul 1943 MTBs *640, 650, 651* sank Italian *Flutto* at 37-34 N, 15-43E.[1368]

On 12 July 1943 *MTB-81* sank German *U-561* in the Strait of Messina.[1369]

BY US PT BOATS

On 9 Dec 1942 *PT-59* sank Japanese *I-3 (Togami)* at 09-13S,159-40E.[1370]

On 14 May 1943 PTs *150* and *152* sank Japanese *RO-102 (Kamemoto), at* 05-55 S,147-34 E.[1371]

BY ITALIAN MAS BOATS

On 19 June 1942 MAS-571 Soviet sub sank *ShCh 214*,off Sevastopol.[1372]

BY GERMAN S-BOATS

on June 23 and 24, 1941 *S-35* and 60 sank Soviet Sub *S-3* in the Irben Strait, Gulf of Riga, Baltic Sea.[1373]

BY SOVIET TKAS

On 19 April 1942, *TKA 13* and *14* sank German *U-454* near Varde, west coast of Jutland, southwest Denmark.[1374]

On 13 June 1942, a TKA force sank Italian *CB-5* near Yalta, Black Sea.[1375]

[1367] *United States Submarine Losses World War II*, Naval Historical Division, 1963, at 179.
[1368] *Ibid.* at 179.
[1369] *Ibid.* at 161.
[1370] *Ibid.* at 175.
[1371] *Ibid.* at 175.
[1372] Internet correspondence with Richard Worth, July 12, 2007.
[1373] Frank, *German S-boats in the Second World War*, at 45.
[1374] Internet correspondence with Richard Worth, July 12, 2007.
[1375] Internet correspondence with Richard Worth, July 12, 2007.

APPENDIX T. SHIPS SUNK OR DAMAGED BY MTBS IN WW II

Date	Event
1 Sep 1939	ORP *Lloyd Bydgoskii (*pilot vessel, sunk, S-23)
19 Apr 1940	HNoMS *Sael* (torpedo boat, sunk, 2nd SFltl.)
19 Apr 1940	HNoMS *Vale* (minelayer, captured, 2nd SFltl)
30 Apr 1940	HMS *Dunoon* (J-52) (minesweeper, sunk, S-boat mine)
10 May 1940	HMS *Kelly* (destroyer, damaged, *S-31*, 2nd SFltl.)
21 May 1940	HMS *Corburn* (destroyed, sunk, *S-19, 21, 22, 23, 24)*
23 May 1940	French *Jaguar* (destroyer, sunk, S-*24, 23 ,21*, 1st SFltl.)
26 May 1940	HMS *Vega* (destroyer, attacked but not hit, 1st SFltl., *S-19, 21, 22, 23, 24)*
28 May 1940	HMS *Grafton* (H-89) (destroyer, sunk, *S-38*)
29 May 1940	HMS *Wakeful* (i) (H-88) (destroyer, sunk, *S-30*)
30 May 1940	French *Cyclone* (destroyer, damaged, captured, *S-24)*
31 May 1940	French *Sirocco* (destroyer, sunk, *S-23, 26)*
1 Jun 1940	HMS *Argyllshire* (armed trawler, sunk, *S-34)*
1 Jun 1940	HMS *Stella Dorado* (*FY-131)* (armed trawler, sunk, S-*34*)
19 Jun 1940	SS *Roseburn* (British, sunk*, S-19, 26)*
24 Jun 1940	SS *Kingfisher* (British, motor vessel, sunk, *S-19, 36*)
24 Jun 1940	SS *Albuera* (British, tanker, sunk, *S-19, 36)*
4 Jul 1940	SS *Elmcrest* (British, sunk, *S-14, 19, 20, 26)*
11 Jul 1940	SS *Mallard* (British, sunk, SFltl.)
Jul 24 41	SS *Sydney Star* (British, sunk, *MAS-532, 533)*
26 Jul 1940	SS *Mekenes* (Fr, sunk, *1st* SFltl.)
26 Jul 1940	SS *Lulonga* (British, sunk, SFltl.)
26 Jul 1940	SS *Broadhurst* (British, sunk, SFltl.))
26 Jul 1940	SS *London Trader* (British, sunk, SFltl.))
16 Oct 1940	HMS *Dundalk (*J 69) (minesweeper, sunk, S-boat mine)
8 Aug 1940	SS *Holme Force* (British, sunk, SFltl.)
8 Aug 1940	SS *Fife Coast* (British, motor vessel, sunk, SFltl.)

Date	Event
4 Sep 1940	SS *Corbrook* (British, sunk, *S-21*)
4 Sep 1940	SS *New Lambton* (British, sunk, *S-21)*
4 Sep 1940	SS *Joseph Swan* (British, sunk, *S-18*)
4 Sep 1940	SS *Nieuwland* (Netherlands, sunk, *S-18)*
4 Sep 1940	SS *Fullham V*, (British, sunk, *S-22)*
7 Sep 1940	SS *Stad Alkmaar* (Netherlands, sunk, *S-33)*
24 Sep 1940	SS *Continental Coaster* (British, sunk, SFltl.)
17 Oct 1940	SS *Hauxley* (British, sunk, *S-18*)
15 Dec 1940	SS *N.C. Monberg* (Danish, sunk, *S-58)*
23 Dec 1940	SS *Stad Maastricht* (Netherlands, sunk, *S-59, 101*)
23 Dec 1940	HMS Pelton (trawler, sunk, *S-28*)
18 Aug 1940	USSR minesweeper *T-51/Pirmunlas* (sunk, *S-58*)
20 Aug 1940	SS *Czenstochowa* (Polish, sunk*, S-48*)
6 Feb 1941	SS *Angularity* (British, motor vessel, sunk, *S-30*)
19 Feb 1941	SS *Algarve* (British, sunk, SFltl.)
26 Feb 1941	SS *Minorca* (British, sunk, SFltl.)
7 Mar 1941	SS *Dotterel* (British, sunk, SFltl.)
7 Mar 1941	SS *Kenton* (British, sunk, SFltl.)
7 Mar 1941	SS *Corduff* (British, sunk, SFltl.)
7 Mar 1941	SS *Boulderpool* (British, sunk, SFltl.)
7 Mar 1941	SS *Rye* (British, sunk, SFltl.)
8 Mar 1941	SS *Togston* (British, sunk, SFltl.)
8 Mar 1941	SS *Norman Queen* (British, sunk, SFltl.)
11 Mar 1941	SS *Trevethoe* (British, motor vessel, sunk, SFltl.)
18 Mar 1941	SS *Daphne II* (British, sunk, SFltl.)
25 Mar 1941	HMS *Pericles* (sunk, M.T.M.)
25 Mar 1941	HMS *York* (cruiser, damaged by Italian MTM)
17 Apr 1941	SS *Effra* (British, sunk, SFltl.)
28 Apr 1941	SS *Ambrose Fleming* (British, sunk, *S-26, 29)*

Date	Event
10 Jun 1941	HMS *Pintail* (corvette, S-boat mine)
28 Jul 1941	Estonian *Lacplesis* (sunk, 3rd SFltl.)
11 Aug 1941	SS *Sir Russell* (British, sunk, *S-49)*
19 Sep 1941	SS *Fiona Shell* (British, tanker, sunk by *Siluro a lenta corsa* [SLC] low-speed manned torpedoes launched from Italian sub *Scirè)*
19 Sep 1941	SS *Durham* (British, tanker, sunk by Italian SS Scirè launching *Siluro a lenta corsa* [SLC] low-speed manned torpedoes
19 Sep 1941	SS *Denbydale* (British, tanker, sunk by Italian SS Scirè launching *Siluro a lenta corsa* [SLC] low-speed manned torpedoes
7 Sep 1941	SS *Duncarron* (British, sunk, SFltl.)
9 Sep 1941	SS *Trifels* (German, sunk, *MTB-202, 208*)
13 Sep 1941	*V-308/Oscar Neynaber* (German, sunk, TKA)
17 Sep 1941	SS *Teddington* (British, sunk, *S-50, 51, 52)*
4 Oct 1941	SS *Mv Borgny* (Norwegian, sunk, *MTB-56*). Oct 4
12 Oct 1941	SS *Chevington* (British, sunk, SFltl.)
19 Nov 1941	SS *Aruba* (British, sunk, SFltl.)
19 Nov 1941	SS *Waldinge* (British, sunk, SFltl.)
23 Nov 1941	SS *Virgilla* (British, tanker, sunk, *S-109*)
23 Nov 1941	SS *Groenlo* (Ne, sunk, *S-109*)
29 Nov 1941	SS *Asperity* (British, tanker, sunk, 4th SFltl.)
29 Nov 1941	SS *Cormarsh* (British, sunk, SFltl.)
30 Nov 1941	SS *Empire Newcomen* (British, sunk, SFltl.)
19 Dec 1941	HMS *Valiant* (battleship, sunk by *Siluro a lenta corsa* [SLC] low-speed manned torpedoes launched from Italian sub *Scirè;* later salved)
19 Dec 1941	HMS *Queen Elizabeth* (battleship, sunk by *Siluro a lenta corsa* [SLC] low-speed manned torpedoes launched from Italian sub *Scirè;* later salved)
19 Dec 1941	SS *Sagona* (tanker, sunk by *Siluro a lenta corsa* [SLC] low-speed manned torpedoes launched from Italian sub *Scirè;)*
19 Dec 1941	HMS *Jervis* (destroyer, damaged by *Siluro a lenta corsa* [SLC] low-speed manned torpedoes launched from Italian sub *Scirè;)*
9 Jan 1942	HMS *Vimiera* (L-29) (destroyer, sunk, S-boat mine)

Date	Event
9 Feb 1942	FNFL *Alysse* (K-58) (ex-HMS *Aconite*, Lost)
11 Mar 1942	SS *Horseferry* (British, sunk, 2nd SFltl.)
14 Mar 1942	HMS *Vortigern* (sunk, *S-104*)
24 Mar 1942	HMS *Southwold* (sunk, S-boat mine)
9 Apr 1942	USS *Capella* (AK-13) (damaged, friendly fire, *PT-59*)
6 May 1942	HMS *Urge* (sub) (sunk, S-boat mine)
7 May 1942	HMS *Olympus* (sub, sunk, S-boat mine)
10 May 1942	HMS *Moor* (armed trawler, sunk, S-boat gunfire)
11 May 1942	HMS *C-308* (tug) (sunk, S-boat mine)
12 May 1942	German *Seeadker* (sunk, *MTB-221, 219)*
12 May 1942	German *Iltis* (sunk, *MTB-221, 219)*
26 May 1942	HMS *Eddy* (armed trawler, sunk, S-boat mine)
27 May 1942	HMS *Fitzroy* (i) (J-03) (minesweeper, S-boat mine)
30 May 1942	HMS St. *Angelo* (tug, sunk, S-boat mine)
4 Jun 1942	HMS *Cocker* (armed trawler, sunk, 3rd SFltl.)
10 Jun 1943	HMS *Trusty Star* (drifter, sunk, S-boat mines)
14 Jun 1942	HMS *Newcastle* (light cruiser, sunk, *S-56)*
15 Jun 1942	HMS *Hasty* (destroyer, sunk, 3rd SFltl.)
16 Jun 1942	HMS *Drifter Justified* (sunk, S-boat mine, 3rd SFltl.)
16 Jun 1942	ORP *Kujawiak (*Polish, ex-HMS *Oakley,* destroyer, sunk, S-boat mine)
21 Jun 1942	HMS *Parktown* (minesweeper, (sunk, S-boat gunfire, 3rd SFltl.)
21 Jun 1942	HMS *J.T.A.I.* (tug) (sunk, S-boat gunfire, 3rd SFltl.)
21 Jun 1942	HMS *Alaisia* (tug, sunk, S-boat gunfire, 3rd SFltl.)
21 Jun 1942	HMS *Highland Queen* (armed trawler, sunk, S-boat gunfire)
21 Jun 1942	HMS *LCM-110* (sunk, S-boat gunfire, 3rd SFltl.)
21 Jun 1942	HMS *LCM-113* (sunk, S-boat gunfire, 3rd SFltl.)
21 Jun 1942	HMS *LCM-145* (sunk, S-boat gunfire, 3rd SFltl.)
21 Jun 1942	HMS *LCM-119* (captured, 3rd SFltl.)
21 Jun 1942	HMS *LCT-146* (captured, 3rd SFltl.)

Date	Event
21 Jun 1942	HMS *LCA-193* (captured, 3rd SFltl.)
21 Jun 1942	HMS *LCP-64* (captured, 3rd SFltl.)
9 Jul 1942	SS *Bokn* (Norway, sunk, *S-70*)
9 Jul 1942	SS *Kongshaug* (Norway, sunk, 3rd SFltl.)
9 Jul 1942	SS *Rosten* (Norway, sunk, 3rd SFltl.)
16 Jul 1942	SS *Aramis* (Norway ,damaged by German *Leichte Schnellboote LS-4-Esau* and armed merchant raider SS *Michel HSK-9* and sunk by *Michel* the next day.) These light attack craft were designed specifically for service aboard commerce raiders.[1376] [1377] Eleven were built; *Esau* was typical with an armament of one 20-mm cannon and two 450-cm torpedo tubes, a speed of 40.9 knots, a displacement of 11.9 tons, and a crew of seven.[1378] Pictures at page 632.
9 Jul 1942	HMS *Manor* (FY-333) (trawler, sunk, SFltl.))
9 Jul 1942	SS *Pomella* (British, motor tanker, sunk, *S-67)*
2 Aug 1942	USSR *Kharkov* (destroyer, damaged, MAS 568, 568, 573)
2 Aug 1942	USSR *Molotov* (cruiser, damaged)
12 Aug 1942	SS *Almeria Lykes* (Am, sunk, SFltl.)
13 Aug 1942	HMS *Manchester* (15) (light cruiser, sunk, *Ms 16, Ms 22)*
13 Aug 1942	SS *Glenorchy* (British, motor vessel, sunk, SFltl.)
13 Aug 1942	SS *Wairangi* (British, motor vessel, sunk, SFltl.)
15 Aug 1942	German *MFP F-138A* (gun ferry, sunk, TKA)
29 Aug 1942	HMS *Eridge* (destroyer, constructive total loss, Italian MTM)
2 Oct 1942	HMS *Lord Stonehaven* (trawler, sunk, S-*65, 77, 82, 112)*
6 Oct 1942	German *MFP F-131A* (gun ferry lighter, sunk, TKA)
7 Oct 1942	HMS *Caroline Moller* (W 09) (tug, sunk, SFltl.)
7 Oct 1942	SS *Sheaf Water* (British, sunk, SFltl.)
7 Oct 1942	SS *Ilse* (British, sunk, SFltl.)
7 Oct 1942	SS *Jessie Maersk* (British, sunk, SFltl.)

1376 http://www.navypedia.org/ships/germany/ger_mb_ls2.htm

1377 "Auxiliary Cruisers Raiding Tactics." http://www.oocities.org/pentagon/2833/general/tactics/raiding/raiding.html#launch. Archived from Geocities, no other publishing information available.

1378 http://www.navypedia.org/ships/germany/ger_mb_ls2.htm

Date	Event
14 Oct 1942	German SS *Komet* (raider, sunk, *MTB-236)*
28 Oct 1942	German *MFP-F-470C* (gun ferry lighter, sunk by TKAs, later salvaged)
19 Nov 1942	HMS *Ullswater* (trawler, sunk, 5th SFltl.)
19 Nov 1942	SS *Yewforest* (British, sunk, SFltl.)
19 Nov 1942	SS *Birgitte* (British, sunk, SFltl.)
23 Nov 1942	SS *Numidia* (lighter, sunk, damaged, aircraft) [seized by Germans September 1943]
1 Dec 1942	HMS *Jasper* (armed trawler, sunk, *S-81)*
3 Dec 1942	HMS *Penylan* (destroyer, sunk, *S-115)*
3 Dec 1942	SS *Gatinais* (British, sunk, SFltl.)
11 Dec 1942	IJN *Teruzuki* (sunk, *PT-37* and *40*)
12 Dec 1942	SS *Avonwood* (British, sunk, *S-48*)
12 Dec 1942	SS *Glen Tilt* (British, sunk, SFltl.)
12 Dec 1942	SS *Lindisfarne* (British, sunk, SFltl.)
12 Dec 1942	SS *Knitsley* (British, sunk, SFltl.)
7 Jan 1943	HMS *Horatio* (minesweeper, sunk, *S-58*).
10 Jan 1943	IJN *Hatsukaze* (damaged, sunk, *PT-36, 39,40, 43, 46, 59,112, 115*)
19 Jan 1943	RS *Santorre Santarosa* (sub, sunk, *MTB-260, 264, 313*)
1 Feb 1943	IJN *Makigumo* (destroyer, damaged, scuttled, PT)
5 Feb 1943	HMS *Stronsay* (armed trawler, sunk, S-boat mine)
5 Feb 1943	HMS *Harstad* (armed whaler, sunk, 5th SFltl., *S-65, 68, 81, 85)*
27 Feb 1943	SS *Modavia* (British, motor vessel, sunk, 5th SFltl S-65,68, 81, 85)
27 Feb 1943	HMS *LCT No.381* (MKII, sunk, 5th SFltl., *S-65, 68, 81, 85)*
2 Mar 1943	IJN SS *Ogawa Maru* (damaged, sunk, *PT-143, 159*)
12 Mar 1943	HMS *Lightning* (destroyer, sunk, *S-55*)
12 Mar 1943	German SS *Liege* (sunk, *MTB-617, 624, 628)*
12 Mar 1943	German SS *Hermond* (sunk, *MTB-617, 624, 628)*
26 Mar 1943	SS *Beckenham* (British, damaged, 3rd S-boat mine).
10 Apr 1943	RN *Trieste* (heavy cruiser, sunk, USAAF)
10 Apr 1943	RN *Gorzia* (heavy cruiser, damaged, USAAF)

Date	Event
14 Apr 1943	HMS *Eskdale* (destroyer, sunk, *S-82, 90, 112).*
14 Apr 1943	HMS *Adonis* (armed trawler, sunk, 2nd, 6th SFltl.)
14 Apr 1943	SS *Stanlake* (British, sunk, SFltl.)
28 Apr 1943	RN SS *Teranio* (damaged, *MTB-633, 637, 639)*
28 Apr 1943	RN *Sagittario* (destroyer, damaged, *MTB-633, 637, 639)*
12 May 1943	HMS *MMS-89* (minesweeper, sunk, S-boat mine)
12 May 1943	HMS *M.L. 1154* (minelayer, sunk, S-boat mine)
14 May 1943	IJN *RO-102 Kanemoto* (sub, sunk, *PT-150, 152*)
4 Jun 1943	German SS *Altenfels* (sunk, *MTB-620, 626)*
27 Jun 1943	German *MFP F-325A* (damaged, TKA)
30 Jun 1943	USN *McCawley* (sunk, *PT-59)*
12 Jul 1943	*U-561* (sunk, *MTB-81*)
13 Jul 1943	RS *Flutto* (destroyed, *MTB-640, 651, 670*)
5 Aug 1943	HMS *Red Gauntlet* (armed trawler, sunk, 2nd,6th SFltl., *S-86*)
23 Aug 1943	FN *Riilahti* (minelayer, sunk, TKA)
9 Sep 1943	SS *Larchbank* (British, sunk, U-boat)
9 Sep 1943	RN *Sella* (destroyer, sunk*, S-54, 61*)
10 Sep 1943	HMS *Abdiel* (cruiser/minelayer, sunk, S-boat mine)
11 Sep 1943	RN *Aurora* (Great Britain, sunk*, S-54, 61*)
11 Sep 1943	RN *Leopardi* (troop trans, captured, SFltl.)
11 Sep 1943	USS *Rowan* (DD-405) (destroyer, sunk*, S-151, 152, 158)*
21 Sep 1943	German SS *Antje Fritzen* (sunk, *TKA-15*)
24 Sep 1943	HMS *Franc Tireur* (armed trawler, sunk, *S-96*)
24 Sep 1943	HMS *Donna Nook* (armed trawler, sunk, *S-96*)
26 Sep 1943	German *V-15019* (sunk, *MTB-202, 204, 231)*
26 Sep 1943	German *M-534/Jungingen* (sunk, *MTB-202, 204, 231*)
8 Oct 1943	German *FB-477* (sunk, USSR sub)
8 Oct 1943	German *FB-474* (sunk, USSR sub)
8 Oct 1943	German assault boat *229* (sunk, USSR air)

Date	Event
23 Oct 1943	German *Juminda*, ex-Italian RN *Elbano Gasperi* (minelayer, sunk, MTB)
23 Oct 1943	HMS *Charybdis* (88) (light cruiser, sunk by German *T-23,T-27*)
23 Oct 1943	HMS *Limbourne* (destroyer, badly damaged by German *T-22*; sctl)
24 Oct 1943	HMS *William Stephen* (armed trawler, sunk, 6th SFltl., *S-74*)
2 Nov 1943	SS *Dona Isabel* (British, sunk*, S-146*)
2 Nov 1943	SS *Foam Queen* (British, sunk, *S-100, 138)*
2 Nov 1943	SS *Storaa* (British, sunk, *S-100, 138)*
4 Nov 1943	German *Main* (buoy-laying ship, *MTB-606)*
5 Nov 1943	SS *British Progress* (British, steam tanker, constructive total loss*, S-62*)
21 Nov 1943	German *MFP F-472* (sunk, TKA)
1 Dec 1943	HMS *Avanturine* (armed trawler, sunk, 5th SFltl.)
12 Dec 1943	German Siebel Ferry *(SF-193)* (sunk, *MTB-297, 637*)
18 Dec 1943	HMS *Felixstowe* (Bangor minesweeper,sunk, S-boat mine)
19 Dec 1943	German *Niobe* (cruiser, ex-RS *Cattaro*, ex-Yugo. *Dalmacija*, ex-German *Niobe*, grounded, destroyed)
22 Dec 1943	German SS *Vardo* (damaged, TKA)
22 Dec 1943	German SS *Kirkenes* (damaged, TKA)
22 Dec 1943	German SS *Petsamo* (damaged, TKA)
22 Dec 1943	German *V-6106* (armed trawler, sunk, TKA)
26 Dec 1943	IJN *Kiyoshimo* (destroyer, sunk, *PT-223*)
6 Jan 1944	SS *Polperro* (British, sunk, 5th SFltl.)
6 Jan 1944	SS *Underwood* (British, sunk, 5th SFltl.) c
6 Jan 1944	SS *Solstad* (Swedish, sunk, 5th SFltl.)
6 Jan 1944	HMS *Wallasea* (T-345) (armed trawler, sunk, 5th SFltl.)
31 Jan 1944	HMS *Pine* (T-101) (tug, sunk, *S-142*)
31 Jan 1944	SS *Emerald* (British, sunk, 5th SFltl.)
31 Jan 1944	SS *Caleb Sprague* (British, sunk, 5th SFltl.)
12 Feb 1944	German *Irma* (coaster, sunk)
13 Feb 1944	HMS *Cap d' Antifer* (FY-350) (armed trawler, sunk, SFltl.)
22 Feb 1945	SS *Goodwood* (British, sunk, SFltl.)

Date	Event
24 Feb 1944	SS *Philipp M* (British, sunk, 2nd SFltl.)
6 Mar 1944	German *V 1306* (sunk, 53rd MTB Fltl.)
15 Mar 1944	German *M-3630* (sunk, *MTB-353, 417)*
15 Mar 1944	German *M-10* (sunk, *MTB-353)*
17 Mar 1944	HMS *MTB-266* (ex-*PT-17*) (sunk, RN)
26 Mar 1944	HMS *MTB-352* (collision, sunk)
26 Mar 1944	German V *1416* (destroyed, RAF)
8 Apr 1944	German *M-37* (sunk)
18 Apr 1944	F.N.F.L *L'Escarmouche* (K 267) (frigate, collision constructive total loss, *MTB-707)*
24 Apr 1944	HMS *Roode Zee* (W-162) (tug, sunk, *S-100*)
25 Apr 1944	German *Tebessa* (tug, sunk, British MTB)
26 Apr 1944	USS *LST-531 (*sunk, 5th, 9th SFltl., Exercise Tiger)
26 Apr 1944	USS *LST-507* (sunk, 5th,9th SFltl., Exercise Tiger)
26 Apr 1944	USS *LST-289* (damaged, 5th,9th SFltl., Exercise Tiger)
7 May 1944	German M/K *Moder II* (sunk, *TKA-215, 218, 219)*
8 May 1944	German *Uj-104* (sunk, TKA Fltl.)
8 May 1944	German *Elbe*-5 (gun ferry lighter, sunk, TKA Fltl.)
8 May 1944	German *G-3102/Drache* (sunk, TKA Fltl.)
8 May 1944	German SS *Cornelis Anni* (sdk,TKA Fltl.)
18 May 1944	HNMS *Marken* (*MMS-227,* sunk, *S-83, 87*, 8th SFltl.)
19 May 1944	German *Vp-5531* (sunk, *MTB-709, 712*)
19 May 1944	German *Vp-5525* (sunk, *MTB-709, 712)*
20 May 1944	HMS *Wyoming* (FY1862) (minesweeper, sunk, *S-83, 87,* 8th SFltl.)
24 May 1944	German *Uj-2223* (ex-RS Marangone (constructive total loss, *PT-202, 213, 218)*
24 May 1944	German *Uj-2222* (ex-RS Tuffeto (constructive total loss, *PT-202, 213, 218)*
28 May 1944	British *MTB-732* (sunk, friendly fire from HMS *La Combattante)*
31 May 1944	German *T-29* (damaged, *PT-304, 306, 307*)
31 May 1944	German *T-30* (damaged, *PT-304, 306, 307)*
4 Jun 1944	German *M-37* (sunk, *TKA-101*)

Date	Event
6 Jun 1944	HNoMS *Svenner* (G-03) ex-HMS *Shark* (destroyer, sunk, by three German T-boats)
6 Jun 1944	USS *Meredith* (DD-726) (destroyer, damaged, sunk, S-boat mine)
6 Jun 1944	USS *Osprey* (AM-56) (minesweeper, sunk, S-boat mine)
6 Jun 1944	USS *Corry II* (DD-463) (sunk, S-boat mine)
7 Jun 1944	USS *Tide* (AM-125) (minesweeper, sunk, S-boat mine)
7 Jun 1944	USS *Susan B. Anthony* (AP-72 (sunk, 2,S-boat mines)
7 Jun 1944	USS *PT-505* (damaged, S-boat mine)
7 Jun 1944	USS *LCT-875* (sunk, 9th SFltl.)
7 Jun 1944	USS *LC(I) 105* (sunk, 9th SFltl.)
7 Jun 1944	USS *LST-314 (*sunk, 4th SFltl.)
7 Jun 1944	USS *LST-376* (sunk, 4th SFltl.)
7 Jun 1944	USS *ECM-1* (sunk, 4th SFltl.)
9 Jun 1944	SS *Ezra Weston* (Liberty, damaged, shell/fire)
8 Jun 1944	USS *LST-499* (sunk, S-Mine)
8 Jun 1944	USS *Rich* (DE-695) (destroyer escort, sunk by three S-boat mines)
9 Jun 1944	USS *Glennon* (DD-620) (destroyer, sunk, shore batteries)
9 Jun 1944	SS *Hannibal Hamlin* (Liberty, sunk, Normandy)
9 Jun 1944	FNFL *Courbet* (battleship, scuttled, Mulberry Harbor, Normandy)
10 Jun 1944	SS *Charles Morgan* (sunk, A/C)
10 Jun 1944	SS *Dungrange* (sunk, *S-177,178,* 2nd SFltl.)
10 Jun 1944	SS *Eleazar Wheelock* (US, War Shipping Administration) (damaged, A/C)
10 Jun, 1944	SS *Ashanti* (British, sunk, *S-177, 178*, 2nd SFltl.)
11 Jun 1944	SS *Brackenfield* (British, sunk, *S-189*)
11 Jun 1944	HMS *Halsted* (frigate, constructive total loss, 5th SFltl.)
10 Jun 1944	USS *Partridge* (tug, sunk, 9th SFltl.),
10 Jun 1944	USS *LST-538* (damaged, 9th SFltl.)
11 Jun 1944	SS *Nohno* (scuttled, Normandy)
11 Jun 1944	USS *LST-496* (sunk, S-boat mine)
11 Jun 1944	HMS *Sesame* (W-144) (tug, sunk, SFltl.)

Date	Event
12 Jun 1944	USS *Nelson* (DD-623) (destroyer, damaged, *S-138)*
12 Jun 1944	HMS *Onslow* (destroyer, damaged, 4th SFltl.)
12 Jun 1944	HMS *Offa* (destroyer, damaged, 4th SFltl.)
12 Jun 1944	HMS *Oribi* (destroyer, damaged, 4th SFltl.)
13 Jun 1944	British *MS No.229* (Motor Minesweeper) (sunk, mine, Normandy)
15 Jun 1944	*TKA-239* (ex-*PT-412)* (sunk, sub chaser *UJ 1211)*
15 Jun 1944	German *T-31* (torpedo boat, sunk, TKA Fltl.)
15 Jun 1944	German SS *Hugin* (drifter, sunk, *TKA-239)*
15 Jun 1944	German *TA-26* (sunk, *PT-552, 558, 559)*
15 Jun 1944	German *TA-30* (sunk, *PT-552, 558, 559)*
16 Jun 1944	German *M-133* (minesweeper, scuttled, sunk *MTB-748*)
19 Jun 1944	German *V-1314 (*sunk, *MTB-681)*
19 Jun 1944	German *V-2020* (sunk, *MTB-681)*
18 Jun 1944	German *V-2021* (sunk, *MTB-681)*
19 Jun 1944	USS *LST-496* (sunk, 9th SFltl.)
19 Jun 1944	*USS LST-523* (sunk, S-boat mine)
19 Jun 1944	SS *Powder River* (T2-SE-A1) (tanker, sunk, SFltl.)
20 Jun 1944	HMS *Solitaire* (tug, capsized)
21 Jun 1944	HMS *Fury* (H-76, destroyer, sunk, SFltl.)
21 Jun 1944	SS German *Memelland* ex-Estonian *Julia* (sunk, *S-59, 60,* 3rd SFltl.)
22 Jun 1944	German SS *Hydra* (sunk, 65th MTB Fltl.)
22 Jun 1944	SS *Cyrus H.K. Curtis* (Liberty, damaged, S-boat mine)
24 Jun 1944	German *TA-34* (ex-Yugo. *T-7*) ((grounded, *MTB-670*)
24 Jun 1944	HMS *Swift* (G-46) (destroyer, sunk, S-boat mine)
24 Jun 1944	Motor Minesweeper *No. 8 (*sunk, S-boat mine)
25 Jun 1944	USS *LCH-185* (sunk, S-boat mine or S-boat torpedo)
28 Jun 1944	USAT (United States Army Transport) *Charles W. Eliot* (Liberty, sunk, S-boat mine)
28 Jun 1944	SS *Nerissa* (sunk by *TKA 239, 241)*
29 Jun 1944	SS *Edward M. House* (Liberty, damaged, S-boat torpedo)

Date	Event
9 Aug 1944	USN *PT-509* (destroyed, German minesweeper)
14 Aug 1944	HMS LCI(L) *No.99* (sunk, S-boat mine, S-boat torpedo)
16 Aug 1944	USN *PT-202 (*sunk, S-boat mine)
16 Aug 1944	USN *PT-218* (sunk, S-boat mine)
19 Aug 1944	German *V-6102/ Köln* (sunk, *TKA-205, 206)*
19 Aug 1944	German SS Co*lmar* (ex-*Volgoles*) (sunk, *TKA-205, 206)*
22 Aug 1944	HMS *Rattler* (J-217) (renamed Loyalty, June 1943, sunk, U-480)
23 Aug 1944	*V-716* (damaged, MTB Fltl, HMS *Thornbourgh, Talybont, Melbreak, Retalick*)
23 Aug 1944	*R-229* (damaged, MTB Fltl , HMS *Thornbourgh, Talybont, Melbreak, Retalick)*
23 Aug 1944	*R-219* (sunk, MTB-Fltl, HMS *Thornbourgh, Talybont, Melbreak, Retalick)*
25 Aug 1944	German *M-3857* (sunk, MTB)
25 Aug 1944	German *V-243* (damaged, MTB)
25 Aug 1944	German *AF-103* (sunk, MTB)
25 Aug 1944	German *AF-98* (gun ferry, sunk, *PT-513, 516, 519*)
25 Aug 1944	German *AF-108* (gun ferry, sunk, *PT-513, 516, 519*)
27 Aug 1944	German *UJ-1433* (sunk, MTB)
27 Aug 1944	German *KUJ-9* (sunk, MTB)
27 Aug 1944	German *R-231* (damaged, constructive total loss, MTB)
29 Aug 1944	HMS *Retalick* (destroyer, damaged, 8th SFltl.)
29 Aug 1944	HMS *Cattistock* (destroyer, damaged, 8th SFltl.)
8 Sep 1944	USN *PT-555* (sunk, S-boat mine)
14 Sep 1944	German *V-6101* (V, sunk, TKA)
18 Sep 1944	HMS *Stayner* (*MTB-724, 728,* damaged*, S-83, 200, 702*)
8 Oct 1944	German *Freikoll* (coaster, grounded, *MTB-711, 712, 722)*
Oct 12 1944	German *M-303* (minesweeper, sunk, *TKA 205, 219*)
18 Oct 1944	USN *PT-311 (*sunk, surface craft)
21 Oct 1944	German *R-311* (mine hunter, damaged, *TKA-205, 215, 219*)
21 Oct 1944	German *M-31* (large mine hunter, sunk, *TKA-215)*
26 Oct 1944	IJN *Abukuma* (light cruiser, sunk, *PT-137, 146, 151, 190, 493)*

Date	Event
26 Oct 1944	IJN *Nachi* (cruiser, sunk *PT-194)*
1 Nov 1944	German *V-5525 (*Patrol Boat, German *Vorspostenboot* (V), sunk *MTB-709, 712*)
1 Nov 1944	German *V-5531* (Patrol Boat, German *Vorspostenboot* (V), sunk *MTB-709 , 712*)
2 Nov 1944	SS *Rio Bravo* (tanker, sunk, *S-167, 175, 207*)
2 Nov 1944	HMS *Colsay* (armed trawler, sunk, *S-167,175, 207*)
9 Nov 1944	IJN *Okinawa* (frigate, sunk, *PT-492, 497, 524, 525)*
9 Nov 1944	IJN *Shimushu* (frigate, sunk, *PT-492, 497, 524, 525)*
28 Nov 1944	IJN *SC-53* (sub chaser\r, sunk, *PT-229, 331*)
29 Nov 1944	IJN P-*105* (PV, ex-Philippine *Arayat*, sunk *PT-127, 128, 191)*
12 Dec 1944	IJN *Uzuki* (destroyer, sunk, *PT-490, 492)*
17 Dec 1944	HMS *Kellett* (J-05) (minesweeper, damaged, SFltl.)
22 Jan 1945	SS *Halo* (sunk, SFltl.)
22 Feb 1945	SS *Blacktoft* (British, sunk, SFltl.)
23 Feb 1945	SS *Aquarius* (British FV (Fishing Vessel), sunk, 5th SFltl.) February 25 mine
26 Feb 1945	SS *Auretta* (British, sunk, S-boat mine)
19 Mar 1945	SS *Rogate* (British, sunk, SFltl.)
19 Mar 1945	SS *Crichtoun* (British, sunk, SFltl.)
9 Apr 1945	German SS *Neuwerk* (sunk, 1st SFltl.)
27 Apr 1945	USS *Hutchins* (DD-476) (constructive total loss, Japanese suicide boat)

Figure 460. Top: *Leichte Schnellboote* LS-2, *Meteorit,* being hauled out from the raider *Komet.* Bottom: LS-4, *Esau,* being put aboard raider *Michel.*[1379] Although these vessels were small, they were as fast as *Schnellboote* and carried torpedoes that could sink any ship.

1379 "Auxiliary Cruisers Raiding Tactics," supra.

THE SHIP KILLERS

THE DEFINITIVE ILLUSTRATED HISTORY OF THE TORPEDO BOAT

VOLUME XI: INDEXES

CONTENTS

BY JOE HINDS

NIMBLE BOOKS LLC

CONTENTS

DEDICATION

This volume is dedicated to all of the people who make up the web site, Coastal Forces. There must be thirty different people who have responded positively to me when I have posted questions about MTBs. It has always amazed me that this group is so readily forthcoming in assisting me to find answers to my questions and many times a note to cheer me on. I am forever pleased to be a part of this group of fine people. To you all, I dedicate this volume in your honor.

APPENDIX U. SOVIET MOTOR TORPEDO BOATS (RICHARD WORTH)

***STAL'NOI* type (2 boats) 1930-1931**

Displacement: 28 tons

Dimensions: 23.5 x 3.8 x 1.4 m

Machinery: 2250 hp = 26 knots

Bunkerage: 170 nm

Armament: two 7.62-mm machine guns, three 450-mm torpedo tubes

Complement: 6

***SH-4* type (59 boats) 1930-1933**

Displacement: 10 tons

Dimensions: 16.8 x 3.3 x 0.8 m

Machinery: 1200 hp = 50.5 knots

Bunkerage: 300 nm

Armament: one 7.62-mm machine gun, two 450-mm torpedo tubes, two mines

Complement: 5

***ANT-5* type 1931-1933**

Displacement: 14.5 tons

Dimensions: 19.1 x 3.4 x 1.2 m

Machinery: 2000 hp = 58 knots

Bunkerage: n/a

Armament: two 7.62-mm machine guns, two 533-mm torpedo tubes, four mines

Complement: 6

Used for training

_G-6_ type 1931-1939

Displacement: 86 tons

Dimensions: 36.5 x 6.6 x 1.9 m

Machinery: 7760 hp = 49.8 knots

Bunkerage: 435 nm at 28.5 knots

Armament: one 45-mm gun, one 12.7-mm machine gun, four 7.62-mm machine guns (1 x 4), six 533-mm torpedo tubes, 3 mines

Complement: 30

_L-5_ type (4 boats) 1936-1939

Displacement: 11.3 tons

Dimensions: 24 x 5.4 x 0* m

Machinery: 2000 hp = 70 knots

Bunkerage: 250 nm

Armament: two 12.7-mm machine guns, two 450-mm torpedo tubes

Complement: 5

*Air-cushion vehicles, not operational

Figure 461. The Levkov *L-5* air cushion torpedo boat.[1105]

[1105] http://jrc-1138.blogspot.com/2008/09/ground-effect-vehicles.html

_G-5_ type (329 boats) 1936-1945

Displacement: 17.8 tons

Dimensions: 19.1 x 3.3 x 1.2 m

Machinery: 2000 hp = 52 knots

Bunkerage: 220 nm

Armament: one 7.62-mm machine gun, two 533-mm torpedo tubes

Complement: 6

_G-8_ type 1937-1940

Displacement: 31.3 tons

Dimensions: 24.2 x 3.8 x 1.5 m

Machinery: 4000 hp = 32 knots

Bunkerage: 350nm at 27 knots

Armament: three 12.7-mm machine guns, two 533-mm torpedo tubes, three mines

Complement: 10

SM-3 type 1938-1940

Displacement: 34 tons

Dimensions: 20.8 x 3.9 x 1.5 m

Machinery: 3600 hp = 30 knots

Bunkerage: 380 nm at 24 knots

Armament: two 12.7-mm machine guns, two 533-mm torpedo tubes

Complement: 8

SM-4 type 1939-1941

Displacement: 42 tons

Dimensions: 22 x 4.1 x 1.8 m

Machinery: 4000 hp = 30 knots

Bunkerage: n/a

Armament: three 12.7-mm machine guns (x)

Complement: 9

Used as mine layer

D-3 type (73 boats) 1939-45

Displacement: 36.1 tons

Dimensions: 22.4 x 4 x 1.9 m

Machinery: 2550 hp = 32 knots

Bunkerage: 500 nm at 16 knots

Armament: two 12.7-mm machine guns, two 533-mm torpedo tubes

Complement: 9

56 others converted as sub chasers

***YUNGA* class (5 boats) 1943-45**

Displacement: 46.9 tons

Dimensions: 23.4 x 4 x 1.7 m

Machinery: 3-shaft petrol engines, 3600 hp = 30.7 knots

Bunkerage: 488 nm at 9.5 knots

Armament: three 12.7-mm machine guns, two 533-mm torpedo tubes

Complement: 11

***A-1* (83 boats)**

Displacement: 43.5 tons

Dimensions: 22.1 x 6.1 x 1.6 m

Machinery: 3600 hp = 39.3 knots

Bunkerage: 450 nm at 21.8 knots

Armament: one 20-mm Oerlikon, four 12.7-mm machine guns (2 x 2), two 533-mm torpedo tubes, two mines

Complement: 11

Vosper Lend-Lease 1944

***A-2* type (52 boats)**

Displacement: 49 tons

Dimensions: 23.8 x 6.1 x 1.7 m

Machinery: 4050 hp = 36.3 knots

Bunkerage: 420 nm at 21.4 knots

Armament: one 20-mm Oerlikon, four 12.7-mm machine guns (2 x 2), two 533-mm torpedo tubes, two mines

Complement: 11

Higgins Lend-Lease 1943

***A-3* type (31 boats)**

Displacement: 51.1 tons

Dimensions: 24.48 x 6.28 x 1.54 m

Machinery: 3600 hp = 38.4 knots

Bunkerage: unknown nm at 22.2 knots

Armament: one 40-mm Bofors, one 20-mm Oerlikon, four 12.7-mm machine guns (2 x 2), two 533-mm torpedo tubes, 2 mines

Complement: 14

Elco Lend-Lease 1945

STK DD type 1942-1943

Displacement: 50 tons

Dimensions: 25 x 4 x unknown

Machinery: 3000 hp = 34 knots

Bunkerage: 1000 nm

Armament: three 12.70-mm machine guns, four 450-mm torpedo tubes, three mines

Complement: 11

Lürrsen (3 boats)

Displacement: 62 tons

Dimensions: 28 x 4.5 x 1.6 m

Machinery: 2850 hp = 33 knots

Bunkerage: 880 nm at 30 knots

Armament: one 20-mm cannon, two 15-mm machine guns, two 533-mm torpedo tubes

Complement: 14

Captured from Bulgaria in 1944

KOMSOMOLETS prototype 1939-40

Displacement: 16.98 tons

Dimensions: 18.5 x 3.4 x 1.2 m

Machinery: 2400 hp =51.6 knots

Bunkerage: 650 nm at 32 knots

Armament: one 12.7-mm machine gun, two 533-mm torpedo tubes

Complement: 7

***KOMSOMOLETS* class (31 boats) 1944-45**

Displacement: 20 tons

Dimensions: 18.7 x 3.4 x 1.2m

Machinery: two gasoline engines, 2400hp = 46 knots

Bunkerage: 345nm at 28.8 knots

Armament: two 12.7-mm machine guns (1 x 2), two 450-mm torpedo tubes

Complement: 7

APPENDIX V. SUCCESSES OF SOVIET TORPEDO CUTTERS (MINE WARFARE NOT INCLUDED)

WS confirmed at warsailors

Gr confirmed by Gröner

Jo confirmed by Jordan

RH confirmed by Rohwer & Hummelchen

Northern Fleet

6.10.1941: D/S *Bjørnungen* (165 grt[1106]) torpedoed and sunk by *TKA No. 12* near Sture-Ekkerøy. WS

24.04.1942: submarine U-454 seriously damaged by depth charges of *TKA No. 13* and *No. 14* near Vardø.

21.06.1943: M/S *Foula* (109 grt) (previously damaged by German aircraft) sunk by *TKA No. 13* (by machine-gun) off Rybachij peninsula. WS but disputed

21.09.1943: D/S *Antje Fritzen* (4330 grt) (previously damaged on two mines) torpedoed (two hits) and sunk by *TKA No. 15* near Cape Nurmensetty (69.43 N 31.25 E). Jo RH

22.12.1943: D/S *Marie* (200 grt) torpedoed (2 hits) and sunk by *TKA No. 13* in Varangerfjord.

7.05.1944: M/K *Moder II* (124 grt) captured by *TK-215*, *TK-218* and *TK-219*; set on fire and sunk in Varangerfjord. WS

28.06.1944: D/S *Nerissa* (992 grt) torpedoed and sunk by *TK-239* and *TK-241* off Petsamo. RH

15.07.1944: D/S *Hugin* (124 grt) attacked by *TK-239*; shelled by MG, blown up by depth charge and set ablaze. The following day she was towed to Vadsø, where she stayed afloat until September when she sank by lack of attendance. RH

Around the same date *TK-12* torpedoed and sank the Norwegian (?) *Storegga*—unconfirmed

19.08.1944: patrol ship *V-6102 / Köln* (472 grt) and D/S *Colmar* (ex-*Volgoles*, 3992 grt) torpedoed and sunk by the Soviet torpedo cutters Kibergness Cape. RH WS

14.09.1944: minesweeper *M-252* damaged by blind torpedo N of Vardø. Participating in the attack were *TK-211*, *TK-213*, *TK-242* and *TK-243*.

25.09.1944: patrol ship *V-6101 / Gauleiter Bohle* (505 grt, disputed) torpedoed and sunk by the Soviet torpedo cutters in Varangerfjord.

[1106] Gross register tonnage.

12.10.1944: minesweeper *M-303* (775 tons) torpedoed and sunk by the Soviet torpedo cutters near Kibergness Cape (*TK-230, TK-238, TK-241* participated in attack). [RH]

21.10.1944: minesweeper *R-311* attacked by *TK-205* and *TK-230* south of Bussesund; damaged by blind torpedo.

21.10.1944: minesweeper *M-31* (874 t) torpedoed and sunk by *TK-237* and *TK-244* in Persfjord (70.31,8 N 30.55 E). [RH]

Baltic Fleet

14.09.1941: minesweeper *M-1707* (Lüneberg, 473 grt, disputed) (previously damaged by mine and hit by 130-mm shell from Soviet shore battery; abandoned by crew) torpedoed and sunk near Arensburg.

23.09.1941: patrol ship *V-308* (*Oscar Neynaber*, 312 grt) sunk near Gogland Island. [RH]

26.08.1942: submarine hunter *Uj-1216* (*Star XXI*, 299 grt) sunk near Bol'shoj Tuters Island. [RH]

23.08.1943: Finnish minelayer *Riilahti* (311 tons) sunk near Tiiskeri Island. [RH]

4.06.1944: minesweeper *M-37* (717 grt) sunk near Gogland Island. [RH]

20.06.1944: torpedo boat *T-31* (1754 tons) sunk near Nerva Island. [RH]

15.04.1945: destroyer *Z-34* (3691 tons) damaged in Danzig Bay.

26.04.1945: MFP *F-248 A* (220 tons) sunk near Hela.

Black Sea Fleet

13.06.1942: Italian submarine *CB-5* sunk at Yalta.

1.08.1942: MFP *F-334* (220 t) sunk at Ivan-Baba Bay. [RH Gr]

15.08.1942: MFP *F-138A* sunk near Kazantip Cape.

6.10.1942: MFP *F-131A* sunk near Anapa. [Gr?]

28.10.1942: MFP *F-470C*, trying to avoid TKA attack, beached near Anapa (later salvaged).

27.06.1943: MFP *F-325A* damaged by blind torpedo near Chauda Cape.

21.11.1943: MFP *F-472* sunk near Eltigen.

27.04.1944: subhunter *Uj-104* (*KT17*, 834 grt) torpedoed by TKA near Sevastopol. The ship broke in two parts; the bow sank and the stern was towed to Sevastopol where it was finally destroyed by soviet aircraft. [RH]

Black Sea Fleet (continued)

8.05.1944: lighter *Elbe-5* (1188 t, disputed), patrol boats *G-3102 Drache*? and *Cornelis Anni* (disputed) sunk near Sevastopol.

11/12.5.44: remains of wrecked *Romania* sunk by *TK-301, 353* off Chersones.

Pacific Fleet

two Japanese vessels in 1945

INDEX OF BRITISH MGBS

INDEX OF BRITISH MTBs

Boat number	Volume and page
MTB-204	7:365
MTB-205	8:469, 472
MTB-208	8:469
MTB-209	8:472
MTB-210	8:472
MTB-212	8:469
MTB-213	5:234
MTB-214	5:234
MTB-215	5:256
MTB-216	5:234
MTB-217	5:234
MTB-218	6:295
MTB-219	7:344
MTB-220	6:276
MTB-221	7:344
MTB-222	7:371
MTB-225	8:390
MTB-226	7:365
MTB-228	7:365
MTB-230	7:371
MTB-231	7:365
MTB-233	7:337
MTB-234	7:337, 340, 8:390
MTB-235	7:351
MTB-236	6:300
MTB-237	6:296
MTB-240	7:351
MTB-241	7:337, 340,, 8:390, 396

Boat number	Volume and page
MTB-242	8:454
MTB-244	7:340, 8:390, 396
MTB-245	8:396
MTB-248	8:449
MTB-252	8:472
MTB-254	8:472
MTB-255	7:359, 9:520
MTB-257	8:472
MTB-256	8:472
MTB-259	6:281
MTB-260	7:320, 351, 8:381
MTB-261	9:548
MTB-262	7:322
MTB-264	7:320, 340
MTB-265	7:331
MTB-266	7:334, 7:371, 8:392, 399
MTB-267	7:334
MTB-269	5:228
MTB-271	5:228
MTB-272	5:228
MTB-273	4:215
MTB-274	4:215
MTB-276	7:378
MTB-284	*7:357*
MTB-285	*7:357*
MTB-287	8:493
MTB-288	7:353
MTB-297	7:377

Boat number	Volume and page
MTB-298	7:378
MTB-308	6:298
MTB-309	8:381
MTB-310	6:298
MTB-311	7:331, 7:334
MTB-312	6:298
MTB-313	7:320
MTB-314	6:296
MTB-315	7:371
MTB-316	7:351
MTB-338	6:276
MTB-339	6:284
MTB-342	6:284
MTB-344	6:287
MTB-344	6:298
MTB-344	6:287
MTB-347	8:480
MTB-349	8:480
MTB-350	8:396, 480
MTB-351	8:480
MTB-352	8:394
MTB-353	8:392
MTB-354	8:425
MTB-356	7:366
MTB-357	7:378
MTB-360	8:480
MTB-361	8:451
MTB-371	8:493

Boat number	Volume and page
MTB-372	8:454
MTB-394	9:520
MTB-412	8:456
MTB-415	8:390
MTB-416	8:402
MTB-417	8:392
MTB-422	9:520
MTB-430	8:456
MTB-431	8:390
MTB-438	9:520
MTB-439	7:368
MTB-444	8:387
MTB-444	9:520
MTB-447	8:472
MTB-448	8:426, 431
MTB-452	8:472
MTB-453	8:472
MTB-455	8:387, 449
MTB-457	8:449
MTB-458	8:449
MTB-459	9:520
MTB-460 (RCN)	8:443
MTB-461	9:520
MTB-462	9:520
MTB-465	9:520
MTB-466	9:520
MTB-468	8:449

Boat number	Volume and page
MTB-469	8:449
MTB-470	8:449
MTB-471	8:468
MTB-472	8:480
MTB-473	8:468, 480
MTB-474	8:459, 468
MTB-475	8:459, 480
MTB-476	8:449, 480, 468
MTB-477	8:468
MTB-478	8:426
MTB-479	8:468
MTB-480	8:480
MTB-494	9:533
MTB-495	9:513
MTB-601 "D"	6:287
MTB-605 "D"	9:520
MTB-606	7:371
MTB-608 "D"	8:459
MTB-610 "D"	8:387
MTB-612 "D"	8:440
MTB-617 "D"	7:330, 371, 8:402
MTB-618 "D"	7:313
MTB-619 "D"	7:331
MTB-620 "D"	6:302
MTB-620 "D"	6:306
MTB-620 "D"	7:344
MTB-622 "D"	7:330

Boat number	Volume and page
MTB-623 "D" (RNoN)	6:302, 306, 7:313, 8:408,:496, 9:507
MTB-624 "D"	7:330, 8:426
MTB-625 "D"	7:313, 8:384
MTB-626 "D"	7:313, 3447:373
MTB-627 "D" (RNoN)	7:313, 8:387, 493, 496, 500
MTB-628 "D"	7:331, 344
MTB-629 "D"	7:331, 344
MTB-630 "D"	7:340
MTB-631 "D"	7:313, 330
MTB-632 "D"	8:402, 444
MTB-633 "D"	7:337, 337, 8:436
MTB-635 "D"	7:337, 346
MTB-636 "D"	7:366
MTB-637 "D"	7:3337, 377
MTB-639 "D"	7:337
MTB-640 "D"	7:336, 345,351, 8:436, 443
MTB-649 "D"	7:377
MTB-650 "D"	8:444
MTB-651"D'	7:345, 351
MTB-653 "D" (RNoN)	8:387,:484, 496
MTB-654 "D"	7:344
MTB-655 "D"	8:436
MTB-655 "D"	9:533
MTB-656 "D"	7:345
MTB-656 "D"	9:527
MTB-657 "D"	8:475

Boat number	Volume and page
MTB-658 "D"	8:443
MTB-660 "D"	9:520
MTB-663 "D"	8:480
MTB-665 "D"	7:357
MTB-666 "D"	8:447
MTB-668 "D" (RNoN)	8:493
MTB-669 "D"	7:368
MTB-670 "D"	7:351, 8:440, 9:534
MTB-671 "D"	8:402
MTB-672 "D"	8:420
MTB-675 "D"	9:531
MTB-681 "D"	8:431
MTB-682 "D"	8:426
MTB-686 "D"	7:373
MTB-688 "D" (RNoN)	8:484, 493
MTB-690 "D"	8:387, 9:513
MTB-692 "D"	8:469
MTB-694 "D"	8:469
MTB-695 "D"	8:469
MTB-697 "D"	7:344, 9:534
MTB-703 "D"	9:527
MTB-705 "D"	9:527, 533
MTB-707 "D"	8:399
MTB-709 "D" (RNoN)	8:417, 451, 489, 9:542
MTB-710 "D"	9:520, 533

Boat number	Volume and page
MTB-711 "D" (RNoN)	8:411, 480, 484
MTB-712 "D" (RNoN)	8:417, 431, 432, 480,:489, 500, 9:507
MTB-715 "D" (RNoN)	8:496, 9:513, 542
MTB-716 "D" (RNoN)	9:507
MTB-717 "D" (RNoN)	8:394, 408,496, 8:500
MTB-718 "D"	8:475
MTB-720 "D"	9:538
MTB-722 "D" ' (RNoN)	8:451, 480, 500, 9:507
MTB-723 "D" (RNoN)	8:496
MTB-724 "D"	8:452, 477
MTB-728 "D"	8:477
MTB-729 "D"	8:447
MTB-732 "D" (RNoN)	8:420
MTB-734 "D"	8:440
MTB-748 "D"	8:436
MTB-776 "D"	9:520
MTB-782 "D"	8:500
MTB-789 "D"	9:520, 538
MTB-791 "D"	9:520
MTB-798 "D"	9:520
MTB-5001 "D"	9:533

INDEX OF COMMERCIAL VESSELS SUNK BY MTB, TKA, PT, MAS AND S-BOOTE

INDEX OF CONVOYS

INDEX OF GERMAN S-BOOTE

Boat number	Volume and page
S-64	5:251
S-65	6:298, 7:322, 336, 8:387
S-67	6:283, 9:507
S-68	6:302, 7:322, 353, 365
S-69	6:300
S-70	6:283, 7:327
S-71	6:300, 7:322
S-72	6:297, 7:336, 342, 8:465
S-73	7:342
S-74	7:327, 357, 368, 371
S-75	6:300, 7:327
S-77	6:298, 302, 7:353
S-79	8:451, 459
S-80	6:309, 7:357, 8:475
S-81	7:322, 336
S-82	6:298, 302, 7:336
S-83	7:336, 357, 765, 8:417, 436
S-84	8:381, 426, 432, 436
S-85	7:322
S-86	7:357
S-87	8:390, 417
S-88	7:368
S-89	7:357, 8:387
S-90	7:336, 8:451, 456
S-91	8:451, 456, 459
S-93	7:365, 8:394
S-94	6:262, 7:322, 357, 8:387
S-96	7:365

Boat number	Volume and page
S-97	8:456
S-98	9:507
S-99	8:387
S-100	7:368, 8:381, 402, 426, 432, 436
S-101	4:222, 5:230
S-102	5:230, 6:281, 7:336, 351
S-103	6:296, 8:438, 9:538
S-104	5:250, 6:265, 283, 7:316
S-105	5:250
S-107	5:245
S-109	5:251, 251, 6:283
S-110	5:251, 6:309, 7:353, 357
S-111	6:265
S-112	6:295, 298, 302, 7:336, 357, 8:381, 418, 440
S-114	7:327, 371, 8:451, 456, 459
S-115	6:302, 306
S-116	6:302, 7:336, 371
S-117	6:309, 7:357
S-119	7:327
S-121	7:336
S-123	7:342
S-124	7:342
S-125	7:342
S-126	7:342
S-128	7:322, 8:387
S-129	8:394
S-131	8:468

Boat number	Volume and page
S-133	8:387, 436
S-134	7:342
S-134	7:342
S-135	8:451
S-136	8:381, 390, 431
S-138	7:368, 8:381, 426, 432
S-139	8:381, 426
S-140	8:381, 426
S-141	8:381, 415
S-142	8:381, 390, 426, 432
S-143	8:432
S-144	8:432
S-145	8:426, 428, 440
S-146	7:368, 8:432
S-147	8:381
S-148	8:468
S-149	8:468
S-150	8:426, 432
S-151	8:413, 9:538
S-152	8:413, 9:538
S-153	8:413
S-154	8:413
S-155	8:413, 9:538
S-156	8:413, 9:538
S-157	8:413
S-158	8:413
S-167	8:402, 459, 489
S-168	8:426, 9:513

Boat number	Volume and page
S-169	8:432
S-171	8:432
S-172	8:431, 432
S-173	8:432
S-175	8:489
S-175	9:513
S-177	8:431
S-178	8:431
S-178	8:432
S-179	8:432
S-180	8:431, 459, 9:507
S-181	8:431, 432, 459
S-182	8:456
S-183	8:477
S-184	8:475
S-185	8:477
S-186	8:477
S-187	8:431, 432
S-188	8:431, 432
S-189	8:432
S-190	8:431, 436, 438
S-191	8:477, 9:542
S-192	8:477
S-197	7:365
S-199	9:513
S-200	8:477
S-201	9:538
S-202	9:533

INDEX OF ITALIAN MAS BOATS

INDEX OF ITEMS, ENGINES, RADARS, BATTLES, AND OTHER ODDS AND ENDS

INDEX OF OPERATIONS (IN CHRONOLOGICAL ORDER)

(

INDEX OF PEOPLE

INDEX OF PLACES

INDEX OF SOVIET TORPEDO CUTTERS (TKAS)

INDEX OF WARSHIPS